Knowledge Generation in STEM and STEAM Education

Gillian Kidman · Hazel Tan ·
Roland Gesthuizen · Dominador D. Mangao

Knowledge Generation in STEM and STEAM Education

Integrative Thinking and Agency

palgrave
macmillan

Gillian Kidman
Faculty of Education
School of Curriculum, Teaching and Inclusive Education
Monash University
Clayton, VIC, Australia

Hazel Tan
Faculty of Education
School of Curriculum, Teaching and Inclusive Education
Monash University
Clayton, VIC, Australia

Roland Gesthuizen
Department of Education Victoria, Faculty of Education
School of Curriculum, Teaching and Inclusive Education
Monash University
Clayton, VIC, Australia

Dominador D. Mangao
College of Alternative and Lifelong Learning
Philippine Normal University
Manila, Philippines

ISBN 978-3-032-03246-1   ISBN 978-3-032-03247-8 (eBook)
https://doi.org/10.1007/978-3-032-03247-8

This Palgrave Macmillan imprint is published by the registered company Springer Nature Switzerland AG
The registered company address is: Gewerbestrasse 11, 6330 Cham, Switzerland

To my husband, Sid, whose patience and unwavering support for all these years have been a foundation for this work. Your encouragement has been my constant source of strength and perseverance. To our grandson, Felix, whose future is yet unwritten but filled with infinite possibilities. May you grow into a world where curiosity, creativity, and knowledge shape your path, and may the joy of discovery always guide you. And to your future teachers, who will nurture your potential and inspire your love for learning. This book is for you—the educators who shape minds, ignite imaginations and lay the neurological foundations for a lifetime of inquiry and innovation.
In some small way, may this work contribute to a future where STEA$^+$M *education transforms lives and expands horizons.*

Gillian Kidman

To my family —the ones who shaped who I am, my identity, my curiosity, and my life.
To my parents, Louis and Lily, for your unwavering love, wisdom, and the values you have instilled in me.
To my siblings, Grace and Philip, for our infinite camaraderie, exponential laughter, and enduring bond that forms the perfect equation in my life.
And to my niece, Victoria, and nephews, Matthew, Perry, and Preston, for inspiring me to dream of a future filled with innovation, creativity, and endless possibilities.

This book reflects the world I hope you will help create—one where STEA⁺M integrative thinking bridges disciplines, solves challenges, and builds a brighter tomorrow.
With all my love and gratitude,

Hazel Tan

To my father, Johannes, whose ingenuity, imagination, and inventiveness shaped my own. My research and writing exist because of you. To my mother, Hendrika, who taught me a mother tongue, how to read, and how to dance in clogs.
To those who inspired me to think, write, and reflect deeply: to friends Gary and Rob, who built cubby houses, spun yarns, and debated our shared neurodiversity; to students and scouts, who let me pose big questions to nudge them out of trees and their comfort zones.
To my wife, Linda, who harboured boundless patience as she lovingly unpacked my recursive thinking. To my children Erika, Karl and Niels, who rekindled my love of learning as we counted stars and watched ants walk around the house.

Roland Gesthuizen

To my parents, brothers and sisters, relatives, teachers, and friends who have stood by me, providing care, protection, and refuge amidst life's challenges since my formative years.
I owe my collaborators my constant inspiration, collegiality, and contagious passion for transformative pedagogies, which awakened my curiosity and excitement to explore innovative ideas and develop a passion for writing, research, and publication.
To the educators in the Southeast Asian region and beyond who selflessly devote themselves to instilling universal values and propagating 21st-century literacies and skills to find solutions for today's global issues. This book hopes to inspire educators to develop their teaching repertoires and innovate approaches to solving world problems. Arts, creativity, culture, economy, environment, ethics, global citizenship, politics, and the power of technology have to be interwoven for holistic learning and a sustainable future.

Dominador D. Mangao

Preface

Our Starting Point: The Philosophical Foundations of STEM Education

The effective integration of Science, Technology, Engineering, and Mathematics (and the associated STEM derivatives) results in integrative STEM/STEAM education—a progressive educational paradigm that fosters creativity, critical thinking, and interdisciplinary collaboration. The philosophical foundations draw significantly on the work of Dewey, Kolb, Vygotsky, and Papert, whose theories collectively offer a framework for implementing STEM/STEAM practices. These frameworks emphasise inquiry, experiential learning, social interaction, and the construction of meaningful artefacts as essential elements of education.

At its heart, STEM/STEAM education embodies Dewey's philosophy of learning through inquiry and experience. Dewey regarded education as active engagement with the world to solve real-life and personally relevant problems. He argued that learning should be participatory, requiring learners to construct knowledge through interactions with their environment, followed by reflective thinking. In STEM/STEAM, this principle materialises through project-based learning, where learners address open-ended, real-world challenges that necessitate integrating diverse disciplinary knowledge, skills, and attitudes. By doing so, learners cultivate critical thinking and a curiosity-driven mindset, essential for fostering innovation in an increasingly complex world.

We can further interpret Dewey's ideas through Kolb's experiential learning theory, which explains how experience shapes learning in STEM/STEAM education contexts. Kolb's four-stage cycle—concrete experience, reflective observation, abstract conceptualisation, and active experimentation—emphasises the iterative nature of learning. In STEM/STEAM education, learners immerse themselves in creative processes such as designing artefacts, conducting experiments, and engaging in interdisciplinary collaboration. These collaborative experiences encourage reflection, enabling learners to abstract and apply key concepts in novel contexts. The iterative ethos of Kolb's model aligns with STEAM education's exploratory nature, where experimentation, failure, and revision are integral to deep learning. For example, learners co-designing and co-refining a holographic image viewer (as we consider in Chapter 9) might navigate Kolb's cycle, enhancing adaptability and fostering innovative solutions.

STEM/STEAM education's collaborative dimension focuses on social interaction and cooperative learning. Vygotsky's social constructivist theory underscores the importance of cultural and social contexts in shaping cognitive development. Central to Vygotsky's philosophy is the zone of proximal development, which posits that learners achieve higher levels of understanding when supported by knowledgeable peers. In STEM/STEAM education, this principle is exemplified in team-based projects, where learners with diverse disciplinary perspectives collaborate to solve complex problems. Such collaboration enhances cognitive development and mirrors the interdisciplinary demands of contemporary challenges. The dialogic exchange intrinsic to collaborative learning (as explored in Chapters 4 and 9) fosters critical reflection and collective creativity, enabling learners to approach innovation from multiple viewpoints.

Papert's constructionism theory emphasises designing and creating artefacts as tangible outcomes as evidence for learning. Extending Vygotsky's constructivist ideas, Papert argues that learners internalise knowledge when creating meaningful, tangible artefacts (as outlined in Chapter 4). This principle aligns with STEM/STEAM education, where learners engage in hands-on team-based projects integrating artistic creativity with scientific and mathematical principles. For instance, designing a holographic image viewer involves synthesising concepts from optics, engineering, and visual arts, culminating in an interdisciplinary learning artefact. The collaborative team-based nature of these creations encourages learners to take ownership of their learning, engage in feedback-driven refinement, and develop resilience and adaptability.

What unites these philosophical perspectives is their shared focus on active, student-centred learning that prioritises the learning process over rote outcomes. Dewey's inquiry, Kolb's experiential learning cycle, Vygotsky's collaborative scaffolding, and Papert's construction of artefacts converge to present learning as a dynamic, context-driven activity. These philosophies challenge traditional, discipline-specific approaches to education, advocating for environments where learners are empowered to explore, innovate, and collaborate.

Recent research highlights the relevance of these theoretical foundations in contemporary STEM/STEAM education. Inquiry-based and experiential approaches have been shown to enhance student engagement, creativity, and critical thinking significantly. For example, in this book, we present studies demonstrating how integrating the Arts into STEM curricula equip learners with the skills to consider complex, real-world challenges. Similarly, constructionist methodologies, particularly in educational technologies like coding and robotics, enhance problem-solving abilities and motivation, aligning with Papert's vision of learning through creation. Collaborative learning environments informed by Vygotsky's principles also promote inclusivity and diverse thinking. By engaging learners from varied disciplinary backgrounds, STEM/STEAM projects foster empathy, communication, and teamwork—essential skills for considering global challenges.

A particularly promising application of these principles is in STEM makerspaces, which embody the philosophy of constructionism and social constructivism. STEM makerspaces provide collaborative, hands-on environments where learners draw on their STEM disciplinary backgrounds to undertake a society-based issue (shifting STEM to STEAM as explored in Chapter 3). Learners explore their interests, experiment, and create tangible outcomes. These spaces encourage curiosity, innovation, and problem-solving by offering meaningful opportunities to integrate various disciplines. For example, learners in Central Australia working in makerspaces might design and prototype camel-proof fence gates, blending engineering with creative problem-solving and fostering real-world impact. STEAM makerspace environments support the iterative learning process advocated by Kolb, the collaborative learning championed by Vygotsky, and the artefact-driven knowledge construction central to Papert's ideas.

The philosophical foundations of STEAM education—rooted in the works of Dewey, Kolb, Vygotsky, and Papert—offer a cohesive and transformative framework for modern education. By integrating inquiry, experiential engagement, collaboration, and artefact creation, STEAM transcends traditional disciplinary boundaries and equips learners with the skills and dispositions necessary for navigating an interconnected and complex world. These principles enhance cognitive and creative capacities while fostering adaptability, resilience, and a commitment to lifelong learning and innovation. Makerspaces exemplify these philosophies in practice, illustrating how STEAM education can transform classrooms into hubs of creativity and interdisciplinary exploration. However, in this book, we contend that this is not sufficient. We need STEAM education to go beyond merely equipping learners with the skills and dispositions necessary for navigating an interconnected and complex world. We need to equip learners to have a student agency.

In the future with student agency: The evolving landscape of education demands a departure from traditional, compartmentalised teaching toward integrative, interdisciplinary approaches that align with the complexities of the 21st century. This book emerges from a deep inquiry into how Science, Technology, Engineering, and Mathematics—and most crucially, *A*gency—can be dynamically interwoven to create a learning ecosystem that is both cognitively rigorous and socially transformative. Central to this book lies the argument that STE*A*⁺M education must go beyond disciplinary silos. It must embrace epistemic flexibility, critical thinking, and problem-solving, all while fostering the development of ethical and socially responsible learners ready to take *A*ction. Drawing on neuroscience, disciplinary reasoning, *A*gency development, and interdisciplinary approaches, this book presents a compelling case for a paradigm shift in educational practice.

Part I lays the theoretical foundations by exploring cognitive integration, disciplinary thinking, and student *A*gency within STE*A*⁺M education. It introduces new frameworks, such as the *Cognitive and Emotional Neuroeducation: Principles for STEM* (CEN-STEM) and the *Cognitive, Inclusive, Resilient, Creative Learning Environments* (CIRCLE) Framework, which bridge neuroscience and pedagogy. The *Plexus Model for STEA⁺M Education* is a guiding framework for reimagining STEM, placing student *A*gency at the centre of learning. Understanding the processes involved in integration and creation is essential in STEM education. Part I concludes with two input-output neuroeducation frameworks, the *Cognitive Integration Framework* and the *Crafting Knowledge Framework,* identifying key mechanisms facilitating cognitive growth in STEM education.

Part II expands on these ideas, critically examining 21st century skills, *STEA⁺M Thinking Systems*, and the hidden STEM and STEAM curriculum. The Authors argue that to prepare learners for an increasingly complex world, educators must move beyond traditional skill development toward cultivating thinking systems—integrated cognitive and social competencies that enable *A*gency, adaptability, and innovation.

Part III brings these concepts into practice, exploring design thinking, professional development, and curriculum integration in STE*A*⁺M education. The *Planning and Design Learning (PaDL) Framework* is introduced as a structured yet flexible model for educators to engage students in real-world problem-solving and creativity. The book concludes with case studies and research illustrating how educators can integrate STE*A*⁺M principles within regional and local educational contexts.

This book is written for educators, researchers, and policymakers seeking to transform STE*A*⁺M education into a more inclusive, interdisciplinary, and impactful field. It invites readers to rethink the purpose of education—not just as a means to develop technical expertise but as a transformative process that empowers learners to navigate an uncertain future with confidence, creativity, and *A*gency. By reframing STE*A*⁺M education as an interconnected system of learning, problem-solving, and ethical engagement, we hope to contribute to a more equitable and dynamic educational landscape. This book is both a call to action and a guide for those committed to re-envisioning education for the challenges and possibilities of the 21st century.

Clayton, Australia Gillian Kidman
Clayton, Australia Hazel Tan

Prologue: Boots on the Moon

"Maar waarom vandaag?" [1]

My Dutch mother never let me skip school. Never.

"Omdat school belangrijk is"

She adjusted the shoulder straps on my leather school satchel. We were a first-generation family settled in Melbourne, Australia. She greatly valued school, but I felt that today was even bigger than Grade 2. She hesitated, so I pointed at the television.

"The first man on the Moon! That's history!"

I dropped to the floor in frustration. The satchel was still strapped on my back and contained distinctly separate exercise books for handwriting, spelling, mathematics, science, and history, a reader, a handful of pencils, and my drawings. Everything was in its place.

She gave in and sighed, *"OK, but you stay in your school uniform."*

My learning at school followed a *strict segmentation*: writing in one time-block, mathematics in another, with science, geography and art whenever they were timetabled. *Knowledge was distilled* into subject silos that matched the exercise books in my satchel.

I was alone in making the connections.

The week before, during breakfast, a lunar lander had fallen out of my cereal box, and my teacher had mentioned the Apollo Mission. As we sat

[1] Translation: But why today? Because school is important.

patiently at her feet after a book reading, I recall her briefly holding up a magazine clipping of an astronaut.

But this was different.

I leaned in as a boot pressed into the lunar dust. The words crackled through.

"One small step ..."

I barely breathed. What would it feel like to walk in low gravity? How did they land without tipping over? How thin was their bulky spacesuit? Could they see the stars?

Later, I built my own spacesuit with a bucket and fruit box. Inside, I stuffed a soft toy kangaroo, a Vegemite sandwich, and some How-and-Why books. Mum helped me to strap it onto my back and nudged me outside. I dreamed of a capsule full of compartments, electronics that could perform the mathematics calculations I struggled with, rocket propulsion, and survival. I wondered if I had enough oxygen left to make it back.

Roaming the backyard, I mimicked the crackling squelch of a radio and its beeps. I crafted a claw from sticks, rubber bands and string to pick things up remotely. I walked slowly and deliberately across the sandpit, bending my knees as I moved around in the low gravity. I felt the sand crunch under my feet and looked back at my footprints.

At school, none of that had a place.

The next day, the teacher stood at the front of the classroom, directing and pacing students through the rote learning. We entered the room to the sound of a bell and ran outside to another. There was no tinkering or creative exploration, just explicit instructions with words to hear, copy, and write.

So, I started writing, imagining and learning on my own.

In the blank pages of a spare exercise book, I scribbled orbital diagrams for rocket trajectories and annotated mission logs. I recruited friends to help fill in the gaps. We formed a club named PLOG (People's Loony Overdone Gadgets – you need to contribute an invention to join). We added details of gadgets we could invent and experiments we could run.[2] Our sketches included gadgets, buttons, screens, gears and jargon.

[2] They were probably Rube Goldberg machines, on hindsight.

There was no space or time in class to question, imagine, cut, prototype, build, test and create. The prescriptive curriculum and siloed disciplines prohibited any further convergence.

The children were ready.

The school system was not.

Teaching was more about compliance, with knowledge confined to separate exercise books, fragmented subjects and rigid lesson plans.

Looking forward to today, we can reflect on our observations of Western classrooms and Southeast Asian workshops. In these varied STEM Education settings, we observed a common struggle: balancing traditional structures with the pressing need for, and widely advocated, interdisciplinary, creative, and cognitive learning. All these systems shared one flaw: an excessive focus on structure over curiosity and knowledge over understanding.

This book is more than a call to rethink our education.

It is a roadmap for moving our teaching and learning toward a *cohesive* and *holistic transformation experience*. By tapping into *neuroscience*, *cognitive integration*, *interdisciplinary education*, and *authentic learning experiences*, you will see how students can become challenged to think, inspired to question and make, and empowered to innovate.

As you turn the pages, you will read how disparate learning disciplines can be woven together to create a *rich tapestry of knowledge*, incorporating *cultural relevance* that resonates with every learner. You will understand how curiosity can be nurtured, allowing it to flourish. This will require *complex thinking, problem-solving,* and *adaptability*. You will better understand how learning spaces can be designed to *engage creativity, challenge understandings,* and *inspire*.

The next time a child looks up at you with wide-eyed wonder, your teaching and classroom should be ready. Not just to direct their attention but to stimulate *curiosity*, *creativity*, and *agency* that will equip them to shape the STE*A*⁺M-powered future they will inherit.

Why is that important?

Because the world does not wait.

Clayton, Australia — Roland Gesthuizen

Acknowledgements Much of the research and insights presented in this book were made possible through the then-leadership of Dr. Suhaidah Tahir at the Southeast Asian Ministers of Education Organization (SEAMEO) Regional Centre for Education in Science and Mathematics (RECSAM) and the generous funding support of the SEAMEO Secretariat in Bangkok, Thailand. We are deeply grateful for their contributions. We also sincerely thank the then-Ministers of Education of SEAMEO for their financial support and the science and mathematics curriculum specialists whose expertise and insights greatly enriched our work. Our gratitude extends to the hundreds of workshop attendees whose collaborations, perspectives, and generous exchange of ideas played a vital role in shaping our understanding. This book would not have been possible without the willingness of both creators and learners to share their knowledge and experiences. Finally, we acknowledge Simone Macdonald for her invaluable assistance with data collection. We are also grateful to Karl Gesthuizen for his meticulous work in data collation and archiving, which was instrumental in developing the disciplinary discourse typology presented in Chapter 9.

Competing Interests The authors have no competing interests to declare that are relevant to the content of this manuscript.

Contents

About the Authors

Associate Prof. Dr. Gillian Kidman is from *Monash University's Faculty of Education, School of Curriculum, Teaching and Inclusive Education.* She specialises in science, mathematics, geography education, and integrative STEM/STEAM education. She is the Interdisciplinary Approaches Research and Scholarship Group leader, Co-Chair of the International Geographical Union Commission on Geographical Education and the Editor of International Research in Geographical and Environmental Education. Gillian is a lead writer of the Australian Curriculum: Science. Her curriculum design work has won many Australian awards. Gillian is well-respected throughout Southeast Asia, with extensive experience in STEM/STEAM and numeracy education.

Dr. Hazel Tan is a Senior Lecturer at *Monash University's Faculty of Education, School of Curriculum, Teaching and Inclusive Education.* She specialises in secondary mathematics and integrative STEM/STEAM education. With a strong disciplinary foundation in mathematics and technology, she has served as a teacher, Head of Mathematics, and Educational Technology Officer at the Ministry of Education in Singapore. Since 2018, she has led teacher professional learning initiatives across Southeast Asia. Dr. Tan's research employs quantitative and mixed methods, including innovative uses of technology, such as Facebook, for research recruitment. Her work bridges practice, policy, and research to empower learners with skills and agency to address real-world challenges and shape the future.

Roland Gesthuizen is a technology and science educator with extensive experience in both secondary and tertiary education. As a senior teacher with the *Department of Education Victoria*, he leads VCE Data Analytics and Applied Computing, integrating emerging technologies into dynamic, student-centred learning environments. With a strong background in STEM education, makerspaces, and interdisciplinary collaboration, Roland specialises in digital pedagogy and computational thinking. He has shaped the curriculum, trained teachers, and advanced the use of technology in education. His research explores how technology fosters student engagement, critical thinking, and problem-solving skills. Currently a Ph.D. candidate at Monash University, he studies the pedagogy of inspiration.

Dominador D. Mangao is the former Deputy Dean of the College of Flexible Learning and ePNU, *Philippine Normal University, Manila*. He is currently a faculty member of the College of Alternative and Lifelong Learning at the same institution. Dominader served at the Southeast Asian Ministers of Education Organization (SEAMEO) Regional Centre for Education in Science and Mathematics (RECSAM) in Penang, Malaysia, as Acting Deputy Director for Research & Development and as a Specialist in Science Education in the Training & Programme for 12 years. He was the project leader of STEM Education and co-author of SEAMEO STEM Planning and Design (PaDL) Framework: Towards 21st Century Skills and Design Thinking in collaboration with Monash University, Faculty of Education. While at the Philippines' Department of Education, he served as a Supervising Education Program Specialist at the DepEd Central Office, an Education Supervisor II in Science, and an Elementary School Principal II. Dominador has received Japanese Government Scholarships for Foreign Teachers to pursue Science Education at Toyama University, Japan.

List of Figures

List of Tables

PART I

Foundations of Integrative Thinking in STEA^+M Education

This Part establishes the foundational principles that underpin integrative learning and thinking in STEA^+M education, drawing from neuroscience, disciplinary thinking, agency development, and interdisciplinary approaches. Part I examines the cognitive and theoretical dimensions of learning and teaching through four interconnected chapters, presenting new models and frameworks that inform evidence-based pedagogical strategies. The overarching theme is the necessity of a shift from traditional, compartmentalised disciplinary instruction toward more interconnected, systems-oriented approaches that equip learners with the skills needed to address complex global challenges.

The first chapter explores the neurophysiological principles that shape learning, offering insights from neuroscience into how cognitive processes such as memory, attention, and problem-solving can be optimised for educational contexts. Central to this discussion is the role of neuroplasticity, the brain's ability to rewire itself in response to learning experiences. The chapter introduces two key frameworks, the *Cognitive and Emotional Neuroeducation: Principles for STEM* (CEN-STEM) and the *Cognitive*, *Inclusive*, *Resilient*, *Creative Learning Environments* (CIRCLE) Framework for STEM Neuroeducation, which bridge the gap between neuroscience and educational practice. These frameworks emphasise inclusivity, cognitive load management, and emotional engagement, positioning neuroscience as a crucial tool for enhancing STEAM education. The chapter further highlights the need to dispel neuromyths and foster an evidence-based understanding of how brain research can inform teaching strategies. By advocating for interdisciplinary collaboration between neuroscientists and educators, this chapter presents a compelling argument for integrating neuroeducation into STEAM pedagogical design.

Building on the cognitive foundations established in Chapters 1 and 2 explores the epistemological dimensions of disciplinary thinking in STEM education. It defines disciplinary literacy as engaging in domain-specific reasoning, inquiry, and discourse. The chapter analyses how different STEM disciplines conceptualise knowledge, contrasting the structured methodologies of mathematics and science with the iterative approaches found in engineering and technology. The role of disciplinary literacy is examined in the Australian Curriculum, illustrating how disciplinary processes are codified in educational standards and shaping students' engagement with STEM content. A key argument is that while disciplinary thinking remains essential for deep learning, contemporary educational challenges require a paradigm shift that integrates disciplinary knowledge within interdisciplinary problem-solving frameworks. The chapter critiques the limitations of siloed disciplinary instruction and advocates for a more fluid, cross-disciplinary approach that prepares students to tackle real-world challenges that do not conform to disciplinary boundaries. This chapter lays the groundwork for rethinking STEM education as an integrative, systems-oriented domain by balancing epistemic rigour with interdisciplinary flexibility.

The third chapter extends the discussion by introducing the Plexus Model of STE*A*⁺M Education, which places student *A*gency at the heart of interdisciplinary learning. Traditional STEM education has often been critiqued for prioritising technical expertise while neglecting the development of ethical reasoning, civic engagement, and problem-solving capacities. This chapter argues for a more holistic approach that reframes STE*A*⁺M as an interconnected learning framework where knowledge is actively applied to address societal challenges. The Plexus Model integrates *A*gency, positioning students not as passive recipients of knowledge but as active participants in shaping the world. This conceptual shift is reinforced by an analysis of how the Australian Curriculum supports or limits the development of student *A*gency. The chapter calls for greater integration of action-oriented learning experiences, encouraging students to engage in ethical inquiry, participatory problem-solving, and community-based projects. STE*A*⁺M education is reimagined through this model as a transformative process that empowers learners to become innovators, decision-makers, and ethical leaders in an increasingly complex world.

The final chapter in Part I explores interdisciplinary thinking in STE*A*⁺M education, emphasising cognitive integration and makerspaces as tangible environments that support collaborative, hands-on learning. The discussion begins by differentiating between intradisciplinary, crossdisciplinary, multidisciplinary, interdisciplinary, and transdisciplinary approaches, clarifying how these frameworks facilitate different levels of knowledge synthesis. The chapter then introduces the *Cognitive Integration Framework*, which models how learners develop interdisciplinary understanding through analogical reasoning, conceptual mapping, and metacognitive reflection. These cognitive processes enable students to move beyond rote memorisation and fragmented knowledge acquisition toward a more holistic, interconnected understanding of complex problems. The chapter also explores the role of makerspaces as dynamic learning environments where students engage in iterative design, experimentation, and problem-solving. The *Crafting Knowledge Framework* is introduced to illustrate how hands-on engagement fosters the development of creator identity, reinforcing cognitive and tactile learning processes. This chapter argues that interdisciplinary learning is essential for preparing students for the fluid, interconnected nature of contemporary challenges. Educational institutions must adopt more,integrative pedagogies to support this shift.

The chapters in Part I present a compelling case for rethinking STE*A*⁺M education as an integrative, interdisciplinary, and *A*gency-driven endeavour. The synthesis of neuroscience, disciplinary thinking, agency development, and interdisciplinary pedagogy underscores the necessity of moving beyond traditional compartmentalised instruction. By embracing neuroeducation, epistemic flexibility, student agency, and cognitive integration, educators can design learning experiences that are cognitively rigorous, socially relevant, and transformative. This section serves as the theoretical foundation for the rest of the book, providing the conceptual tools necessary for reimagining STE*A*⁺M education as a dynamic, interconnected, and future-oriented field.

As educational systems evolve in response to global challenges, Part I calls for a paradigm shift that foregrounds the interconnected nature of knowledge, the importance of fostering adaptable learners, and the necessity of aligning teaching practices with contemporary cognitive and societal realities. This section aims to inspire educators, researchers, and policymakers to embrace a more holistic vision of education that integrates the sciences, arts, ethics, and human agency into a cohesive and transformative learning experience.

Gillian Kidman and Hazel Tan

CHAPTER 1

Neuroeducation in STEM: Bridging Cognitive Science and Pedagogy for Effective Learning

1.1 Introduction

The intersection of neuroscience, neurophysiology, and education offers transformative possibilities for understanding and enhancing learning in the STEM (science, technology, engineering, and mathematics) disciplines. As education systems strive to equip learners with the skills necessary for a rapidly evolving world, leveraging insights from neuroscience becomes crucial. Neuroeducation, an interdisciplinary field bridging neuroscience, psychology, and pedagogy, seeks to illuminate the biological underpinnings of learning and thinking (Gola, 2024). Central to this is understanding how neural processes (neuroplasticity, synaptic activity, and large-scale brain networks) influence cognition, memory, and problem-solving. By aligning educational practices with these processes, educators can foster deeper engagement, improve retention, and enhance educational outcomes (Kidman & Tan, 2025).

This chapter introduces two foundational frameworks, Cognitive and Emotional Neuroeducation: Principles for STEM (CEN-STEM) and the CIRCLE Framework for STEM Neuroeducation, that integrate neuroscience insights into educational practice. These frameworks provide actionable principles (Ho & Griffiths, 2022) for designing evidence-based, inclusive, engaging pedagogies in the STEM disciplines. The emphasis is on aligning teaching strategies with cognitive, emotional, and motivational processes to optimise learning outcomes.

G. Kidman et al., *Knowledge Generation in STEM and STEAM Education*, https://doi.org/10.1007/978-3-032-03247-8_1

Ultimately, this chapter positions neuroeducation as a critical tool for bridging the gap between scientific research and classroom application. By translating the complexities of brain science into practical pedagogical strategies, educators can empower learners (Howard-Jones, 2014) to navigate and excel in STEM education. Exploring neural mechanisms, emotional dynamics, and evidence-based teaching methods underscores the potential of neuroeducation to revolutionise educational practices, fostering academic success, lifelong learning, and innovation (Ansari et al., 2011). Through interdisciplinary collaboration and a commitment to evidence-based practices, integrating neuroscience into education promises to unlock new pathways for advancing teaching and learning in STEM fields.

1.2 Overview of Neuroeducation

Integrating neuroscience and neurophysiology into educational practices (known as neuroeducation) (Ansari et al., 2011) has implications for improving teaching and learning. Insights into how we learn and retain information provide a foundation for evidence-based teaching pedagogies and curricula to enhance educational outcomes. Neuroplasticity research highlights the brain's ability to adapt neuronal networks through targeted learning experiences that explicitly consider memory (long-term and short-term), coupled with instructional strategies like retrieval practice and the spacing effect (Yuan, 2022). Such an approach improves long-term retention and understanding.

Neurophysiology complements neuroscience by providing critical insights into how sensory inputs and physiological processes impact learning. For example, understanding the role of sensory systems like vision and hearing in classroom settings can inform the design of environments that optimise learner engagement and comprehension (Flavian, 2024). Neurological research on attention and fatigue highlights the interdisciplinary link to education and the importance of managing classroom cognitive load and implementing breaks during learning sessions to sustain focus and performance (Baratali & Zardeini, 2023). Research into stress and its physiological effects also has practical implications for education, as high-stress environments are known to impair memory and learning. Creating supportive and emotionally safe classroom settings can mitigate these effects, facilitating better educational outcomes (Baratali & Zardeini, 2023).

In addition to the impact on pedagogical selection, applying neuroscience and neurophysiology in education ensures teachers possess adequate neuroscience literacy (Torrijos-Muelas et al., 2021). This is vital to prevent the inadvertent adoption of neuromyths, misconceptions about the brain and learning that lack scientific support (Torrijos-Muelas et al., 2021). Common examples include the belief in learning styles or the idea that individuals only use 10% of their brains. Torrijos-Muelas and colleagues emphasise that educators with limited neuroeducation knowledge are more likely to accept such myths, which can detract from implementing effective teaching strategies.

Neuroeducation offers a robust framework for improving teaching methods, supporting diverse learners, and creating effective learning environments (Ansari et al., 2011). However, the success of these efforts depends on fostering neuroscience literacy among educators to ensure that their practices are grounded in scientific evidence and free from misconceptions. This alignment between science and practice can significantly enhance the quality of STEM education and its outcomes for learners.

1.2.1 Neuroeducation and STEM and STEAM Education

Neuroscience focuses on the biological processes of the nervous system, while neuropsychology examines behaviour and mental processes. Their intersection provides an understanding of how individuals learn, process information, and develop cognitively (Gola, 2024). Neuroeducation bridges the neuroscience and neuropsychology disciplines, offering insights into the neural mechanisms that underlie learning and cognition. This interdisciplinary approach enables educators to design teaching strategies that align with how the brain naturally acquires and processes information. This interdisciplinary field seeks to apply insights from brain science to educational practice, leading to an understanding of the cognitive processes behind learning (Gola, 2024).

Understanding how we learn is central to educational research and, indeed, this book as it underpins the development of good teaching strategies and curriculum design. The neuroeducation behind cognition (obtaining, understanding, retaining, and recalling information) (Yuan, 2022) that underlies learning and problem-solving in STEM education is important to enhance educational outcomes. The intricate web of neural connections and circuits in the human brain forms the biological foundation for cognitive and mental processes, influencing how we learn and think and informing strategies to optimise teaching methods and learning environments. For example, the spacing effect—the phenomenon where information is better retained when learning is distributed over time, aligns with findings on synaptic consolidation and neural plasticity (Yuan, 2022). Neurons communicate via synapses, where electrical signals are converted into chemical messages by releasing neurotransmitters. These synaptic connections form vast networks that enable the brain to process and transmit information. The plasticity of these synapses, their ability to strengthen, weaken, or form anew in response to stimuli, is crucial for learning and memory and is influenced by the learner's experiences. The landmark study by Bliss and Lømo (1973) on long-term potentiation demonstrated how repetitive stimulation of synapses strengthens neural pathways, a process fundamental to learning and memory consolidation. Neural circuits, comprising interconnected groups of neurons, govern specific cognitive functions and behaviours. The prefrontal cortex, for example, plays a central role in executive functions such as decision-making, attention, and working memory. Prefrontal circuits integrate sensory inputs with stored information to guide behaviour (Zhang & Xu, 2022). These prefrontal circuits balance excitatory and inhibitory signals (Attinger et al., 2017; Yuan, 2022), mediated by neurotransmitters like glutamate and gamma-aminobutyric acid to maintain cognitive stability and flexibility (Rodríguez-Nieto et al., 2024). Incorporating spaced repetition in lesson plans for science and mathematics can enhance long-term retention and application of knowledge. Furthermore, promoting metacognitive strategies, such as self-reflection and goal-setting, aligns with the role of the prefrontal cortex and supports the development of independent learning skills.

As discussed in the Torrijos-Muelas et al. (2021) systematic review, neuroeducation integrates insights from the allied disciplines of educational psychology, educational neuroscience, and cognitive neuroscience. We need to understand how the mind processes information, how we solve problems, how our memories are formed, and how we acquire skills and knowledge (Kidman & Tan, 2025). Our research contends that educators can enrich educational approaches by integrating these allied disciplines into education, specifically STEM and STEAM (science, technology, engineering, arts, and mathematics) education. While this integrative approach is still under development and remains a challenge (Gola, 2024), the interdisciplinary convergence involving the study of the neurophysiology behind cognitive processes and behaviour in the context of education lays the foundation for more informed and scientifically grounded pedagogical practices and approaches. With the progress in technologies and methods for studying brain functions, we can now investigate how brain research can enhance educational practices. Insights from cognitive neuroscience are being transformed into practical strategies for teaching and learning, to improve academic results by matching teaching approaches with the brain's inherent learning mechanisms (Esterman et al., 2013).

Understanding neural mechanisms offers insights into effective teaching strategies in STEAM education. For instance, problem-based learning and inquiry-driven instruction synchronise with the brain's inherent active engagement and exploration tendencies (Amjad et al., 2023). As demonstrated through synaptic changes, neural plasticity underscores the importance of iterative learning processes in mathematics and science education. Repeated exposure to complex problems and collaborative problem-solving help reinforce neural connections, deepening conceptual understanding (Abaniel, 2021). The role of neural circuits in shaping learning also extends to motivation and reward, which are critical factors in educational settings. The mesolimbic dopamine system, encompassing the ventral tegmental area and nucleus accumbens, mediates reward processing and reinforcement learning. Dopaminergic signalling reinforces behaviours associated with positive outcomes, facilitating the acquisition of new skills and habits. Lerner et al. (2021) demonstrated how dopaminergic neurons encode reward prediction errors, the difference between expected and actual outcomes, a mechanism that fine-tunes learning. In classrooms, providing timely feedback and celebrating small successes (reward prediction) can engage these reward circuits, promoting persistence and enthusiasm for learning. Beyond individual neural circuits, large-scale neural networks underpin higher-order cognitive functions. The default mode network, active during introspective tasks such as reflection or recalling memories, contrasts with the task-oriented network, which engages during goal-directed activities (Seeburger et al., 2024). The interplay between these networks reflects the brain's capacity to shift between modes of cognition, a process integral to adaptive thinking and problem-solving. This is explored further in Chapter 4. Studies using functional magnetic resonance imaging (fMRI) have shown that efficient communication between networks correlates with higher cognitive performance (Esterman et al., 2013). Pilcher

(2015) informs us that fMRI studies indicate ah ha moments or flashes of insight are stimulated in the amygdala.

The field of neurodidactics links neuroscience and education and has been researched in terms of active methodologies that centre the learner in the learning process (da Costa et al., 2024). Stimulating brain areas related to attention, memory, and cognitive processing through collaboration, problem-solving, and the practical application of knowledge activates neural circuits responsible for retention, deep understanding, and metacognitive skill development essential for academic and professional success. da Costa and colleagues (2024) promote neurodidactic practices in the sciences, specifically to improve pedagogical strategies for developing critical thinking skills and using problem-based learning in STEM/STEAM education to make teaching more engaging and effective. Research has shown that targeted interventions, such as scaffolded problem-solving and differentiated instruction, can stimulate neuroplasticity, enabling learners to overcome difficulties and achieve mastery (Abaniel, 2021; Amjad et al., 2023; Baratali & Zardeini, 2023; Kleim & Jones, 2008). Additionally, using technology in classrooms, from interactive simulations to virtual laboratories to makerspaces, leverages multisensory engagement to activate multiple neural pathways, enhancing comprehension and retention.

Emotions are pivotal in shaping engagement, learning, and thinking, and they are mediated by neural circuits involving the amygdala, hippocampus, and prefrontal cortex. The amygdala's role in emotional processing influences the salience of memories, as demonstrated in studies on emotional memory enhancement (Paré & Headley, 2023). Interactions between the amygdala and hippocampus modulate how emotionally charged events are encoded and consolidated into long-term memory in mathematics (Amjad et al., 2023). In education, fostering positive emotional experiences and creating safe, supportive environments enhance engagement and retention of material, particularly in subjects often perceived as challenging, such as mathematics and science (Baratali & Zardeini, 2023). By grasping the principles of cognition, educators can craft instructional strategies that enhance cognitive resources and promote effective learning experiences. For example, utilising cognitive load theory enables educators to develop materials and prompts that alleviate unnecessary cognitive burden, thus boosting student engagement and comprehension and minimising emotional stress.

In summary, the neural connections and circuits underpinning cognitive and mental processes are fundamental to learning and thinking and directly relevant to STEM/STEAM education. From the microscopic dynamics of synaptic plasticity to the macroscopic interactions of large-scale networks, these biological systems shape our ability to process information, adapt to new challenges, and innovate. Advances in neuroscience continue to unravel the complexities of these systems, offering insights that inform educational practices. As research progresses, the integration of neuroscience with pedagogy holds the potential to unlock new pathways for enhancing student learning and achievement.

1.2.2 *Implications for teaching and learning in the STEM Subjects*

Neuroeducation has transformative implications for STEM education subjects. Educators can develop evidence-based pedagogies and curricula that optimise teaching and learning outcomes by leveraging insights into the neural mechanisms underlying learning to create evidence-based, student-centred practices (Kidman & Tan, 2025). By emphasising neuroeducation literacy among educators, fostering supportive learning environments, and leveraging innovative teaching strategies, this interdisciplinary approach can enhance educational outcomes and prepare learners for the complexities of a rapidly evolving world (Pradeep et al., 2024; Torrijos-Muelas et al., 2021). As allied-field research continues to illuminate the neural mechanisms underlying learning, the potential for neuroeducation to revolutionise the teaching and learning of STEM subjects remains vast, promising transformative impacts on teaching and learning practices. To truly transform neuroeducation, we must distil the field into practical guidelines that improve our understanding, efficiency, consistency, and innovation in applying neuroeducation to STEM subjects (Kidman & Tan, 2025). To this end, we identified the neurocognitive processes involved in STEM teaching and learning and converted these into 10 cognitive principles of neuroeducation (see Table 1.1). Each of the 10 principles fits the definition of a 'principle'—a fundamental idea that clarifies how something operates. These principles are both generalisable, meaning they apply broadly to the understanding of STEM education, and actionable, providing clear guidance for implementation (Ho & Griffiths, 2022). The *Cognitive and Emotional Neuroeducation: Principles for STEM* (CEN-STEM) balances academic rigour and accessibility while emphasising neuroeducation's cognitive and emotional aspects in the STEM context. CEN-STEM includes a range of cognitive processes such as retrieval, attention, perception, and metacognition. Gaining insights into these cognitive mechanisms equips educators with knowledge on how learners acquire, process, and retain information, which aids in crafting effective teaching strategies and instructional materials (Chew & Cerbin, 2021). CEN-STEM also encompasses the interaction between emotions and cognitive processes in STEM learning to sustain motivation and nurture the learner's creative potential.

CEN-STEM aims to clarify how cognition influences STEM/STEAM neuroeducation. It is built on recognising the interplay between cognition, learning contexts, and outcomes related to skills, attitudes, and expectations. These principles encourage interdisciplinary methods within STEM/STEAM education and are examined throughout this book. They improve our understanding of learning across disciplines and promote strong reasoning, knowledge development, and skill improvement. Therefore, these principles are suggested as foundational guidelines for creating interdisciplinary STEM/STEAM education. As described in the following section, the 10 principles are categorised into four sets based on their underlying objectives.

Our research connects theoretical neuroeducation with practical applications in STEM/STEAM problem-solving. By clarifying the cognitive and emotional underpinnings of STEM/STEAM teaching and learning, we equip educators with actionable insights to improve their instructional methods and enhance learning experiences in STEM/STEAM classrooms (Kidman & Tan, 2025). This work supports the ideas of Chew and Cerbin (2021), advocating that teaching approaches should resonate with the brain's natural learning and developmental processes, ultimately resulting in better educational outcomes.

The CEN-STEM principles are evidence-based and derived from the intersection of cognitive science, neuroscience, and education. They explain how the brain learns, processes, retains, and applies information. The principles optimise STEM/STEAM teaching and learning strategies by aligning them with the natural mechanisms of cognition, including memory, attention, problem-solving, and emotional engagement. They aim to enhance educational practices, particularly in fields like STEM/STEAM, by providing actionable pedagogical insights into creating more effective, inclusive, and scientifically grounded learning environments.

Table 1.1 Cognitive and emotional neuroeducation: principles for STEM (CEN-STEM)

Principle	*Basic description*
Set 1—Educator knowledge and cognitive skill development	
1. Addressing Neuromyths through Educator Literacy	Neuromyths lead to ineffective teaching practices. Integrate neuroscience literacy into teacher education programs and provide professional development opportunities that focus on evidence-based neuroeducational strategies and avoid counterproductive practices
2. Metacognition and Executive Function Support	The prefrontal cortex governs executive functions such as decision-making, goal-setting, and self-regulation, which are vital for independent learning. Teach metacognitive strategies, such as self-reflection and goal-setting exercises, to support learners in planning and monitoring their learning. Use tools like checklists and guided prompts to develop executive function skills
Set 2—Emotional and motivational foundations	
3. Reward and Motivation	The brain's reward system, mediated by dopamine, reinforces learning behaviours and promotes persistence. Provide timely feedback, celebrate incremental achievements, and design STEM/STEAM challenges that offer immediate rewards or visible progress to maintain motivation and engagement
4. Emotional Safety in Learning	Stress and negative emotions impair memory and learning, while positive emotional experiences enhance engagement and retention. Foster supportive and inclusive classroom environments. Use positive reinforcement, celebrate small successes, and design collaborative STEM/STEAM activities that promote a sense of achievement and belonging

(continued)

Table 1.1 (continued)

Principle	*Basic description*
5. Emotional Engagement in Complex Subjects	Emotional salience enhances memory encoding and retrieval, particularly in challenging STEM/STEAM subjects. Use storytelling, real-world applications, and emotionally engaging scenarios to make STEM/STEAM topics relatable and memorable. Please encourage learners to connect STEM/STEAM concepts to their interests and aspirations
Set 3—Neural adaptation and information processing	
6. Neuroplasticity and Targeted Learning	Neuroplasticity adapts and reorganises neural connections, enabling learning. Targeted, repetitive learning experiences strengthen these connections. Incorporate spaced repetition and retrieval practice in STEM/STE*A*⁺M curricula to enhance long-term retention—design lesson plans with incremental challenges to build on prior knowledge and strengthen neural pathways
7. Cognitive Load Management	Excessive cognitive load impedes learning by overwhelming the brain's working memory capacity (short-term and long-term memory). Reduce extraneous cognitive load, break STEM/STEAM concepts into manageable chunks, provide scaffolding, and use visual aids. Plan regular breaks during lessons to prevent fatigue and maintain attention
8. Large-Scale Neural Network Interplay	The brain's ability to shift between introspective and goal-directed cognitive states supports adaptive thinking and problem-solving—alternate STEM/STEAM activities between focused tasks and reflective exercises. Use journaling, group discussions, and individual problem-solving to engage reflective and task-oriented cognitive modes
Set 4—Engagement and active learning	

(continued)

Table 1.1 (continued)

Principle	*Basic description*
9. Active Engagement	Active, problem-based, and inquiry-driven learning stimulates neural circuits responsible for retention, critical thinking, and deep understanding. Implement hands-on STEM/STEAM activities such as experiments, real-world problem-solving, and collaborative projects. Encourage exploration and iteration
10. Multisensory and Technology-Enhanced Learning	Engaging multiple sensory systems activates diverse neural pathways, enhancing comprehension and retention. Technology such as interactive simulations and virtual labs is leveraged in STEM/STEAM education. Multisensory tools, like augmented reality or hands-on manipulatives, can cater to diverse learning preferences

By explicitly integrating neuroscience and cognitive science insights into education, CEN-STEM provides a set of scientifically validated principles to address challenges like cognitive overload, learning disparities, and the complexity of STEM concepts. These principles bridge theoretical research and practical teaching, enabling educators to design pedagogies that align with how learners learn best and foster innovation and lifelong learning.

1.3 Cognitive and Emotional Neuroeducation: Principles for STEM (CEN-STEM)

1.3.1 Set 1: Educator Knowledge and Cognitive Skill Development

Set 1 is educator-centred, involving *Principle 1: Addressing Neuromyths through STEM Educator Literacy* and *Principle 2: Metacognition and Executive Function Support*, both critical in bridging neuroscience and education. Together, they enhance the teaching and learning experience, particularly in the challenging context of STEM education. Set 1 highlights the importance of preparing educators with scientifically correct neuroscience knowledge and equipping learners with metacognitive tools

(Pradeep et al., 2024; Torrijos-Muelas et al., 2021). These are foundational for fostering independent learners who can navigate STEM/STEAM education challenges. Educators can ensure their instructional strategies are accurate and aligned with cognitive science by addressing neuromyths and promoting neuroscience literacy. Meanwhile, fostering metacognitive skills and executive function prepares learners to become independent, lifelong learners capable of tackling the unique challenges of STEM/STEAM fields (Baratali & Zardeini, 2023). Together, these principles create a foundation for STEM/STEAM education that not only disseminates knowledge but also cultivates the skills and mindset necessary for innovation and problem-solving in a rapidly evolving world.

Principle 1: Addressing Neuromyths through STEM/STEAM Educator Literacy

This principle emphasises the need to dispel widespread neuromyths—that can undermine effective teaching practices. For example, beliefs such as the "10% brain usage myth" or the notion of rigid "learning styles" have been discredited by neuroscience yet persist in educational discourse (Torrijos-Muelas et al., 2021). These myths, when unchallenged, facilitate ineffective pedagogies. Educator literacy in neuroscience enables STEM/STEAM teachers to critically evaluate misconceptions and integrate evidence-based practices that reflect how the brain learns (Pradeep et al., 2024). Educators can cultivate more inclusive and effective learning environments by developing a stronger grasp of neural processes, including memory, attention, and learning. This focus on neuroscience literacy also cultivates a mindset of inquiry and adaptability, essential for addressing the evolving demands of STEM/STEAM education.

Principle 2: Metacognition and Executive Function Support

Complementing Principle 1, the Metacognition and Executive Function Support Principle develops learners' self-regulation and independent learning skills. Metacognition involves the learner reflecting on and regulating their learning processes, while their executive functions, such as planning, prioritising, and goal-setting, are higher-order cognitive skills essential for problem-solving and adaptability (Abaniel, 2021; Amjad et al., 2023). Supporting these skills equips learners to take ownership of their learning journey, manage complex tasks, and persist through challenges, all of which are critical for success in STEM subjects. Pedagogies involving learners developing skills to set goals, monitor their progress, and adjust their approaches foster a growth mindset and resilience, enabling them to navigate STEM/STEAM education's iterative and problem-solving nature.

1.3.2 Set 2: Emotional and Motivational Foundations

Set 2 emphasises confidence in STEM learning through *Principle 3: Reward and Motivation*, *Principle 4: Emotional Safety in Learning*, and *Principle 5: Emotional Engagement in Complex Subjects.* This set of principles represents the core elements of neuroeducation that emphasise the interactions between emotions, motivation, and cognitive processes in learning (Baratali & Zardeini, 2023; Paré & Headley, 2023). Together, they highlight the importance of addressing education's emotional and motivational dimensions, particularly relevant in STEM fields, where challenges often evoke anxiety or disengagement among learners. The STEM subjects can be intimidating (Baratali & Zardeini, 2023); however, fostering emotional safety, leveraging motivational strategies, and connecting topics to learners' interests create a supportive and engaging learning environment. These principles help sustain confidence and interest and improve retention where intimidation and disengagement are common barriers (Baratali & Zardeini, 2023). Emotional safety, leveraging the brain's reward systems, and making learning emotionally meaningful, educators create STEM/STEAM environments that encourage curiosity and persistence. Together, this enhances the acquisition of knowledge and the motivation and confidence necessary for learners to navigate and excel in the dynamic and often demanding STEM fields.

Principle 3: Reward and Motivation

The principle of Reward and Motivation is grounded in the role of the brain's dopamine-mediated reward system in reinforcing learning behaviours (Lerner et al., 2021). When learners experience a sense of achievement, whether through solving a problem, completing a task, or receiving positive feedback, the release of dopamine strengthens neural pathways associated with the learned behaviour. This neurochemical process reinforces the learning and fosters a positive emotional association with the activity (Baratali & Zardeini, 2023). In STEM/STEAM education, where problem-solving and experimentation are central, incorporating motivational strategies such as immediate feedback, achievable goals, and gamified learning can harness this reward system. This approach sustains learners' intrinsic motivation and encourages perseverance through challenges, ultimately building a foundation for lifelong engagement with STEM disciplines.

Principle 4: Emotional Safety in Learning

Equally important is the Emotional Safety in Learning principle, which focuses on creating supportive environments that reduce stress and promote cognitive and emotional well-being. Research shows that stress and anxiety impair memory consolidation and executive functions, hindering learning (Baratali & Zardeini, 2023). When learners feel emotionally safe, their brains are better equipped to focus, process information, and retain knowledge. In STEM/STEAM education, emotional safety is fostered by building a classroom culture of respect, empathy, and encouragement. Pedagogies for scaffolding complex tasks, normalising mistakes as part of the learning process, and helping equitable access to resources can reduce intimidation and create a space where learners feel confident to take intellectual risks (Baratali & Zardeini, 2023).

Principle 5: Emotional Engagement in Complex Subjects

The Emotional Engagement in Complex Subjects principle highlights the role of emotional salience in improving attention and retention when dealing with challenging or abstract topics. Neuroscience research indicates that emotionally charged experiences activate the amygdala, enhancing memory encoding and increasing focus (Paré & Headley, 2023). In the context of STEM/STEAM education, connecting complex subjects to real-world applications or personal interests evokes curiosity and emotional investment (Baratali & Zardeini, 2023). Planning lesson topics concerning global challenges, such as climate change or medical innovation, makes abstract concepts more tangible and meaningful. This sparks engagement and deepens learners' understanding by linking theoretical knowledge to practical implications.

1.3.3 Set 3: Neural Adaptation and Information Processing

The third set focuses on *Principle 6: Neuroplasticity and Targeted Learning*, *Principle 7: Cognitive Load Management*, and *Principle 8: Large-Scale Neural Network Interplay*, emphasising the importance of designing teaching strategies that align with the brain's adaptive and processing capacities. These principles highlight how targeted interventions and understanding neural mechanisms can enhance learning, particularly in STEM/STEAM education, where mastering complex and abstract concepts requires strategic cognitive engagement. This set emphasises how STEM/STEAM educators can optimise teaching strategies to align with neural mechanisms (Kidman & Tan, 2025). Managing cognitive load, facilitating the strengthening of neural connections, and designing activities that engage various neural networks (Baratali & Zardeini, 2023) are crucial for a deep understanding of complex STEM concepts. These principles emphasise the importance of aligning teaching strategies with the brain's adaptive and processing mechanisms. STEM/STEAM educators can optimise learning outcomes by designing experiences that capitalise on neuroplasticity, manage cognitive load, and engage large-scale neural networks. Doing so enables the learner to grasp, retain, and apply complex concepts. This fosters an understanding of STEM subjects while equipping learners with the cognitive flexibility to address real-world challenges in innovative ways.

Principle 6: Neuroplasticity and Targeted Learning

Neuroplasticity and targeted learning refer to the brain's ability to adapt and reorganise large neural networks by forming and strengthening neural pathways in response to learning experiences (Gola, 2024). Targeted learning involves engaging learners in purposeful and specific activities that challenge their prior knowledge while building new skills and understandings. STEM/STEAM education involves problem-based learning, scaffolded instruction, and iterative practice, allowing learners to refine their understanding incrementally (Amjad et al., 2023). STEM/STEAM educators leverage neuroplasticity to consolidate learning and create durable neural networks by providing repeated and meaningful exposure to concepts, ensuring learners retain and apply knowledge over time.

Principle 7: Cognitive Load Management

Cognitive load management addresses the brain's finite capacity to process information. When the cognitive demands of a task exceed the brain's working memory capacity, learning will break down, leading to frustration and disengagement (Baratali & Zardeini, 2023). This emphasis is for educators to structure learning in ways that optimise mental resources. For STEM/STEAM education, managing cognitive load involves breaking complex problems into smaller, manageable steps, using visual aids to reduce abstraction, and scaffolding lessons to minimise distractions. Chunking information, sequencing tasks, and allowing time for reflection are strategies that assist learners in focusing on core concepts without becoming overwhelmed. By aligning instructional design with the brain's information processing capacity (Howard-Jones, 2014), educators enable learners to engage with the STEM/STEAM material.

Principle 8: Large-Scale Neural Network Interplay

Large-scale neural network interplay concerns how the brain switches between introspective, self-referential processes and goal-directed cognition, depending on the demands of a task. The default mode network facilitates reflection, creativity, and big-picture thinking, while the task-positive network supports focus, problem-solving, and action-oriented tasks (Seeburger et al., 2024). Effective STEM/STEAM education engages both networks by integrating opportunities for reflection and goal-directed learning. Designing STEM/STEAM lessons that encourage learners to alternate between brainstorming innovative solutions and systematically implementing those solutions mirrors how the brain naturally functions. This interplay enhances problem-solving and nurtures creativity and adaptability, key skills for success in STEM disciplines (Mahama, 2024).

1.3.4 Set 4: Engagement and Active Learning

Principle 9: Active Engagement and *Principle 10: Multisensory and Technology-Enhanced Learning* highlight the importance of fostering immersive and interactive learning experiences that actively involve learners in the educational process. These cognitive principles draw on research demonstrating that active participation and multisensory stimulation in makerspaces enhance understanding, retention, and application of knowledge, particularly in STEM/STEAM education, where abstract concepts and problem-solving are central. Active and multisensory learning strategies make abstract STEM concepts more tangible and accessible. Makerspaces promote critical thinking, innovation, and collaborative problem-solving by directly engaging learners in learning (Gesthuizen et al., 2020). Together, these principles emphasise the importance of designing STEM learning experiences that are interactive, dynamic, and multisensory. Educators fostering learner participation significantly enhance cognitive processing, resulting in a better grasp of STEM concepts. Incorporating multisensory tools and technology further enhances this process by making abstract concepts more tangible and accessible, inspiring curiosity and innovation. These strategies ultimately prepare learners not only to excel in STEM subjects but also to become critical thinkers and problem-solvers capable of addressing the complex challenges of the modern world.

Principle 9: Active Engagement

Active engagement emphasises the transformative influence of hands-on, problem-based, and inquiry-driven learning. This principle aligns with the idea that learners learn best when constructing their understanding rather than passively receiving information. STEM/STEAM activities such as conducting experiments, solving real-world problems, and engaging in collaborative projects enable learners to apply theoretical knowledge in practical contexts (Gesthuizen et al., 2020). In STEM/STEAM education, these strategies foster critical thinking, innovation, and the ability to synthesise information across disciplines. For example, when learners design and test prototypes in an engineering challenge, they deepen comprehension and cultivate resilience, adaptability, and teamwork skills vital for success in STEM fields (Gesthuizen et al., 2020).

Principle 10: Multisensory and Technology-Enhanced Learning

Multisensory and technology-enhanced learning focuses on learning that engages multiple senses, such as sight, sound, and touch, to create richer neural connections, leading to improved memory and understanding. For instance, using 3-dimensional models to visualise molecular structures or interactive simulations to explore physics concepts allows learners to experience abstract ideas tangibly and dynamically. Additionally, technologies used in makerspaces, virtual reality, augmented reality, and digital simulations provide the opportunity to be immersed in complex STEM/STEAM concepts that are sometimes challenging to convey through traditional methods. Such learning is more accessible and engaging, and it also helps learners bridge the gap between theory and real-world applications (Gesthuizen et al., 2020).

1.4 The CIRCLE Framework for STEM Neuroeducation

The CIRCLE (Cognitive, Inclusive, Resilient, Creative Learning Environments) Framework for STEM Neuroeducation (see Fig. 1.1) represents the interconnected approach to STEM education, where each CEN-STEM principle supports the others. The 10 cognitive principles' relationships, overlaps, and key ideas are outlined below. The CIRCLE Framework for STEM Neuroeducation showcases the interconnected nature of evidence-based brain-based learning practices, inclusivity, emotional engagement, and the preparation of future innovators in STEM education.

In the *CIRCLE Framework for STEM Neuroeducation*, evidence-based practice (the foundation layer) forms the innermost circle, providing the foundation and ensuring all other layers are grounded in neuroscience research. This layer supports the first two CEN-STEM principles, emphasising **educator knowledge and cognitive skill development.** Evidence-based practice aligns with promoting inclusivity (social-emotional layer) (Principles 3, 4, and 5). The promoting inclusivity layer emphasises the emotional and social factors that impact learning. The alignment with brain function (cognitive layer) (Principles 6, 7, and 8) ensures that social-emotional and cognitive strategies are grounded in science. This layer bridges evidence-based practices with cognitive strategies for enhancing learning efficiency. The tackling complexity in STEM subjects (instructional layer) represents strategies (Principles 9 and 10) that make teaching and learning STEM subjects more engaging and manageable. Preparing future innovators (skills-development layer) depends on and interacts with all other layers, integrating emotional, cognitive, and instructional strategies to produce adaptable, creative, and resilient problem-solvers, focusing on developing critical STEM skills and

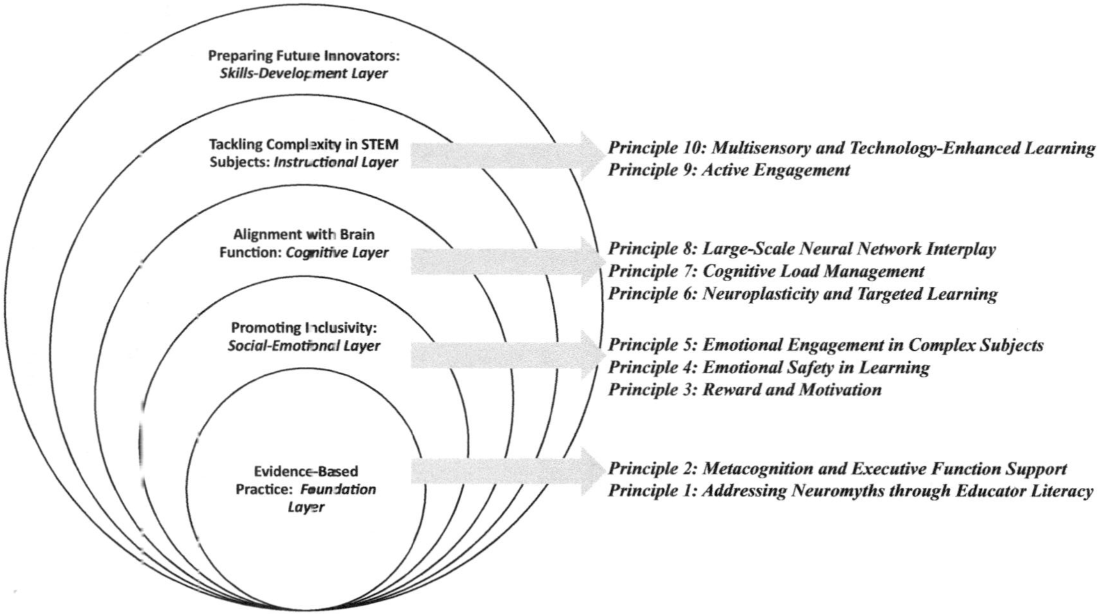

Fig. 1.1 CIRCLE (cognitive, inclusive, resilient, creative learning environments) framework for STEM neuroeducation (*Source* Authors)

preparing learners for real-world problem-solving. The learner experience is central to the CIRCLE Framework for STEM Neuroeducation, symbolising the ultimate goal of creating a supportive, engaging, and effective STEM learning environment.

The CEN-STEM Framework and CIRCLE Framework for STEM Neuroeducation have broad implications. As researchers, we need to investigate the interactions between the layers of the CIRCLE Framework for STEM Neuroeducation in diverse classroom settings to refine and expand the framework's application and the CEN-STEM Framework. As policymakers, the alignment of STEM curriculum design with this holistic framework will ensure resources and professional development for educators to address cognitive, emotional, and practical aspects. As educators, the implications of using the framework to assess lesson plans and teaching practices are to ensure they incorporate all five layers—evidence-based strategies, inclusivity, brain alignment, methods for tackling complexity, and skill development.

1.5 Conclusion

This chapter emphasises the transformative role of neuroscience and neurophysiology in reshaping STEM/STEAM education by exploring the neural foundations of learning and cognition. At the heart of this understanding lies the concept of neuroplasticity, emphasising the brain's capacity to adjust based on learning experiences (Amjad et al., 2023). This understanding lays the groundwork for creating teaching strategies that are evidence-based and resonate with the brain's methods of acquiring, processing, and applying knowledge. The integration of neuroscience into education, or neuroeducation, bridges scientific research with classroom practice (Gola, 2024). By focusing on how neural circuits and brain networks support cognitive functions such as memory, attention, and problem-solving, educators can apply strategies like spaced repetition, retrieval practice, and metacognitive exercises to enhance learning outcomes (Yuan, 2022). Recognising the critical role of emotions and motivation (governed by neural pathways like the mesolimbic dopamine system and the amygdala), effective teaching practices must also address the emotional and motivational dimensions of learning, fostering supportive and engaging environments that sustain interest and resilience.

Frameworks such as the CEN-STEM principles and the CIRCLE Framework for STEM Neuroeducation offer structured methods for applying neuroscience in education. These frameworks emphasise inclusivity, emotional and cognitive alignment, and active, problem-based learning experiences that engage diverse learners. Addressing neuromyths (common misconceptions about the brain) through improved neuroscience literacy among educators is essential for informed, science-based pedagogical practices (Torrijos-Muelas et al., 2021).

The interdisciplinary nature of neuroeducation holds immense potential for advancing education (Kidman & Tan, 2025). Collaboration among neuroscientists, educators, and policymakers, supported by emerging technologies like functional imaging and neurofeedback, can refine our understanding of learning processes and enable more personalised STEM pedagogies. Educators can create inclusive, engaging, and effective STEM/STEAM learning environments by aligning educational practices with biological learning mechanisms. This integration improves teaching methods and gives learners the skills and resilience essential for success in a complex world. The chapter calls for leveraging neuroscience to unlock learners' full potential in the STEM disciplines.

References

Abaniel, A. (2021). Enhanced conceptual understanding, 21st century skills and learning attitudes through an open inquiry learning model in Physics. *Journal of Technology and Science Education, 11*(1), 30–43. https://doi.org/10.3926/joste.1004

Amjad, A. I., Habib, M., Tabbasam, U., Alvi, G. F., Taseer, N. A., & Noreen, I. (2023). The impact of brain-based learning on students' intrinsic motivation to learn and perform in mathematics: A neuroscientific study in school psychology. *International Electronic Journal of Elementary Education, 16*(1), 111–122. https://www.iejee.com/index.php/IEJEE/article/view/2076

Ansari, D., Coch, D., & De Smedt, B. (2011). Connecting education and cognitive neuroscience: Where will the journey take us? *Educational Philosophy and Theory, 43*(1), 37–42. https://doi.org/10.1111/j.1469-5812.2010.00705.x

Attinger, A., Wang, B., & Keller, G. B. (2017). Visuomotor coupling shapes the functional development of mouse visual cortex. *Cell, 169*(7), 1291–1302. https://doi.org/10.1016/j.cell.2017.05.023bliss&Como1973synapse strength

Baratali, M., & Zardeini, A. Z. (2023). Identification of curriculum objectives of brain-based education and positive education. *Journal of Fundamentals of Mental Health, 25*(2), 69–79. https://doi.org/10.22038/jfmh.2023.22253

Bliss, T. V., & Lømo, T. (1973). Long-lasting potentiation of synaptic transmission in the dentate area of the anaesthetized rabbit following stimulation of the perforant path. *Journal of Physiology, 232*(2), 331–356. https://doi.org/10.1113/jphysiol.1973.sp010273

Chew, S. L., & Cerbin, W. J. (2021). The cognitive challenges of effective teaching. *The Journal of Economic Education, 52*(1), 17–40. https://doi.org/10.1080/00220485.2020.1845266

da Costa, H. N., da Silva, F. P., de Almeida Matos, T., dos Santos Cruz, G., de Arruda Lima, S. M., de Mascena Costa, L. A., & Nobre, A. P. M. C. (2024). Active methodologies as a neurodidactic resource in the teaching and learning of science and biology. *Observatório De La Economía Latinoamericana, 22*(12), e8296–e8296.

Esterman, M., Noonan, S. K., Rosenberg, M., & DeGutis, J. (2013). In the zone or zoning out? Tracking Behavioral and Neural Fluctuations during Sustained Attention. *Cerebral Cortex, 23*(11), 2712–2723. https://doi.org/10.1093/cercor/bhs261

Flavian, H. (2024). Transdisciplinary teaching for learners with sensory impairments and challenges. In *Transdisciplinary Teaching in Inclusive Schools. Transdisciplinary Perspectives in Educational Research*, vol 8. Springer, Cham. https://doi.org/10.1007/978-3-031-52509-4_5

Gesthuizen, R., Kidman, G., Tan, H., Mangao, D., & Macdonald, S. (2020). STEM inspiration: A phenomenological investigation exploring beyond the solution. In E. Creely, J. Southcott, K. Carabott, & D. Lyons (Eds.), *Phenomenological Inquiry in Education: Theories, Practices, Provocations and Directions* (1st ed., pp. 225–242). Routledge. https://doi.org/10.4324/9780429285646-15

Gola, G. (2024). *Educational neuroscience in the classroom.* Cambridge Scholars Publishing.

Ho, M. K., & Griffiths, T. L. (2022). Cognitive science as a source of forward and inverse models of human decisions for robotics and control. *Annual Review of Control, Robotics, and Autonomous Systems, 5*(1), 33–53. https://doi.org/10.1146/annurev-control-042920-015547

Howard-Jones, P. A. (2014). Neuroscience and education: Myths and messages. *Nature Reviews Neuroscience, 15*(12), 817–824. https://doi.org/10.1038/nrn3817

Kidman, G. & Tan, H. (2025). Exploring cognitive principles of instructional models and problem-solving processes in the STEM disciplines. In T. Barkatsas & W. Goff (Eds.), *Advances in mathematics and STEM education.*

Kleim, J. A., & Jones, T. A. (2008). Principles of experience-dependent neural plasticity: Implications for rehabilitation after brain damage.

Journal of Speech, Language, and Hearing Research, 51(1), S225–S239. https://www.proquest.com/scholarly-journals/principles-experience-dependent-neural-plasticity/docview/232334661/se-2

Lerner, T. N., Holloway, A. L., & Seiler, J. L. (2021). Dopamine, updated: Reward prediction error and beyond. *Current Opinion in Neurobiology, 67*, 123–130. https://doi.org/10.1016/j.conb.2020.10.012

Mahama, I. (2024). The role of culture and environment in shaping creative thinking. In L. Barlach (Ed.), *Creativity in contemporaneity*. IntechOpen. https://doi.org/10.5772/intechopen.1007481

Paré, D., & Headley, D. B. (2023). The amygdala mediates the facilitating influence of emotions on memory through multiple interacting mechanisms. *Neurobiology of Stress, 24*, https://doi.org/10.1016/j.ynstr.2023.100529

Pilcher, J. (2015). A Modified Delphi Study to Define "Ah Ha" Moments in Education Settings. *Educational Research Quarterly, 38*(4), 51–67. https://www.proquest.com/scholarly-journals/modified-delphi-study-define-ah-ha-moments/docview/1692023351/se-2

Pradeep, K., Sulur Anbalagan, R., Thangavelu, A. P., Aswathy, S., Jisha, V. G., & Vaisakhi, V. S. (2024). Neuroeducation: understanding neural dynamics in learning and teaching. *Frontiers in Education, 9*. https://doi.org/10.3389/feduc.2024.1437418

Rodríguez-Nieto, G., Alvarez-Anacona, D. F., Mantini, D., Edden, R. A. E., Oeltzschner, G., Sunaert, S., Swinnen, S. P. (2024). Association between inhibitory-excitatory balance and brain activity response during cognitive flexibility in young and older individuals. *Journal of Neuroscience, 44*(36), https://doi.org/10.1523/JNEUROSCI.0355-24.2024

Seeburger, D. T., Xu, N., Ma, M., Larson, S., Godwin, C., Keilholz, S. D., & Schumacher, E. H. (2024). Time-varying functional connectivity predicts fluctuations in sustained attention in a serial tapping task. *Cognitive, Affective, & Behavioral Neuroscience, 24*(1), 111–125. https://doi.org/10.3758/s13415-024-01156-1

Torrijos-Muelas, M., González-Víllora, S., & Bodoque-Osma, A. R. (2021). The persistence of neuromyths in the educational settings: A systematic review. *Frontiers in Psychology, 11*, https://doi.org/10.3389/fpsyg.2020.591923

Yuan X. (2022). Evidence of the spacing effect and influences on perceptions of learning and science curricula. *Cureus, 14*(1), https://doi.org/10.7759/cureus.21201

Zhang, N., & Xu, N. L. (2022). Reshaping sensory representations by task-specific brain states: Toward cortical circuit mechanisms. *Current Opinion in Neurobiology, 77*, https://doi.org/10.1016/j.conb.2022.102628

CHAPTER 2

Disciplinary Thinking in STEM: Balancing Epistemologies for Integrated Education

2.1 Introduction

A discipline consists of a body of knowledge and practices for acquiring, confirming, implementing, preserving, and reproducing that knowledge (Post, 2009). Disciplines vary in methodological coherence and subject matter, with their knowledge claims shaped by distinct epistemological constructs (Yoon et al., 2024). To achieve disciplinary literacy, educational experiences must reflect these epistemological constructs, including beliefs about knowledge, inquiry practices, reasoning strategies, key concepts, representational forms, and discourse structures (Goldman et al., 2016). The ability to assess the relevance of information and integrate it into evolving explanations defines disciplinary literacy within a given field (Goldman et al., 2016). In contrast to general content area literacy, disciplinary literacy involves grasping how knowledge is created, shared, and assessed within particular fields (Shanahan & Shanahan, 2017).

Disciplinary literacy and disciplinary thinking are symbiotic. While disciplinary literacy provides the tools and language necessary for engaging deeply within a field, disciplinary thinking informs how this literacy is applied in problem-solving and knowledge construction. Disciplinary thinking emphasises domain-specific approaches to inquiry and analysis, balancing substantive and procedural knowledge (Rantala, 2012). Both are essential for fostering expertise, with educators playing

G. Kidman et al., *Knowledge Generation in STEM and STEAM Education*, https://doi.org/10.1007/978-3-032-03247-8_2

a critical role in developing students' capacity to think and communicate within disciplinary frameworks. Teachers foster disciplinary literacy and disciplinary thinking to prepare students to engage meaningfully with subject matter.

Disciplinary thinking is the intellectual foundation of STEM (science, technology, engineering, and mathematics) education. It shapes how knowledge is structured, applied, and understood. Discipline thinking ensures conceptual depth while supporting interdisciplinary connections crucial for real-world problem-solving. Traditionally, classroom learning has reinforced compartmentalised disciplinary expertise, yet contemporary challenges demand an integrated, systems-oriented approach. STEM education requires educators to bridge disciplinary traditions while preserving the rigour of discipline-specific inquiry (Wudarczyk et al., 2021).

A central theme of this chapter is the tension between the traditional, discipline-bounded nature of knowledge within classroom contexts and the need for integrated, systems-oriented approaches to address real-world challenges. While classroom learning often reinforces compartmentalised subject expertise, the evolving landscape of contemporary education calls for a paradigm shift where disciplines function as interconnected lenses for examining complex societal and environmental issues. With its interdisciplinary emphasis, STEM education requires educators to navigate this shift, facilitating connections between disciplinary traditions while preserving the rigour of domain-specific inquiry (Wudarczyk et al., 2021). Another chapter theme examines the epistemological differences in the STEM subjects. In a comparative exercise, we consider the disciplinary differences in solving wicked problems and developing social agency in students.

2.2 Defining Disciplinary Knowledge: Foundations, Boundaries, and Classroom Contexts

Disciplinary knowledge has formed the foundation of education, encapsulating the principles, methodologies, and epistemologies unique to specific fields of study. Disciplines and sub-disciplines have existed for "as long as humans have been learning from each other" (McElreavy et al., 2016, para. 10). McElreavy et al. indicate that mathematics and music were the earliest recorded academic disciplines, dating back to the Greeks and Socrates. As we know them today, the disciplinary systems date back to the 1800s. During this time, education became more closely related to the "needs of everyday life by teaching disciplines such as history, geography, geometry, algebra, modern languages, navigation and astronomy" (McElreavy et al., 2016, para. 16). Over the years, sciences have evolved into distinct sub-disciplines such as physics, chemistry, mathematics, medicine, biology, and engineering, responding to the increasing demand for scientists trained in applied science. Over the same period, the humanities discipline has remained relatively stable as the home of philosophy, with all languages, history, art history, and religious studies (McElreavy et al., 2016) as sub-disciplines. As some disciplines change as knowledge emerges in response to societal changes, other disciplines remain stable.

In education, disciplines such as science, mathematics, technology, and the humanities retain distinct identities shaped by their evolution, purposes, and practices (Ings, 2019). While these disciplines are deeply interconnected in STEM and STEAM (science, technology, engineering, arts, and mathematics) education, their unique disciplinary characteristics define their epistemologies, boundaries, and classroom applications (Yoon et al., 2024). The following synthesis explores the nature of disciplinary knowledge, the challenges, and opportunities presented by its boundaries, and how it is represented and applied in educational settings.

2.2.1 *The Epistemological Nature of Disciplinary Knowledge in the STEM Subjects*

Disciplinary knowledge is the structured body of concepts, theories, and practices that define a particular field. Yoon et al. (2024) define disciplinary knowledge through an educational lens as "the knowledge base that Shulman ... and various scholars have referred to as content knowledge or subject-matter knowledge ... the skills and concepts generated in disciplinary investigations of theories and practices that members of a discipline would undertake" (p. 4). It encompasses core concepts, methods, tools, and epistemological foundations of knowledge creation and validation. In the STEM subjects, each discipline has a unique identity rooted in its specific goals, methodologies, and approaches to problem-solving. Epistemology, the study of knowledge and how we come to know, is foundational in understanding how different disciplines approach learning and inquiry. Each discipline in STEM education has distinct epistemological underpinnings that shape its pedagogy, methodologies, and knowledge validation. This section explores the individual epistemologies of these subjects and provides a comparative analysis.

If we look at the early work of Habermas (1971), we see that the sciences seek to predict and control the world. Scientific knowledge can be tested against the rules of prediction and control. Science is characterised by its commitment to understanding the natural world through empirical observation and experimentation (Goldman et al., 2016). The scientific method involves hypothesis testing, data analysis, and iterative refinement of theories, which is central to its epistemology. "Thus, inherent in the epistemology of science is its tentativeness: Extant theories, models, and explanations reflect the best accounts given the results of inquiries to date, but these are expected to change with new inquiries" (Goldman et al., 2016, p. 230). Core concepts form the framework for inquiry-based learning. Tools like microscopes, spectrometers, and data analysis software are essential for gathering precise measurements and validating findings. Science education emphasises curiosity, critical thinking, and analytical skills, aiming to explain and predict phenomena to advance technology, address environmental challenges, and improve societal well-being (Bayani et al., 2025). Science education emphasises inquiry-based learning, where students ask questions, design experiments, and analyse data. This approach mirrors scientific practices, fostering critical thinking and problem-solving. For example, students might explore energy concepts through hands-on experiments, such as observing heat transfer. These activities link theory to practical uses, making learning both relevant and engaging.

Mathematics is characterised by abstract reasoning, logic, and precision, built upon a foundation of inductive and deductive reasoning, proofs, axioms, and theorems (Annizar et al., 2020; Khan & Arabia, 2015), essential pillars of the discipline's structure (Suyitno, 2010). Mathematics is distinguished by its abstract objects, reliance on agreements, deductive thinking, and consistency with previously established truths (Suyitno, 2010). These characteristics should be reflected in school mathematics to maintain the integrity of the subject (Suyitno, 2010). Mathematics employs creativity, intuition, and logic to discover new concepts and solve everyday challenges. Mathematicians use inductive reasoning to draw probable conclusions and deductive reasoning to validate conclusions based on accepted hypotheses (Khan & Arabia, 2015). As a universal language, mathematics can connect diverse fields and foster analytical thinking, enabling learners to engage with real-world challenges

(Tsebella et al., 2022). It also needs recognition as a critical discipline in STEM education. In mathematics classrooms, teaching often begins with modelling and practice to develop procedural fluency. Conceptual understanding is then deepened through real-world applications, such as analysing statistical data or solving open-ended problems. Visual aids, manipulatives, and technology, such as graphing calculators and dynamic geometry software, help students grasp abstract concepts and improve problem-solving skills.

Technology is distinguished by its focus on practical application and innovation. Rooted in problem-solving, it integrates theoretical understanding with real-world implementation. Technology is characterised by its practical application and problem-solving approach, integrating theoretical knowledge with real-world implementation. Design thinking, a key aspect of technological innovation, emphasises abductive thinking, iterative refinement (Carlgren & BenMahmoud-Jouini, 2022; Suhaimi et al., 2024), prototyping, and multidisciplinary collaboration to produce innovative solutions. This approach differs from traditional knowledge work, employing an interpretive, emergent, and embodied epistemology rather than a purely rational, analytic one (Rylander, 2009). Design epistemology foregrounds creativity and collaboration in knowledge creation, utilising Popper's three-world ontology to explain how conceptual artefacts emerge from personal thinking and experiences (Kühne, 2024). The distinct epistemology of technology, known as techno-epistemology, is integral to human practice and cannot merely be diminished to a scientific perspective. This perspective challenges the notion of technology as merely applied science and highlights the importance of understanding technology's distinct way of knowing in education (Eijck & Claxton, 2009). Technology education fosters creativity, collaboration, and adaptability, preparing students to design and implement innovative solutions. Technology education relies on project-based learning, where students design and create solutions to real-world challenges. Collaborative tasks like building prototypes or coding applications integrate technical skills with creativity. Teachers act as facilitators, guiding students through design processes and encouraging experimentation and iteration. This hands-on approach reflects technology's practical and innovative nature.

Despite their differences, these disciplines are interdependent. Science and mathematics rely on data analysis and modelling; mathematical tools test scientific hypotheses, while scientific questions inspire mathematical inquiry. Similarly, mathematics underpins technological advancements, while technology enhances mathematical education through visualisation tools. Science and technology collaborate symbiotically as scientific discoveries drive technological innovation and technology expands scientific research capabilities. Understanding these distinctions and connections is crucial for designing curricula that prepare students for interdisciplinary problem-solving in a complex world. We explore the manifestations of disciplinary epistemologies in Chapter 8.

In contrast, the humanities primarily foster "the intersubjectivity of mutual understanding in ordinary-language communication and in action according to common norms" (Habermas, 1971, p. 176). Therefore, unlike the sciences, the humanities do not address problems similarly. Nonetheless, they offer insights into human meaning, understanding, and cooperation, which are becoming more unstable (Post, 2009). Thus, knowledge is more controversial when considered through a humanities lens than a science lens.

Table 2.1 summarises the epistemological aspects of the four STEM disciplines. The epistemologies are distinct but interrelated in their contribution to understanding and problem-solving. While these four disciplines share reasoning and knowledge acquisition aspects, their epistemological foundations, methodologies, validation processes, and objectives differ. This section compares and contrasts these aspects, highlighting each discipline's unique contributions and interconnectedness. We include in the discussion the role of students in developing social agency to help address wicked problems in their communities.

The ***nature of knowledge*** is quite varied among the STEM disciplines. Science is rooted in empirical and theoretical knowledge (Goldman et al., 2016). It seeks to explain natural phenomena through systematic observations and formulating theories to predict and explain occurrences. Technology, by contrast, is more applied and practical, focusing on creating tools, systems, and solutions to improve human life (Rylander, 2009; Suhaimi et al., 2024). Engineering falls between these two, emphasising applied and analytical knowledge to create optimised solutions. Mathematics is a discipline based on abstract and logical knowledge, developing axiomatic structures that underpin scientific and engineering principles (Annizar et al., 2020). While science builds knowledge through systematic observation and hypothesis testing, technology and engineering apply that knowledge to create tangible outcomes (Bayani et al., 2025). Mathematics, meanwhile, provides the logical framework within which science, technology, and engineering operate (Tsebella et al.,

Table 2.1 Epistemological aspects of the STEM disciplines

Aspect	*Science*	*Technology*	*Engineering*	*Mathematics*
Nature of Knowledge	Empirical and theoretical	Applied and practical	Applied and analytical	Abstract and logical
Inquiry Process	Scientific method (experimentation, hypothesis testing)	Design and iteration, hands-on problem-solving	Structured design process, prototyping	Axiomatic-deductive reasoning
Knowledge Validation	Experimentation, peer review, falsifiability	Functionality, usability, ethical considerations	Testing, optimisation, trade-offs	Proof, logical consistency, universal applicability
Objective	Understand natural phenomena	Create useful artefacts and systems	Design and optimise solutions	Establish logical truths and structures
Role of Creativity	Hypothesis generation, innovative experiments	Innovation, product development, adaptation	Design thinking, problem-solving within constraints	Pattern recognition, abstract thinking
Interdisciplinary Connection	Draws from mathematics and engineering	Integrates science, engineering, and social factors	Applies science, mathematics, and technology	Supports all STEM fields with foundational structures

2022). A critical aspect that challenges traditional knowledge structures in the STEM disciplines is the presence of wicked problems. Such problems are complex, multifaceted issues with no definitive solution. Examples include climate change, urban sustainability, global health crises, and poverty. They require interdisciplinary approaches, blending scientific inquiry, technological innovation, engineering design, and mathematical modelling. Unlike traditional problems with clear resolutions, wicked problems involve evolving constraints, competing interests, and uncertain outcomes, necessitating adaptability and creative problem-solving across the STEM disciplines (McCune et al., 2023). To address these issues effectively, students must develop **social agency**, learning that they are responsible as active citizens to contribute to their communities. By engaging with real-world problems through STEM/ STE*A*⁺M , students can foster a sense of civic duty and collective responsibility, using their disciplinary knowledge to drive meaningful change in society (Sarid & Goldman, 2021). We explore social agency further in Chapter 3, where STEM becomes STE*A*⁺M and *A* is *A*gency.

Similarly, the ***inquiry process*** also varies considerably. Science follows the scientific method, which includes hypothesis formation, controlled experimentation, and systematic observation. It relies heavily on empirical evidence to build knowledge and refine theories (Kidman & Casinader, 2017). Conversely, technology employs a design and iteration approach, incorporating hands-on problem-solving, testing, and modification to create functional innovations (Eijck & Claxton, 2009). Engineering shares elements with science and technology but follows a structured design process. Engineers identify problems, create prototypes, and refine their designs based on feasibility, efficiency, and cost considerations. Unlike the other three, mathematics relies on axiomatic-deductive reasoning (Annizar et al., 2020). It derives truths from established axioms through logical proof, meaning its validity depends not on empirical evidence but on logical consistency. This distinction in inquiry processes means that while science is investigative and explanatory, technology and engineering are solution-driven, and mathematics is deductive and abstract. In tackling wicked problems, the inquiry process becomes even more dynamic, as it often requires blending empirical research with applied design thinking and computational modelling (McCune et al., 2023). Science contributes by investigating root causes and evidence-based insights, technology provides iterative and adaptive solutions, engineering designs scalable and feasible implementations, and mathematics formulates predictive models to guide decision-making under uncertainty. Engaging students in these processes allows STEM education, mainly STEA^+M focused on societal wicked problems, to empower students to take charge of their learning and apply their knowledge to socially relevant issues (Mathers et al., 2024). Encouraging project-based learning, community engagement, and problem-solving activities allows students to see themselves as change-makers capable of addressing wicked community problems.

Knowledge validation differs across STEM disciplines. In science, experimentation, peer review, and falsifiability are crucial. Scientific theories must be tested and scrutinised, with the possibility of revision or rejection if new evidence emerges. In technology, knowledge is validated by functionality and usability. A technological innovation must work effectively, meet human needs, and comply with ethical considerations (Eijck & Claxton, 2009). Engineering validation is similar but incorporates testing, optimisation, and trade-offs. Engineers must balance multiple constraints, such as cost, durability, and efficiency when developing a product or system. Mathematics differs fundamentally by relying on proof and logical consistency rather than empirical validation. Mathematical truths are universally applicable and remain valid as long as the axioms they derive hold (Annizar et al., 2020). While scientific theories may be revised, mathematical theorems remain true within their established frameworks. Wicked problems complicate knowledge validation, as solutions are often contingent, context-dependent, and iterative rather than definitive (McCune et al., 2023). For example, scientific models for climate change must be continuously refined, technological responses to crises such as pandemics evolve as new challenges arise, and engineering solutions must be resilient to unforeseen societal and environmental changes. Mathematics aids in structuring these complexities by offering models and simulations that, while not yielding absolute certainty, provide essential guidance for decision-making in ambiguous situations. Education can reinforce the importance of evidence-based reasoning and ethical considerations in problem-solving by integrating student-led inquiry into the knowledge validation process (McCune et al., 2023). Students participating in interdisciplinary STEM/ STEAM projects develop a greater understanding of the iterative process of knowledge validation and the societal implications of their efforts (Sarid & Goldman, 2021).

Each discipline's ***objectives***, or overarching goals, further illustrate epistemological differences. Science seeks to understand and explain natural phenomena through empirical study. Technology aims to create valuable artefacts, systems, and tools that enhance human capability. Engineering focuses on designing and optimising solutions to solve real-world problems. Mathematics, in contrast, is concerned with establishing logical truths and structures that serve as the foundation for other STEM fields. However, when addressing wicked problems, these objectives must often be interwoven (McCune et al., 2023). Scientific understanding alone cannot resolve climate change without technological advancements in renewable energy, engineering solutions for carbon capture, and mathematical modelling for long-term impact analysis. Similarly, global health challenges demand biomedical research, technological innovations, engineering interventions, and epidemiological models to coordinate effective responses. Wicked problems underscore the necessity of interdisciplinary thinking in STEM/STEAM, pushing these disciplines beyond their traditional objectives toward more integrated, adaptive, and systemic solutions (McCune et al., 2023). Developing social agency in students is critical for achieving these objectives (Mathers et al., 2024). Encouraging students to see themselves as contributors to societal progress instils a sense of responsibility and engagement (Marginson et al., 2013). By applying STEM knowledge to real-world contexts, students gain technical competencies and learn to work collaboratively, think critically about ethical implications, and advocate for sustainable and equitable solutions (Wysocki et al., 2013). Despite these differences, the disciplines frequently interact. Scientific discoveries inform technological advances and engineering designs, while mathematics provides the necessary analytical tools for all three.

Creativity is distinct but essential in each field (Wised & Inthanon, 2024). In science, creativity is evident in hypothesis generation and the development of innovative experimental approaches. Scientists must think beyond conventional knowledge to propose new explanations for observed phenomena (Bayani et al., 2025). In technology, creativity is essential for innovation and product development. The iterative nature of technological design requires continuous adaptation and improvement, often requiring novel problem-solving techniques. Engineering creativity is closely related, focusing on design thinking and working within constraints to develop optimal solutions. Mathematics employs creativity in pattern recognition and abstract thinking. While mathematical reasoning follows strict logical structures, creating new theorems, proofs, and problem-solving methods requires deep conceptual insight and originality.

We explore ***interdisciplinary integration*** in greater detail in Chapter 3 as a defining feature of STEM. Science frequently draws from mathematics and engineering for analytical models and practical applications. Technology integrates science, engineering, and social factors, ensuring that innovations are functional and socially relevant. Engineering combines science, mathematics, and technology to develop efficient and optimised solutions to problems. Engineers apply scientific principles, use mathematical models, and incorporate technological advancements to refine their designs. While abstract, mathematics supports all STEM fields by providing the fundamental structures for quantitative reasoning, modelling, and problem-solving. It is the universal language of science and engineering, making it an indispensable tool in all disciplines. Understanding the epistemological differences among science, technology, engineering, and mathematics is crucial for effective STEM education. While science seeks to explain the natural world through empirical investigation, technology applies knowledge to create functional solutions. Engineering combines scientific and mathematical principles in problem-solving, and mathematics provides a foundational logical framework for all STEM disciplines. Recognising these distinctions helps educators design curricula that foster interdisciplinary learning while respecting each field's unique way of knowing.

2.2.2 Teaching Disciplinary Knowledge and Skills in Classrooms

Teachers play a critical role in establishing the balance between the presentation of content knowledge and skill development, shaping how disciplinary knowledge is structured, taught, applied, and represented in the classroom (Popkewitz, 2018). Effective educators convey content knowledge while modelling disciplinary thinking and problem-solving processes (Mason, 2009). This representation of content knowledge and the selected pedagogical approaches tend to align with the epistemology and practices of a discipline. The teacher's disciplinary training manifests in how they plan and conduct each teaching and learning opportunity; however, their professional responsibility and existing knowledge act as filters of disciplinary knowledge (Dai, 2023). Disciplinary biases of the teacher become apparent in their teaching, thus influencing how students engage with disciplinary content and develop transferable skills.

A well-organised, resource-rich environment supports diverse learning styles and promotes engagement and learning. Collaborative group activities and discussions foster peer learning, encouraging learners to co-construct knowledge and develop critical thinking skills (Koretsky et al., 2014). Inquiry-based learning, modelling, and project-based learning resonate with the core principles of science, mathematics, and technology, offering excellent opportunities for integrating STEM disciplines (Starkey et al., 2023). Application-oriented tasks, such as solving mathematical problems or designing technological prototypes, connect theoretical knowledge to practical scenarios (Wensveen & Matthews, 2014), motivate students to engage with disciplinary concepts and skills and develop creativity and critical thinking. Using real-world problems and scenarios, such as climate change or public health, requires integrating scientific principles, mathematical modelling, and technological innovation, thus developing the learners' understanding of disciplinary knowledge (Krajcik & Delen, 2017). As we show in Chapter 4, collaborative and interdisciplinary approaches further enhance the teaching and representation of disciplinary knowledge in the classroom.

In the following section, we explore the STEM disciplines in the Australian Curriculum as a case study. The focus is on the curriculum writers' (disciplinary experts) use of verb clauses to highlight how the Australian Curriculum documents broadly convey the nature of the STEM disciplinary processes and disciplines by mandating inquiry-based practices and application-oriented tasks in Australian classrooms. The premise is that the curriculum materials convey disciplines differently, influencing how teachers and students take up curriculum materials in the classroom (Choppin et al., 2021). Teachers "explicitly or tacitly align themselves with–or resist–the [disciplinary] messages regarding the nature of [disciplinary] activity" (Choppin et al., 2021, p. 1250). Using verb clauses reflects disciplinary processes, evident in disciplinary actions or activities.

2.3 Exploring the STEM Disciplines in Classrooms: An Australian Case Study

This section explores an Australian case study of the three main STEM disciplines to illustrate disciplinary differences in curriculum documents. We use the Australian Curriculum as the case. The data we utilise draw upon in this chapter and throughout the book is from Australia and Southeast Asia, indicating the relevance of Australia as a case study.

The Australian curriculum has a three-dimensional structure with three integrated components: Learning Areas,[1] General Capabilities and Cross-Curriculum Priorities collaborate to specify the knowledge students should acquire (disciplines) and the qualities they should develop (General Capabilities and Cross-Curriculum Priorities) (Lingard, 2018). This case study and its analysis centre on the Australian Curriculum for compulsory schooling, spanning from Foundation to Year 10 (ages 5 to 16). We chose not to analyse the post-compulsory Senior Curricula as this is subject to different state-based approaches to the end-of-secondary assessments, thus creating inconsistencies and making the Learning Area comparisons more challenging.

The Australian Curriculum can be described as a high-performing educational system:

- A "prescriptive curriculum, characterised by disciplinary knowledge, skills and understandings described by content descriptions and achievement standards, supported by general capabilities and cross-curriculum priorities developed across the learning areas" (ACARA, 2019, p. 5).
- "Was challenging in most learning areas/subjects across all curricula, particularly in the secondary years. Science was the most variable, with the AC showing fundamental depth at Year 2 and Year 10 and limited depth at Year 6" (ACARA, 2019, p. 5).
- Goal 2 stipulates, "All young Australians become confident and creative individuals, successful lifelong learners, and active and informed community members". (ACARA, 2020, p. 7).
- Including seven general capabilities that play a significant role "in equipping young Australians to live and work successfully in the twenty-first century" (ACARA, 2017, para 1). Each capability is organised into elements and sub-elements within a continuum focused on the development of skills and attitudes of learners" (ACARA, 2019, p. 7).

[1] 'Learning areas' is the overarching collective term for disciplines of study in the Australian Curriculum A Learning Area is akin to a 'subject' in other jurisdictions.

- Includes a 'traditional' range of Humanities disciplines and the disciplines that underpin STEM. "The competencies/capabilities identified … also include themes/concepts that relate to both Humanities and STEM" (ACARA, 2019, p. 11).
- A web-based format that uses hyperlinks, knowledge, and process strands (ACARA, 2019) that list the mandated content descriptions and optional elaborations.

The content descriptions noted in the last point are used in each learning area to "specify the essential knowledge, understandings and skills that Australian education students are expected to learn across the years of schooling. There is an emphasis on depth rather than breadth (ACARA, 2020). The Learning Areas we consider in the case study relate to the STEM subjects. STEM is not a distinct educational category; it is integrated within science, technology, mathematics, and essential skills, especially numeracy, information and communication technology (ICT) capability, and critical and creative thinking. The Design and Technologies Learning Area addresses engineering through content descriptions focusing on engineering principles and systems. Accompanying the Design and Technologies Learning Area is the Digital Technologies Learning Area. In our analysis, we combine the Design and Digital Technologies Learning Areas to present a Technologies (combined) discussion. Integrating STEM education into the Science, Technology, and Mathematics curriculum guarantees that every student graduates with a solid foundation in STEM knowledge and skills. It encourages students to pursue more advanced STEM subjects and careers after compulsory education. However, timetabling issues made it difficult for teachers in secondary school settings to implement STEM activities that work across learning areas.

In this case study, we examine the nature of the three learning areas as separate entities. The intent is for Chapter 3 to explore their integration into STEM education. This case study aims to establish the nature of the learning areas as separate disciplines. We build on the previous studies by Herbel-Eisenmann and Otten (2011), as well as Choppin et al. (2021), which employ verb clauses to illustrate how curriculum materials reflect the characteristics of disciplinary processes and the disciplines at large. The analysis of verb clauses included within curriculum documents signals the student disciplinary experiences represented in the curriculum documents, thus revealing the meanings that emerge from the text.

Each learning area's content descriptions were individually scrutinised for verb clauses, and a master list of 48 verb clauses was created. We used two categories of verbs: physical and mental verbs. Physical verb clauses were considered those that indicated an action process or procedure. Mental verb clauses indicate a cognitive process that might lead to the communication of ideas. A simple frequency count of science, digital technologies, design technologies, and mathematics content descriptors was systematically made. This analysis is provided in Table 2.2.

Table 2.2 presents the frequency of each verb clause found in the Australian Curriculum documents for the core years of schooling. Two sets of verb clauses emerged in the analysis. The first involves the sharing of verb clauses. Two subsets are evident: a subset of 14 verb clauses that all three learning areas mandate and a subset of 16 verb clauses that two of the three learning areas mandate. The second set consisted of 18 verb clauses mandated by only one of the three learning areas. Fifteen physical verb clauses and 33 mental action verb clauses were identified.

What follows is an exploration of the data patterns in Table 2.1 and a discussion of the implications for educators and curriculum designers. Figure 2.1 visually summarises the multifaceted data from Table 2.1, making patterns, trends, and comparisons more immediately apparent. The differences and commonalities in pedagogical approaches indicate the physical and mental actions required of students.

2.3.1 *Set 1: Shared Verb Clauses Across the Learning Areas*

A subset of 14 verb clauses appears across all three learning areas, reflecting a shared focus on essential cognitive and physical processes in education. Prominent mental verbs such as "evaluate", "investigate", "analyse", and "describe" suggest that critical thinking and intellectual engagement are fundamental in all three disciplines. For example, "evaluate" occurs 16 times in Science, 44 times in Technologies, and 5 times in Mathematics. This variance in frequency indicates a greater emphasis on evaluative thinking in Technologies compared to Science and Mathematics. Physical action verbs also feature in all three areas, though their relative importance varies. "Select" (24 occurrences in Science, 44 in Technologies, and 4 in Mathematics) and "represent" (50 in Science, 27 in Technologies, and 44 in Mathematics) underscore the importance of hands-on activities. Technologies strongly emphasise physically selecting resources and representing concepts, suggesting a

Table 2.2 Verb clause analysis

	Active Verbs Physical or Mental	*Science*	*Technologies Combined*	*Mathematics*
Verb clauses in science, technologies (Combined) and mathematics (n = 14)				
Select	Physical	24	44	4
Collect	Physical	26	12	22
Represent	Physical	50	27	44
Sort	Physical	7	6	3
Record	Physical	18	5	3
Investigate	Mental	1	97	58
Evaluate	Mental	16	44	5
Explore	Mental	7	65	12
Develop	Mental	4	55	10
Analyse	Mental	19	50	3
Plan	Mental	40	44	1
Apply	Mental	8	26	31
Describe	Mental	10	10	58
Identify	Mental	52	22	41
Technologies (combined) and mathematics verb clauses (n = 4)				
Sequence	Physical	0	13	2
Organise	Mental	0	4	2
Recognise	Mental	0	22	27
Interpret	Mental	0	6	31
Science and mathematics verb clauses (n = 7)				
Solve problems	Physical	8	0	46
Construct	Physical	19	0	15
Measure	Physical	14	0	6
Compare	Mental	17	0	52
Reflect	Mental	11	0	6
Decide	Mental	7	0	2
Formulate	Mental	11	0	1
Science and technologies (Combined) verb clauses (n = 5)				
Share	Physical	2	4	0
Communicate	Physical	31	40	0
Address ethics	Mental	14	29	0
Assess risk	Mental	22	21	0
Predict	Mental	40	4	0
Mathematics verb clauses (n = 6)				
Draw	Physical	0	0	2
Trial	Physical	0	0	2
Locate	Mental	0	0	13
Extend	Mental	0	0	4

(continued)

Table 2.2 (continued)

	Active Verbs Physical or Mental	*Science*	*Technologies Combined*	*Mathematics*
Infer	Mental	0	0	4
Conduct	Mental	0	0	2
Technologies (Combined) verb clauses (n = 5)				
Implement	Physical	0	19	0
Combine	Physical	0	13	0
Examine	Mental	0	16	0
Decompose	Mental	0	10	0
Visualise	Mental	0	18	0
Science verb clauses (n = 7)				
Participate	Physical	8	0	0
Respond	Physical	4	0	0
Control variables	Mental	12	0	0
Engage	Mental	2	0	0
Summarise	Mental	7	0	0
Draw conclusions	Mental	18	0	0
Observe	Mental	7	0	0

design-oriented approach integrating physical creation and visualisation. Mathematics, in contrast, emphasises representation in abstract forms, such as graphs or equations.

In the second subset, 16 verb clauses are shared between two learning areas, highlighting specific intersections in their objectives. For instance, Science and Mathematics both emphasise "solve problems", "compare", and "construct", indicating their shared focus on logical reasoning. Notably, "solve problems" appears 46 times in Mathematics but only 8 times in Science, reflecting Mathematics' stronger association with problem-solving frameworks. In contrast, Science and Technologies share verbs such as "communicate", "address ethics", and "assess risk", indicating their joint emphasis on collaboration and ethical considerations. The frequency of "address ethics" is higher in Technologies (29) compared to Science (14), possibly due to the ethical implications of technological development. Mathematics and Technologies overlap in verbs like "sequence" and "interpret", reflecting their shared focus on organising and making sense of data. However, the relative frequencies suggest that these processes are more emphasised in Technologies

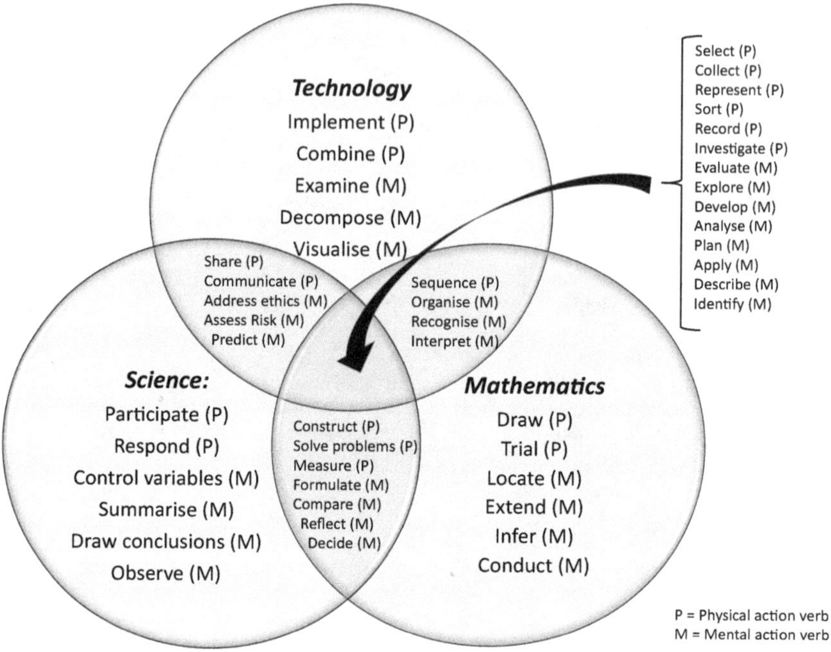

Fig. 2.1 Verb clause analysis for science, technologies (combined) and mathematics (*Source* Authors)

where sequencing is integral to prototyping and the design processes, and interpretation is crucial for analysing digital data.

2.3.2 *Set 2: Distinctive Features of Each Learning Area*

Science employs unique verb clauses such as "control variables", "observe", and "draw conclusions", which underscore its experimental and inquiry-based nature. "Control variables" appear 12 times, emphasising the precision required in scientific investigations. Similarly, "observe" and "draw conclusions" are critical for fostering empirical reasoning. Technologies feature unique verbs such as "implement", "decompose", and "visualise", reflecting its focus on project-based learning and iterative design. For example, "implement" (19 occurrences) highlights the application of theoretical knowledge in creating tangible outputs, while "decompose" (10 occurrences) emphasises breaking down complex problems into manageable components. Mathematics, on the other hand, employs unique verbs such as "locate", "extend", and "infer", demonstrating its abstract and logical nature. "Locate" (13 occurrences) and "infer" (4 occurrences) point to its focus on spatial reasoning and deductive logic, both of which are essential for advanced mathematical thinking.

2.3.3 *Physical and Mental Verb Clauses*

The distribution of physical and mental verb clauses reveals no distinct pedagogical priorities. The prevalence of mental verbs in all three learning areas suggests that the Australian Curriculum, at least for these learning areas, strongly emphasises cognitive development. Science and mathematics favour mental verbs, possibly because both disciplines primarily involve cognitive processes such as reasoning, analysing, hypothesising, and calculating. Science, however, also incorporates physical verbs, particularly when hands-on investigations are involved. This includes measuring, constructing, and recording activities essential to empirical investigation and laboratory work. Science, by its investigative nature, inherently involves thinking (mental verbs) and doing (physical verbs). Mathematics follows a similar pattern, heavily relying on mental verbs due to its abstract nature. Students engage in activities such as inferring, formulating, evaluating, and applying, which are fundamentally cognitive. However, when mathematics is applied in real-world problem-solving—such as drawing geometric figures, measuring, or sorting physical objects—it incorporates some physical verbs.

As represented in the Australian Curriculum, the technologies (combined) primarily emphasise mental verb clauses of a conceptual nature. For example, design, planning, and digital problem-solving, are inherently cognitive processes. This is why many of the verbs uniquely associated with technologies (combined) are mental, such as visualise, examine, and decompose. Physical verbs in the technologies (combined) curriculum are those shared with science and mathematics, usually because these disciplines intersect in applied fields. For example, engineering-related technological activities involve communicating, sequencing, collecting, and representing during prototyping, actions that usually appear when linked to practical applications in STEM/ STE*A*$^{+}$M. When technologies share verbs with science and mathematics, it is likely because those aspects involve applied, hands-on elements (e.g., coding in technologies shares analytical thinking with math, while robotics in technologies shares prototyping with science).

2.3.4 *Neuroeducation and Disciplinary Thinking Strengths*

The differences in verb clauses across the disciplines reflect distinct cognitive strengths that align with neuroeducation findings on brain function and learning processes (Pradeep et al., 2024). Technologies' emphasis on verbs such as "develop", "implement", and "visualise" suggests a focus on creativity and problem-solving, which neuroeducation links to activation in the prefrontal cortex and parietal lobes (Aldous, 2007) (see Chapter 1). These neural areas are responsible for planning, decision-making, and spatial reasoning, skills critical for design and innovation. Mathematics' reliance on abstract reasoning and mental verbs such as "solve problems", "compare", and "interpret" aligns with the left hemisphere's dominance in logical and analytical processing. This disciplinary focus fosters neural pathways related to precision, sequential thinking, and symbolic representation. Neuroeducation further suggests that activities like interpreting data and solving complex problems engage working memory and reinforce neural connections in the prefrontal cortex. Science's balance of physical and mental verbs, such as "control variables", "observe", and "draw conclusions", reflects an integrated cognitive approach. Neuroeducation indicates that combining hands-on activities with critical thinking engages multiple brain regions, including the motor cortex, sensory areas, and higher-order cognitive areas (Pradeep et al., 2024). This integration supports inquiry-based learning by fostering neural plasticity and promoting long-term memory retention through experiential learning.

2.3.5 *Implications of Disciplinary Verb Processes in Curriculum Documents*

The variation in verb frequencies across learning areas indicates how each learning area or discipline approaches skill development. Technologies' high emphasis on verbs like "develop" (55 occurrences) and "implement" (19 occurrences) highlights its focus on creativity and application, preparing students for real-world problem-solving. Integrating ethical considerations through verbs like "address ethics" and "assess risk" ensures students consider broader societal implications. Mathematics' focus on mental processes like "solving problems" and "interpreting" emphasises abstract reasoning, logic, and precision. The scarcity of physical verbs aligns with its theoretical nature but may suggest opportunities to incorporate more practical, hands-on activities to enhance engagement. With its balance of physical and mental verbs, science offers a comprehensive approach to learning through experimentation and critical thinking. However, its lower frequency of design-oriented verbs like "develop" and "implement" suggests less emphasis on applied innovation than Technologies. The distinctions and overlaps also highlight opportunities for interdisciplinary learning. For example, the shared focus on "analyse" and "evaluate" across all three areas could form the basis for integrated STEM/STE*A*$^{+}$M activities, as discussed in Chapter 3. Similarly, Science and Technologies' emphasis on "communicate" and "assess risk" could foster collaborative projects that explore the ethical dimensions of scientific and technological advancements, as explored in other chapters in this book.

Neuroeducation findings reinforce the importance of leveraging these disciplinary strengths. For example, incorporating collaborative, hands-on projects that require ethical decision-making and creative problem-solving could stimulate the prefrontal cortex and enhance interdisciplinary learning (Wysocki et al., 2013). Similarly, integrating abstract reasoning tasks in science and technology could foster connections to the cognitive strengths developed in mathematics, promoting well-rounded cognitive growth (Ablin, 2008).

Analysing verb clauses in the Australian Curriculum for Science, Technologies, and Mathematics reveals shared priorities and distinct pedagogical approaches. While all three learning areas emphasise critical thinking and intellectual engagement, their specific foci reflect their unique disciplinary goals. Technologies prioritise practical application and ethical considerations, Mathematics emphasises abstract reasoning, and Science balances inquiry-based learning with theoretical understanding. Neuroeducation findings highlight how these disciplinary approaches align with cognitive strengths, providing a basis for designing interdisciplinary learning experiences, as discussed in Chapter 3. By leveraging these insights, educators can create dynamic, engaging curricula that harness the unique strengths of each discipline while fostering holistic cognitive development.

These patterns provide insights into curriculum design and suggest avenues for interdisciplinary integration, fostering a holistic approach to STEM education. By leveraging these insights, educators can design learning experiences that harness the strengths of each discipline while addressing their unique challenges.

2.4 The Integration Challenge: Moving Toward Interdisciplinary Thinking

This chapter outlines the nature of disciplinary knowledge that guides educators' classroom planning and lesson enactment approaches. Disciplinary knowledge is present in each curriculum regarding content knowledge, inquiry processes, and skills advocated in curriculum documents (Choppin et al., 2021). Traditional school structures emphasise disciplinary boundaries, with subjects taught in isolation and assessed independently. While this model ensures a degree of depth of knowledge within specific domains, it fails to equip learners with the ability to integrate learning across disciplines (Krajcik & Delen, 2017; Popkewitz, 2018). Foundational assumptions and methodologies guide problem-solving but can lead to blind spots. For instance, science's reliance on empirical evidence may overlook subjective or qualitative experiences better addressed by the humanities (Mazzocchi, 2019). Mathematics's emphasis on deductive reasoning might neglect real-world uncertainties. While offering clarity, these paradigms risk limiting creativity and innovation (Ings, 2019).

This fragmentation hinders learners from tackling complex, real-world issues that demand interdisciplinary thinking, like climate change, public health, and technological innovation. The continued dominance of single-discipline teaching raises concerns about its effectiveness in preparing learners for the interconnected challenges of modern workplaces and global issues (Reich, 2022; Vladova et al., 2024). A limitation of discipline-based teaching is its constrained approach to problem-solving. Knowledge is presented as compartmentalised, preventing students from recognising connections across subjects (Ings, 2019). This separation inhibits their ability to synthesise knowledge and develop holistic understandings of complex issues that require multiple perspectives. The rigid structure of curricula reinforces this fragmentation by prioritising subject-specific content and limiting cross-disciplinary exploration (Tsebella et al., 2022; Vladova et al., 2024). Science education, for example, often treats biology, physics, and chemistry as distinct areas rather than integrating them to address broader scientific inquiries. This structure can make learning feel disconnected from practical applications, reducing students' ability to see the relevance of their education beyond the classroom. Motivating students to link their subjects and utilise their knowledge and skills meaningfully will enhance their readiness for future challenges, where integrative thinking and interdisciplinary problem-solving are becoming more crucial (Reich, 2022; Vladova et al., 2024).

Assessment practices further entrench these disciplinary divisions. Standardised testing typically evaluates proficiency within isolated subjects, reinforcing the perception that knowledge exists in separate domains. This discourages interdisciplinary thinking and prevents students from developing the skills to apply their learning in complex, real-world scenarios (Tsebella et al., 2022). As a result, students may achieve high marks in individual subjects yet struggle to integrate their knowledge when confronted with multifaceted problems that demand broader, more connected approaches.

2.4.1 Bridging the Gap: Moving Toward Interdisciplinary Thinking

Addressing the challenges of disciplinary fragmentation in education requires a shift toward interdisciplinary and systems-oriented thinking (Barile & Saviano, 2021). While we describe this further in Chapters 4 and 6, we consider it a necessary link between disciplinary and interdisciplinary teaching and learning, so we briefly introduce the notion here. Rather than replacing disciplinary teaching and learning, the systems-oriented thinking approach emphasises meaningful integration between subjects, allowing students to connect knowledge and skills across different fields. We see the goal of contemporary STEM/STE*A*⁺M schooling as enabling learners to see the relevance of their learning beyond isolated school subjects, equipping them with the ability to apply their knowledge across domains. Identifying opportunities for interdisciplinary learning within existing curricula provides a starting point. Such an approach strengthens students' grasp of individual subjects while fostering cognitive flexibility, enabling them to tackle complex problems from multiple perspectives.

A significant challenge in implementing a systems-oriented interdisciplinary education is balancing disciplinary depth with interdisciplinary breadth. We share the concern that interdisciplinary learning, if not carefully structured, could lead to a superficial understanding of concepts rather than deep expertise (Reich, 2022). To mitigate this risk, schools should adopt strategies that ensure strong foundational learning within individual disciplines while providing structured interdisciplinary engagement opportunities. One effective approach is "disciplinary anchoring", where students understand key concepts within a subject before applying them in broader interdisciplinary contexts (Barile & Saviano, 2021). For example, in a STE*A*⁺M-focused curriculum, students might first build proficiency in mathematical modelling before applying those skills to analyse engineering challenges in renewable energy or environmental sustainability. This would ensure disciplinary knowledge and skills are still on the school agenda; however, so is the recognition that learners need to develop the ability to synthesise and apply information across multiple domains. Education jurisdictions continue to structure the STEM offerings as distinct subjects to align with tertiary prerequisite entry requirements. However, most school graduates do not undertake further STEM studies. Consideration needs to be made to keep STEM subject offerings relevant so that all learners learn to think in ways that can contribute to solving contemporary global concerns. By engaging in interdisciplinary learning, learners gain practical experience in applying interdisciplinary knowledge and skills, thus preparing them for the complexities of whatever future professional work they undertake and in addressing societal challenges.

To achieve this, we should consider the evolving role of disciplinary knowledge and skills within interdisciplinary education. Rather than treating subjects as separate and independent, educators should present them as interconnected components of a broader knowledge system. This shift enables learners to recognise the interdependencies between disciplines, fostering a more holistic and systems-orientated approach to learning (Barile & Saviano, 2021; Wysocki et al., 2013). Scientific inquiry, for instance, frequently relies on mathematical precision, technological applications, and ethical considerations (Wysocki et al., 2013). By adopting an interdisciplinary perspective, we can better understand how different fields interact, making learning more relevant and applicable to real-world challenges (Barile & Saviano, 2021).

Integrating interdisciplinary education requires rethinking curriculum design, pedagogical approaches, and assessment practices. Traditional curricula are usually structured around rigid subject boundaries, limiting opportunities for the teaching and learning of cross-disciplinary exploration. However, schools must create environments that encourage collaboration between subjects and provide opportunities for teachers to develop interdisciplinary teaching strategies. Assessment methods must also evolve to support a systems-orientated approach to interdisciplinary education. Traditional assessments often reinforce disciplinary divisions by evaluating students based on isolated subject knowledge rather than their ability to synthesise and apply learning across fields. Alternative assessment models, such as interdisciplinary capstone projects and performance-based tasks, can better measure students' ability to integrate knowledge from multiple disciplines. These evaluations promote essential skills like critical thinking, creativity, and problem-solving, vital for success in our interconnected world.

The role of the educator in fostering a systems-orientated approach to interdisciplinary education is critical. Nevertheless, due to their disciplinary training, many educators may feel unprepared to implement interdisciplinary teaching strategies. Professional development programs must be designed to support teachers/educators in integrating interdisciplinary approaches into their instruction. Collaborative training sessions that bring together teachers/educators from different subject areas can provide opportunities to share best practices and co-design interdisciplinary projects. An exciting initiative along these lines occurred in 2024 when the then-Indonesian Ministry of Education, Culture, Research and Technology (MOECRT) partnered with Monash University to offer a micro-credential for teachers from all disciplines alongside university teacher educators from all disciplines. The teachers and lecturers explored the systems-orientated approach to developing numeracy skills across the curriculum. The MOECRT encouraged cross-disciplinary partnerships between teachers and lecturers, enabling them to collaborate to develop lessons that span multiple subjects. Fostering a culture of collaboration among educators, schools, and providers of initial teacher education can create cohesive and integrated learning experiences.

Beyond preparing students for academic success, systems-orientated interdisciplinary education fosters the skills needed for the future workforce. Many of society's most pressing challenges—climate change, cybersecurity, healthcare innovation, and artificial intelligence—require expertise that spans multiple disciplines (Barile & Saviano, 2021). Employers increasingly seek individuals who can navigate complex problems, synthesise diverse perspectives, and adapt to rapidly evolving industries. By exposing learners to systems-orientated interdisciplinary learning, schools can equip learners with the adaptability and problem-solving skills needed for careers that require cross-disciplinary expertise (Barile &

Saviano, 2021). Additionally, systems-orientated interdisciplinary education promotes innovation by encouraging students to think creatively and draw insights from different fields. Many of the most incredible technological and scientific advancements have emerged at the intersections of disciplines. Promoting such collaboration requires a shift in educational and institutional priorities to foster integrative approaches (Wudarczyk et al., 2021). Schools can inspire students to explore new ways of approaching problems and contribute to advancements in their chosen fields.

2.4.2 Future Directions: Preparing Students for Real-World Challenges

The transition to interdisciplinary education will not be without its challenges (Barile & Saviano, 2021). Resistance to change, institutional constraints, and a lack of resources will hinder efforts to integrate subjects. To support such change, a mind shift is needed. Schools must invest in curriculum redesign, teacher education, and developing interdisciplinary learning materials and pedagogical approaches. Additionally, policymakers should recognise the value of interdisciplinary education and provide the necessary support for schools to implement these approaches effectively. Interestingly, in recent workshops with Indonesian civil engineering lecturers, the lecturers commented on first-year students' lack of systems-orientated thinking. There is an expectation among the tertiary lecturers that school graduates entering the training for the engineering profession would have entry-level basic skills from school—but they do not.

Promoting a systems-oriented, interdisciplinary approach (Barile & Saviano, 2021) in STEM/STEAM education is crucial for preparing students to tackle the complexities of the modern world. Schools should empower students with the necessary skills, knowledge, and mindsets to confront real-world challenges by dismantling disciplinary silos and fostering collaborative efforts across disciplines. This approach enhances learning outcomes and ensures that learners are prepared to contribute meaningfully to society in academia, industry, or global problem-solving efforts. As STEM/STEAM education evolves, systems-oriented interdisciplinary learning must be seen not as an alternative but as a necessary and integral part of a contemporary, future-focused education.

Aligning systems-oriented disciplinary teaching with interdisciplinary goals is essential to preparing students for future challenges (Barile &

Saviano, 2021). Schools should design learning experiences that encourage students to think beyond the confines of individual STEM subjects, fostering critical thinking, collaboration, and adaptability skills and creating environments where knowledge is fluid, interconnected, and relevant to the world students will enter beyond graduation. Future directions in education must prioritise interdisciplinary and systems-oriented learning. Employers increasingly seek individuals who can navigate complexity, synthesise diverse perspectives, and adapt to rapidly evolving industries. Interdisciplinary learning prepares students for these demands and fosters a more engaging and meaningful educational experience. Ultimately, the integration challenge in education is not about diminishing disciplinary expertise but leveraging it to build a more interconnected, real-world learning experience. By promoting interdisciplinary systems thinking, educators empower learners with essential skills to tackle the complex challenges of the twenty-first century, preparing them to make valuable contributions in a swiftly evolving world. We elaborate on this in Chapter 4.

2.5 Conclusion

Disciplinary thinking is both an anchor and a catalyst within STEM/ STEAM education, providing essential structures for knowledge acquisition and serving as a gateway to interdisciplinary problem-solving (Barile & Saviano, 2021). This chapter has outlined how disciplinary literacy and disciplinary thinking shape educational experiences, emphasising how knowledge is organised, processed, and communicated across different domains. While traditional classroom practices have long reinforced discrete subject boundaries, the increasing complexity of contemporary challenges necessitates a shift toward more integrative and systems-oriented educational approaches (Krajcik & Delen, 2017). STEM education provides a compelling framework for this shift, requiring educators to navigate the dual imperatives of disciplinary depth and interdisciplinary breadth.

The discussion around the epistemological contributions of the STEM subjects informs us that individual disciplines are insufficient to solve wicked problems and underscores the necessity of multiple disciplines interacting beyond their traditional objectives toward more integrated, adaptive, and systemic solutions. Developing social agency in students is critical for achieving these objectives. The Australian curriculum case study analysis further reinforced the role of disciplinary verb clauses in shaping how students experience knowledge within formal education, highlighting commonalities and distinctions among STEM subjects. These insights underscore the necessity of designing curricula and professional learning experiences that foster disciplinary expertise and interdisciplinary fluency.

As STEM/STEAM education continues to evolve, educators and policymakers must construct learning environments that maintain disciplinary rigour while embracing the interconnectedness of knowledge. Systems-oriented interdisciplinary learning offers a promising pathway for achieving this balance, equipping students with the cognitive flexibility required to address global challenges such as climate change, technological innovation, and ethical decision-making (Barile & Saviano, 2021). Future directions in STEM education must focus on integrating disciplinary strengths within interdisciplinary frameworks, ensuring students develop the analytical, creative, and collaborative skills essential for navigating an increasingly complex and interconnected world.

References

Ablin, J. L. (2008). Learning as problem design versus problem solving: Making the connection between cognitive neuroscience research and educational practice. *Mind, Brain, and Education, 2*, 52–54. https://doi.org/10.1111/J.1751-228X.2008.00030.X

Aldous, C. R. (2007). Creativity, problem solving and innovative science: insights from history, cognitive psychology and neuroscience. *International Education Journal, 8*(2), 176–187. http://iej.com.au

Annizar, A. M. R., Jakaria, M. H. D., Mukhlis, M., & Apriyono, F. (2020). Problem solving analysis of rational inequality based on IDEAL model. *Journal of Physics: Conference Series, 1465*(1), Article 012033. https://doi.org/10.1088/1742-6596/1465/1/012033

Australian Curriculum, Assessment and Reporting Authority (ACARA). (2017). *F–10 curriculum: General capabilities*. Retrieved from https://www.australiancurriculum.edu.au/f-10-curriculum/general-capabilities/

Australian Curriculum, Assessment and Reporting Authority (ACARA). (2019). *Program of research: Key findings from four international comparative studies: The British Columbia Curriculum, The Finnish National Core Curriculum, The Singapore Curriculum, The New Zealand Curriculum*. Retrieved from https://www.australiancurriculum.edu.au/media/4394/ac-bcc-fncc-sc-nzc-key-findings.pdf

Australian Curriculum, Assessment and Reporting Authority (ACARA). (2020). *The shape of the Australian Curriculum (Version 5.0)*. Retrieved from https://www.acara.edu.au/docs/default-source/curriculum/the_shape_of_the_australian_curriculum_version5_for-website.pdf

Barile, S., & Saviano, M. (2021). Interdisciplinary systems thinking for a new scientific paradigm: toward a re-founding of human values. In G. Minati (Ed.), *Multiplicity and interdisciplinarity: Essays in honor of Eliano Pessa* (pp. 17–39). Springer. https://doi.org/10.1007/978-3-030-71877-0_3

Bayani, F., Rokhmat, J., Hakim, A., & Sukarso, A. (2025). Research trends in analytical thinking skills for science education: insights, pedagogical approaches, and future directions. *International Journal of Ethnoscience and Technology in Education*, *2*(1), 129–157. https://doi.org/10.33394/ijete.v2i1.14142

Carlgren, L., & BenMahmoud-Jouini, S. (2022). When cultures collide: What can we learn from frictions in the implementation of design thinking? *Journal of Product Innovation Management, 39*(1), 44–65. https://doi.org/10.1111/jpim.12603

Choppin, J., Davis, J., McDuffie, A. R., & Drake, C. (2021). Influence of features of curriculum materials on the planned curriculum. *ZDM, 53*, 1249–1263. https://doi.org/10.1007/s11858-021-01305-7

Dai, Y. (2023). Negotiation of epistemological understandings and teaching practices between primary teachers and scientists about artificial intelligence in professional development. *Research in Science Education, 53*(3), 577–591. https://doi.org/10.1007/s11165-022-10072-8

Eijck, M. V., & Claxton, N. X. (2009). Rethinking the notion of technology in education: Techno-epistemology as a feature inherent to human praxis. *Science Education, 93*(2), 218–232. https://doi.org/10.1002/sce.20308

Goldman S. R., Britt A. M., Brown W., Cribb G., George M., Greenleaf C., Lee, C., Shanahan C. & Project READI. (2016). Disciplinary literacies and learning to read for understanding: A conceptual framework for disciplinary literacy. *Educational Psychologist, 51*(2), 219–246. https://doi.org/10.1080/00461520.2016.1168741

Habermas, J. (1971). *Knowledge and human interests* (J. J. Shapiro, Trans). Beacon Press.

Herbel-Eisenmann, B. A., & Otten, S. (2011). Mapping mathematics in classroom discourse. *Journal for Research in Mathematics Education, 42*(5), 451–485. https://doi.org/10.5951/jresematheduc.42.5.0451

Ings, W. (2019). Undisciplined thinking: Disobedience and the nature of design. In T. Pernecky (Ed.), *Postdisciplinary Knowledge* (pp. 48–65). Routledge. https://doi.org/10.4324/9780429058561-3

Khan, L. A. & Arabia, S. (2015). What is mathematics—An overview. *International Journal of Mathematics and Computational Science, 1*(3), 98–101. https://doi.org/10.13140/RG.2.1.3626.8967

Kidman, G. & Casinader, N. (2017). *Inquiry-based teaching and learning across disciplines: comparative theory and practice in schools*. London Palgrave Macmillan.

Koretsky, M., Nolen, S. B., Gilbuena, D. M., Lehtinen, E., Vauras, M., Tierney, G., Violet, S. E. (2014). Studying & supporting productive disciplinary engagement in STEM learning environments. Paper presented at *2014 ASEE Annual Conference & Exposition*, Indianapolis, Indiana. https://doi.org/10.18260/1-2--23070

Krajcik, J. & Delen, I. (2017). Engaging learners in STEM education. *Eesti Haridusteaduste Ajakiri. Estonian Journal of Education, 5*(1), 35–58. https://doi.org/10.12697/eha.2017.5.1.02b

Kühne, O. (2024). *Redescribing horizontal geographies: A neopragmatist approach to spatial contingency, complexity, and relationships*. Springer. https://doi.org/10.1007/978-3-031-59124-2_3

Lingard, B. (2018). The Australian Curriculum: A critical interrogation of why, what and where to? *Curriculum Perspectives, 38*(1), 55–65. https://doi.org/10.1007/s41297-017-0033-7

Marginson, S., Tytler, R., Freeman, B., & Roberts, K. (2013). *STEM: Country comparisons: International comparisons of science, technology, engineering and mathematics (STEM) education*. Australian Council of Learned Academies. https://acola.org.au/wp/project-2/

Mason, J. (2009). Teaching as disciplined enquiry. *Teachers and Teaching, 15*(2), 205–223. https://doi.org/10.1080/13540600902875308

Mathers, B., Johnson, J., Kaufmann, A., Sinni, N., Louis, E., & Henneman, E. (2024). Developing student agency through authentic application of socio-scientific issues in STEM classrooms. *Turkish Journal of Education, 13*(5), 508–534. https://doi.org/10.19128/turje.1507933

Mazzocchi, F. (2019). Scientific research across and beyond disciplines: Challenges and Opportunities of Interdisciplinarity, *EMBO Reports, 20*(6), e47682. https://doi.org/10.15252/embr.201947682

McCune, V., Tauritz, R., Boyd, S., Cross, A., Higgins, P., & Scoles, J. (2023). Teaching wicked problems in higher education: Ways of thinking and practising. *Teaching in Higher Education, 28*(7), 1518–1533. https://doi.org/10.1080/13562517.2021.1911986

McElreavy, C., Tobin, V., Martin, T., Damon, M. B., Crate, N., Godinez, A., & Bennett, K. (2016). The history of the academy and the disciplines. In R. DeRosa (Ed.), *Interdisciplinary studies: A connected learning approach*. Rebus Community. https://press.rebus.community/idsconnect/chapter/the-history-of-the-academy-and-the-disciplines/

Popkewitz, T. S. (2018). What is 'really' taught as the content of school subjects? Teaching school subjects as an alchemy. *The High School Journal, 101*(2), 77–89. https://doi.org/10.1353/hsj.2018.0001

Post, R. (2009). Debating disciplinarity. *Critical Inquiry, 35*(4), 749–770. https://doi.org/10.1086/599580

Pradeep, K., Sulur Anbalagan, R., Thangavelu, A. P., Aswathy, S., Jisha, V. G., & Vaisakhi, V. S. (2024). Neuroeducation: Understanding neural dynamics in learning and teaching. *Frontiers in Education, 9*, 1437418. https://doi.org/10.3389/feduc.2024.1437418

Rantala, J. (2012). How Finnish adolescents understand history: Disciplinary thinking in history and its assessment among 16-year-old Finns. *Education Sciences, 2*(4), 193–207. https://doi.org/10.3390/educsci2040193

Reich, Y. (2022). We cannot play 20 questions with creativity and innovation and win: The necessity of practice-based integrative research. *International Journal of Design Creativity and Innovation, 10*(2), 69–74. https://doi.org/10.1080/21650349.2022.2041889

Rylander, A. (2009). Design thinking as knowledge work: Epistemological foundations and practical implications. *Design Management Journal, 4*(1), 7–19. https://doi.org/10.1111/j.1942-5074.2009.00003.x

Sarid, A., & Goldman, D. (2021). A value-based framework connecting environmental citizenship and change agents for sustainability—Implications for education for environmental citizenship. *Sustainability, 13*(8), 4338. https://doi.org/10.3390/su13084338

Shanahan, C., & Shanahan, T. (2017). Disciplinary literacy. In D. Lapp & D. Fisher (Eds.), *Handbook of research on teaching the English language arts* (4th ed., pp. 281–308). Routledge. https://doi.org/10.4324/9781315650555

Starkey, L., Yates, A., de Roiste, M., Lundqvist, K., Ormond, A., Randal, J., & Sylvester, A. (2023). Each discipline is different: Teacher capabilities for future-focussed digitally infused undergraduate programmes. *Educational Technology Research and Development, 71*(1), 117–136. https://doi.org/10.1007/s11423-023-10196-2

Suhaimi, S. N., Walters, A., & Ward, J. (2024). Design thinking mindset: A user-centred approach toward innovation in the Welsh creative industries. *International Journal of Design Creativity and Innovation, 12*(4), 238–257. https://doi.org/10.1080/21650349.2024.2383410

Suyitno, A. (2010). Sistem deduktif aksiomatis dalam matematika dan matematika sekolah, *AKSIOMA: Jurnal Matematika dan Pendidikan Matematika, 1*(2), https://journal.upgris.ac.id/index.php/aksioma/article/view/54

Tsebella, A. T., W/Yesus, K. M. & Kassa, S. A. (2022). Early Grade Mathematics in Ethiopia: Alignment of Contents in the Textbooks with the Corresponding Syllabuses. *Asian Journal of Education and Training, 8*(2), 34–42. https://doi.org/10.20448/edu.v8i2.3866

Vladova, G., Haase, J. & Friesike, S. (2024). Why, with whom, and how to conduct interdisciplinary research? A review from a researcher's perspective. *Science and Public Policy*, scae070, https://doi.org/10.1093/scipol/scae070

Wensveen, S. & Matthews, B. (2014). Prototypes and prototyping in design research. In P. A. Rodgers, & J. Yee (Eds.), *Routledge companion to design research* (pp. 262–276). Routledge. https://doi.org/10.4324/9781315758466-25

Wised, S., & Inthanon, W. (2024). The evolution of STE*A*+M-based programs: Fostering critical thinking, collaboration, and real-world application. *Journal of Education and Learning Reviews, 1*(4), 13–22. https://doi.org/10.60027/jelr.2024.780

Wudarczyk, O. A., Kirtay M., Kuhlen, A. K., Abdel Rahman, R., Haynes, J-D., H., V. V. & Pischedda, D. (2021). Bringing together robotics, neuroscience, and psychology: Lessons learned from an interdisciplinary project. *Frontiers in Human Neuroscience, 15*, 630789. https://doi.org/10.3389/fnhum.2021.630789

Wysocki, B. T., McDonald, N., Fanto, M., & McEwen, T. (2013). Designing STEM activities to complement neural development in children. In *Proceedings of the 2013 IEEE Integrated STEM Education Conference (ISEC)*, 1–5. https://doi.org/10.1109/ISECON.2013.6525234

Yoon, S. A., Miller, K. M., Richman, T., Noushad, N., Hageman, G., Liu, Y., Zhang, W., & Cottone, A. M. (2024). Developing teachers' disciplinary knowledge for high school STEM integration: A review of a decade of educational research. *Review of Educational Research*. https://doi.org/10.3102/00346543241289566

CHAPTER 3

The Plexus Model of STE*A*$^{+}$M Education: Fostering *A*gency, Innovation, and Societal Engagement

3.1 Introduction

The rapid pace of technological, environmental, and social change has necessitated a curriculum shift from content knowledge mastery to interdisciplinary problem-solving. Traditional STEM (science, technology, engineering, and mathematics) education, which prioritises technical expertise within separate subject areas, is no longer sufficient in preparing students for an increasingly complex world. Integrating additional learnings within STEM, forming STE*A*$^{+}$M (science, technology, engineering, *A*gency, and mathematics), offers a more holistic educational approach that encourages students to think critically and engage ethically (Wysocki et al., 2013; Yilmaz & Ayaz, 2021), and apply their knowledge in meaningful, real-world contexts (Bevan et al., 2019). At the heart of this transformation is *A*gency and innovation, which positions students as active participants in shaping the world around them rather than passive recipients of knowledge.

The interdisciplinary nature of STE*A*$^{+}$M education the inclusion of *A*gency and innovation (*A*$^{+}$) challenges conventional notions of subject boundaries. Rather than treating science, mathematics, and technology as isolated fields, a plexus approach emphasises the interconnectedness of these disciplines, weaving them into socially and culturally relevant scenarios (Wood & Puttick, 2019). This approach enables students to

G. Kidman et al., *Knowledge Generation in STEM and STEAM Education*, https://doi.org/10.1007/978-3-032-03247-8_3

understand their work's ethical, civic, and societal implications, fostering responsible *A*gency, innovation and sustainable problem-solving.

A key argument in this chapter is that *A*gency and innovation should be central to STEAM education creating STE*A*⁺M education. *A*gency refers to students' capacity to take initiative, engage in decision-making, and apply their learning to shape their communities and the broader world (Kidman et al., 2024). While many education policies acknowledge the importance of student *A*gency, its integration within STEM education remains limited. Traditionally, STEM subjects have emphasised technical proficiency and problem-solving skills (Goldenkoff, 2020). However, without explicit connections to ethical reasoning, civic engagement, and social responsibility, students may struggle to see the broader impact of their work. By embedding action-oriented learning experiences into the STEAM curriculum, students develop the confidence and ability to apply their knowledge meaningfully. This shift requires an educational framework that integrates ethical inquiry, communication, and interdisciplinary collaboration, ensuring that students are equipped with technical skills and the ability to think critically about the implications of their work.

This chapter introduces the ***Plexus Model of STEA⁺M Education*** to conceptualise the integration of *A*gency, problem-solving, and design thinking innovation. This model frames STE*A*⁺M as an interactive and evolving framework where students are encouraged to develop problem-solving skills, ethical reasoning, and societal engagement through interdisciplinary inquiry. The model operates on the principle that design thinking, *A*gency, and problem-solving must function interdependently, reinforcing one another to cultivate innovative, ethically responsible learners.

The chapter explores how this model can be applied within the Australian Curriculum, examining existing Content Descriptions, General Capabilities, and Cross-curriculum Priorities that support the development of student *A*gency. Through this analysis, it becomes clear that while some curriculum elements encourage interdisciplinary engagement, there is significant potential for STE*A*⁺M education to be further enriched with action-oriented, ethical, and culturally relevant learning experiences.

3.2 Defining STEM and the Shift to STEA⁺M Education

STEM is an acronym for Science, Technology, Engineering, and Mathematics and is widely recognised as a strategic priority in education policies across many governments (Kidman & Tan, 2025; Lyons, 2020). While STEM is often framed as essential for national economic and technological advancement, the conventional understanding of STEM education remains primarily confined to discrete disciplinary silos (Liu & Li, 2023) and some form of integration, but integration is rarely defined. Among these disciplines, science and mathematics typically receive the most emphasis, as they are traditionally regarded as high-status academic subjects (English, 2016). As described in Chapter 2, the Australian Curriculum categorises the 'E' as engineering under "Design & Technology" and is grouped with "Digital Technologies" within the Technologies learning area. However, digital technologies often receive greater priority, leaving engineering and design technology underrepresented despite their inclusion in the STEM acronym (English, 2016).

Beyond this disciplinary framework, Australia's National STEM Education Strategy aims to address several key objectives (Education Council, 2015). These include revitalising student engagement in mathematics and science due to declining participation rates (Marginson et al., 2013), strengthening links between STEM education and professional industry practices (Australian Education Council, 2018), and fostering twenty-first century skills necessary for an evolving global landscape (Prinsley & Baranyai, 2015). However, despite these objectives, the implementation of STEM education in schools faces significant challenges that hinder its effectiveness.

Forbes et al. (2021) highlight several challenges that impede the realisation of STEM education goals in Australian classrooms. At the policy level, the curriculum's rigid structure separates learning areas and does not effectively integrate STEM learning progressions across various school years. In schools, the emphasis is primarily on mathematics and science, which often sidelines technology and engineering. Additionally, STEM is frequently treated as an extracurricular activity instead of being thoroughly incorporated into the curriculum. Other systemic challenges include restrictive assessment practices that limit opportunities for inquiry-based learning, a shortage of STEM-qualified teachers, and inadequate guidance for implementing engineering concepts within the curriculum. Moreover, there is a significant ambiguity surrounding the definition of STEM and its effective implementation in education.

A fundamental limitation of current STEM education models is the persistent emphasis on subject content rather than on interconnected, applied learning experiences and outcomes. While STEM education equips students with technical expertise, it often neglects human and ethical dimensions of knowledge application. This prevailing approach reinforces disciplinary silos (Garibay, 2015), thereby restricting interdisciplinary engagement and limiting students' ability to perceive STEM as a cohesive and dynamic field. While 'integration' is frequently cited as a key goal, it is seldom meaningfully conceptualised or executed (Kidman et al., 2024). As a result, knowledge is often delivered in fragmented ways, with interconnections between STEM fields left implicit. This approach diminishes the potential of STEM education to develop students' ability to address complex, real-world challenges (Goldenkoff, 2020). The societal link is underutilised, especially when an ethical and socially responsible purpose is called for (Yilmaz & Ayaz, 2021). Kidman et al. (2024) and Wood and Puttick (2019) argue that we need a curriculum that lays the foundations for disciplinary learning. They argue that we must engage students with pressing global issues and their roles as active citizens. A plexus curriculum (Kidman et al., 2024; Wood & Puttick, 2019) prioritises the interconnections between theories, methodologies, and societal issues, fostering holistic and networked learning experiences. Applying this concept to STEM education transforms how students engage with knowledge, shifting away from a reductive, content-driven model toward one that encourages agency and complex problem-solving.

A plexus approach to STEM curriculum foregrounds the interwoven nature of scientific, technological, engineering, and mathematical concepts, explicitly connecting students' lived experiences and societal challenges (Kidman et al., 2024). The plexus approach would move beyond simply teaching STEM disciplines in isolation or a form of integrated STEM education, drawing on content from the four STEM disciplines and none other. STEM education is advocated to foster an applied, hands-on learning approach that mirrors the real-world contexts in which these fields operate (Lyons, 2020; Sutaphan & Yuenyong, 2019). However, a plexus approach to the STEM curriculum would further position students as active agents in their learning. By embedding STEM education within meaningful social contexts and drawing on various other disciplines, students are better equipped to engage with contemporary issues such as climate change, technological ethics, and sustainable development (Karahan & Roehrig, 2016). Crucially, this plexus approach would empower students to become agents of change, fostering disciplinary expertise and the ability to apply their knowledge in transformative ways.

As a result, a shift from STEM to STEA^{+}M is necessary. This will ensure that students understand how to create technology and, importantly, why and how they should use it responsibly (Yilmaz & Ayaz, 2021). To appreciate what the 'A' in STEA^{+}M means for a plexus STEM curriculum, we turn to Lyons (2020) to discuss the STEM acronym. Lyons "looks beyond the distractions of the acronym" (p. 225) and suggests we must consider the relationships between the disciplinary areas. Lyons refers to a 'STEM palette' as a way to blend various disciplines for innovative solutions to real-world challenges. Lyons, like many other STEM education researchers, describes several modifications to the STEM acronym as a way of considering the inclusion of different disciplines, "most prominently STEAM (STEM and the Arts), but also STREAM (STEM, Reading and the Arts), and STEMM (STEM and Medicine)" (Lyons, 2020, p. 231). Lyons explains that the conceptual understandings underlying the new acronym may be of lesser concern than its pronounceability. Thus, it is challenging to locate a single definition of STEAM "or by extension of what constitutes STEAM practices (Liao, 2019, p. 44). Liao discusses the STEAM movement as:

- Driving progress that is good for society
- Needing to include the study of ethical use of new technology and scientific advances
- Requiring a diversity and gender equality perspective

From this perspective, STEAM must be seen as going beyond the simplistic inclusion of the Arts. It needs to include the epistemologies of the Arts, going beyond the disciplinary expansion.

STEAM education is far more than a disciplinary expansion. In this chapter, we conceive it to represent a paradigmatic shift in educational purpose and practice. STEAM education offers a robust framework for rethinking how knowledge is produced, whose voices are legitimised, and what purposes education serves. At its core, STEAM invites an epistemological rebalancing, where creative, ethical, and cultural dimensions are placed alongside scientific and technical knowledge. This is particularly significant in contexts of decolonisation, where STEAM can serve to challenge Eurocentric dominance in knowledge systems by valuing Indigenous, local, and alternative ways of knowing (O'Donoghue et al., 2024). In doing so, STEAM becomes a conduit for recognising the relational, holistic, and land-based knowledge traditions often excluded from conventional STEM pedagogies. Through transdisciplinary inquiry, students can engage with global and local challenges, such as climate change, food security, or urban inequality, through lenses that foreground justice, cultural meaning, and collective agency. The inclusion of the Arts does not merely make STEM more engaging; it enables education to address complex socio-ecological problems by integrating emotional intelligence, critical reflection, and ethical responsibility. This positions STEAM as a transformative approach that promotes sustainability, not as a technical problem to be solved, but as a profoundly human question of how we live, relate, and care for one another and the planet. In this way, STEAM is aligned with social justice imperatives, providing educators and learners with the tools to imagine and enact more equitable, inclusive, and regenerative futures.

STEA^{+}M education, as developed in this chapter and book, is housed in the makerspace and promotes sustainability. We do not follow the ambiguous, both narrow and wide, conceptualisations of the A in STEAM (Herranen et al., 2021), where the fifth letter 'A' is seen to be the Arts, Architecture or Aesthetics (Herranen et al., 2021). Instead, we promote the '*A*' in STEA^{+}M to signify *Agency* (Kidman et al., 2024) and *A*ll students (White, 2014). STEA^{+}M learning outcomes, therefore, ensure that students not only acquire content knowledge, skills, and technical expertise but also develop the ability to apply their knowledge for ethical and socially responsible purposes. When STEA^{+}M fosters positive attitude development and social awareness, allowing individuals to apply their knowledge for the benefit of everyone, the Agency application proves innovative. Thus, we consider the '*A*' to be 'A^{+}', giving rise to STEA^{+}M. The innovation is an important learning outcome from the design thinking in the technology's 'T' of STEA^{+}M.

3.3 STE*A*⁺M Education: A Plexus Curriculum

Our *Plexus Model of STEA⁺M Education* (see Fig. 3.1) is a trialectic model integrating three foundational elements: (1) STE*A*⁺M as a problem-solving framework for engaging with societal issues; (2) *A*gency as the foundational core of student-driven learning, and (3) Design Thinking as the process of inquiry and innovation. These elements function interdependently, reinforcing one another to cultivate students as active problem-solvers, ethical decision-makers, and change agents. The model is structured around the principle of circular causality, where each domain strengthens and evolves through interaction with the others (Roos, 2024).

In the Plexus Model of STE*A*⁺M Education, learners are positioned as active participants in shaping their futures rather than passive recipients of knowledge. As *A*gents of change, they engage with disciplinary knowledge not as static content to be memorised but as a dynamic tool

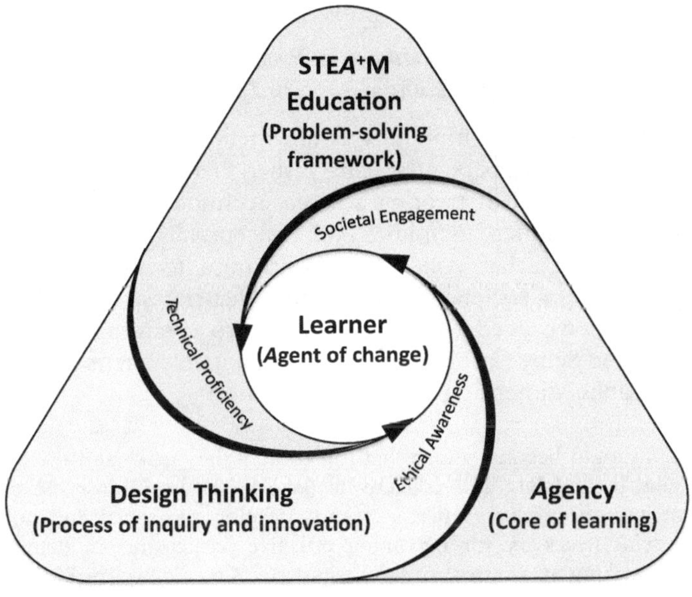

Fig. 3.1 Plexus model of STE*A*⁺M education (*Source* Modified from Kidman et al., 2024)

for inquiry, innovation, and problem-solving. This requires an education prioritising adaptability, ethical reasoning, and the ability to navigate uncertainty (Gouëdard et al., 2020; Yilmaz & Ayaz, 2021).

Rather than confining learning to pre-defined disciplinary silos, students develop the capacity to synthesise insights across fields, applying their knowledge to address societal and environmental challenges. The learner-centred approach fosters curiosity, critical engagement, and collaborative problem-solving, enabling students to question assumptions, design solutions, and contribute meaningfully to their communities. Therefore, the teacher's role extends beyond content delivery to facilitating opportunities for students to apply their learning in real-world contexts (Karahan & Roehrig, 2016). By embedding *A*gency within the curriculum, education transforms from a system of knowledge transmission into a platform for action, empowering students to participate in shaping the world they will inherit. Below, each of the three foundational elements is discussed regarding their influence on the learner as an *A*gent of change.

3.3.1 STEA⁺M Education as a Problem-solving Framework for Engaging with Societal Issues

The STE*A*⁺M approach functions as an intellectual tool that facilitates inquiry, analysis, and problem-solving rather than existing as an isolated field of study. Instead of treating science, technology, engineering, and mathematics as distinct disciplines, we conceptualise them as interconnected perspectives that collectively contribute to understanding and addressing complex societal problems and challenges. It has long been recognised that we need to educate people to create a sense of social action. Pike and Selby (1988) advocated for an involvement consciousness and preparedness dimension to global education:

> Students should become aware that the choices they make and the actions they take individually and collectively have repercussions for the global present and the global future. ... [and] develop the social and political action skills necessary for becoming effective participants in democratic decision making at a variety of levels, grassroots to global. (p. 35)

Through the Plexus Model of STE*A*⁺M Education, students develop a holistic approach to societal problem-solving that transcends traditional subject boundaries. They integrate these disciplines with a notion of *A*gency. Research increasingly highlights the potential for students to develop *A*gency through education (Sarid & Goldman, 2021; von Braun, 2017). While many curricula and international educational charters emphasise fostering student *A*gency, its integration into STEM education remains limited. To fully realise the potential of STEM learning, *A*gency must be explicitly embedded within teaching practices, ensuring that students acquire technical knowledge and learn how to apply it in meaningful, real-world contexts. Geography, from the Humanities, provides a strong example of how STEM education can be linked to student *A*gency. The 2016 International Charter on Geographical Education (International Geographical Union Commission on Geographical Education [CGE], 2016) promotes a teaching approach from students' real-life experiences. It encourages them to inquire, enhance their critical skills, and interact with significant global and local challenges. Geography education naturally cultivates twenty-first century skills, such as problem-solving, inquiry, and critical thinking, by emphasising fieldwork and real-world social contexts. These competencies prepare students to address complex societal and environmental challenges (Garibay, 2015). The approach taken in geography education demonstrates how STEM subjects can be taught in ways that empower learners to engage actively in their education and their communities. By incorporating inquiry-based learning, real-world problem-solving, and interdisciplinary thinking, STEM education can move beyond content delivery to actively foster student *A*gency (Garibay, 2015). This transition is crucial for equipping young learners to engage with and contribute to a complex, interconnected world.

A problem-centred approach to STEA^+M education encourages students to identify and engage with real-world challenges that require interdisciplinary understanding. This method creates a learning atmosphere in which students are not just passive recipients of information but become active problem-solvers who apply their knowledge to significant and relevant challenges (Lyons, 2020; Sutaphan & Yuenyong, 2019). Through this approach, students cultivate the ability to dynamically apply STEM competencies, selecting and integrating appropriate methodologies, technologies, and analytical frameworks based on the specific demands of a problem. The Plexus Model of STEA^+M Education thus emphasises adaptability and the capacity to think across disciplines, preparing students to navigate the complexities of contemporary global issues (Wood & Puttick, 2019).

The Plexus Model of STE*A*⁺M Education emphasises the importance of considering broader social, cultural, and ethical dimensions of STEM applications. Technological and scientific advancements do not exist in a vacuum; they have profound implications for individuals, communities, and societies. By embedding STEM learning within socially and culturally relevant contexts, this model ensures that students engage with knowledge not as a collection of abstract facts but as a set of transformative resources that can be harnessed for the common good (Guyotte, 2019). Ethical considerations, cultural sensitivity, and societal needs become integral to problem-solving, fostering learners' sense of responsibility and civic engagement (Yilmaz & Ayaz, 2021).

In practical terms, the Plexus Model of STE*A*⁺M Education encourages the development of interdisciplinary learnings that challenge students to synthesise knowledge from multiple fields. For example, addressing environmental sustainability through a problem-solving approach requires an understanding of ecological systems (science and geography), the development of sustainable technologies (engineering), the application of mathematical modelling (mathematics), the use of digital innovations (technology), and an awareness of social and policy implications (arts, humanities, and social sciences). By engaging in such problem-solving, students learn to navigate the intersections between disciplines, develop critical thinking skills, and cultivate an appreciation for the complexity of real-world challenges.

In the context of a Plexus Model of STE*A*⁺M Education and a problem-solving framework, the student emerges as a central agent of change by engaging with complex societal issues through interdisciplinary inquiry. The Plexus Model of STE*A*⁺M Education prepares students for the demands of a rapidly evolving world by situating STEM education within a broader, socially responsive framework and ultimately shifting the focus of education from passive knowledge acquisition to active problem-solving, equipping students with the intellectual tools necessary to make meaningful contributions to their communities and beyond. The Plexus Model of STE*A*⁺M Education cultivates technical proficiency, ethical awareness, interdisciplinary thinking, and the ability to address pressing societal challenges through problem-solving that has *A*gency at the core of learning.

3.3.2 STEA⁺M Education as Agency: The Core of Learning

*A*gency is the second central force in the plexus approach to STE*A*⁺M education, equipping students with the skills and motivation to apply their learning for the betterment of society. In the evolving education landscape, fostering student agency has become a central concern in preparing learners for a rapidly changing world. Horn and Gresalfi (2021) argue that students must be empowered to identify social dilemmas, critically assess information, and apply their learning to enact change. Within STE*A*⁺M education, this empowerment is positioned at the heart of learning, transforming students from passive recipients of knowledge into active participants who shape their educational journeys. This approach aligns with Roberts and Bybee's (2014) conceptualisation of STEM literacy, which emphasises that students must develop knowledge, attitudes, and skills that enable them to act in ways that responsibly shape their environments.

*A*gency within the educational context manifests through students' decision-making abilities, self-regulation, and ownership of their learning processes (Moses et al., 2020). It is about developing technical competence and fostering the confidence and motivation necessary for students to engage in meaningful problem-solving and societal transformation. Fostering student *A*gency has become a central concern in preparing learners for a rapidly changing world (Kidman et al., 2024). When students take ownership of their learning, they engage more deeply with content and develop greater confidence in their ability to effect change. Ke et al. (2020) found that students who acquire systems thinking skills and epistemic tools demonstrate a higher capacity for autonomous learning and problem-solving. Thus, an *A*gency-centred approach enhances students' academic skills and broader critical thinking and decision-making abilities.

One of the most effective ways to foster *A*gency in STE*A*⁺M education is to ensure that learning is relevant to students' lives. Research shows that when students see clear connections between their studies and real-world issues, they are more likely to be engaged and to take an active role in their education (Karahan & Roehrig, 2016). Grounding STE*A*⁺M content in local environmental and social challenges has increased student motivation and engagement (Mathers et al., 2024). For example, Morales-Doyle (2017) found that participation in justice-oriented science education activities assisted in developing students' ability to see themselves as able to analyse both scientific and societal issues critically. This approach enhances students' understanding of complex problems and prepares them for civic and scientific decision-making (Lombardi et al., 2022). By embedding real-world contexts into curricula, as advocated by a Plexus Model of STE*A*⁺M Education, educators can cultivate a sense of responsibility and *A*gency in students, motivating them to apply their learning meaningfully (Mathers et al., 2024). Teachers thus play a critical role in designing learning tasks that empower students to take risks, explore new ideas, and engage in interdisciplinary inquiry. The role of the teacher is explored further in Chapter 8.

Research indicates that many educators lack confidence in implementing student-driven pedagogies (Johnson et al., 2020). Professional development focused on facilitating *A*gency can help teachers transition from traditional instructional roles to ones that encourage student ownership of learning. We explore this further in Chapters 8 and 9. Effective professional development programs should provide educators with pedagogical strategies for incorporating inquiry-based learning, collaborative problem-solving, and real-world applications into their teaching practices.

3.3.3 STEA$^+$M Education as a Design Thinking Process of Inquiry and Innovation

Shifting toward an *A*gency-centred STE*A*$^+$M approach transforms classrooms into dynamic environments where students see themselves as active contributors to knowledge creation and problem-solving. By integrating real-world challenges, fostering student-led inquiry, and supporting teachers in adopting facilitative roles, educators can create an engaging and transformative learning experience. Developing agency in STE*A*$^+$M education requires a fundamental pedagogy and shift in classroom dynamics. Traditional teaching models, which position educators as the primary source of knowledge, often restrict students' ability to take initiative, think independently, and make meaningful decisions. However, Karahan and Roehrig's (2017) research indicates that when teachers transition into facilitative roles, agency flourishes, empowering students to navigate complex, interdisciplinary problems with greater autonomy and confidence. This shift necessitates rethinking instructional strategies to create an environment where students are encouraged to question, experiment, and collaborate in pursuing knowledge.

Kidman et al. (2024) emphasise that design thinking, where the teacher is a facilitator, is a pedagogical approach that provides a critical link between STEM education and societal agency, highlighting its role in equipping students with problem-solving skills applicable beyond the classroom. Micheli et al. (2019) characterise design thinking as a cyclical process consisting of three main phases: comprehending users' needs and pinpointing obstacles; participating in creative exploration, ideation, iterative experimentation, implementation, and prototyping; and assessing and refining prototypes to transform them into functional, practical solutions. This structured yet flexible approach encourages students to take ownership of their learning by emphasising real-world problem-solving and human-centred innovation.

By incorporating design thinking into a plexus STE*A*$^{+}$M curriculum, students have opportunities to engage in inquiry-driven learning that is both meaningful and relevant (Marginson et al., 2013). Rather than passively receiving information, students actively participate in learning, critically analysing problems, proposing solutions, and refining their ideas through continuous feedback and iteration (Kidman et al., 2024). This approach nurtures essential twenty-first century skills such as creativity, adaptability, collaboration, and ethical decision-making (Wysocki et al., 2013; Yilmaz & Ayaz, 2021). Additionally, it encourages students to embrace uncertainty and complexity—an essential mindset for tackling global challenges and advancing scientific and technological progress (Gouëdard et al., 2020).

Applying design thinking within a plexus STE*A*$^{+}$M curriculum is particularly effective in fostering a sense of *A*gency because it mirrors how professionals in STEM fields approach problem-solving. Engineers, scientists, and designers do not simply apply static knowledge; they engage in iterative inquiry, experimentation, and refinement cycles to develop innovative solutions (Kidman et al., 2024). Students who participate in similar processes acquire technical knowledge and develop the confidence and autonomy to apply their skills in diverse contexts. This experiential approach to learning makes science, technology, engineering, and mathematics more accessible and engaging, particularly for students who may not initially see themselves as belonging in these fields (Marginson et al., 2013).

Ultimately, integrating *A*gency into STE*A*$^{+}$M education through frameworks such as design thinking offers a powerful way to prepare students for the complexities of the modern world. Education can move beyond passive knowledge transfer to cultivate innovation, adaptability, and a genuine passion for learning by positioning students as active problem-solvers, encouraging interdisciplinary collaboration, and supporting teachers in adopting facilitative roles. As schools evolve in response to societal and technological changes, embracing an agency-centred approach will be essential in fostering the next generation of creative thinkers, ethical decision-makers, and engaged citizens.

3.4 Exploring Agency in the Curriculum: An Australian Case Study

This section explores the Australian Curriculum as a case study, where the compulsory years of education, Foundation to Year 10 (5 to 16 years of age), are the case. The Australian Curriculum has three integrated components: Learning Areas,[1] General Capabilities, and Cross-Curriculum Priorities that the knowledge students need to know (disciplines) and what sort of people they ought to become (General Capabilities and Cross-Curriculum Priorities) (Lingard, 2018) are the outcomes after 10 years of compulsory learning. Depth is generally emphasised rather than breadth (ACARA, 2020). This case study focuses on analysing the content descriptions (the essential knowledge, understandings, and skills of the curriculum) and their potential for developing *Agency*.

Table 3.1 summarises the occurrence of *Agency* (expressed as verb clauses, e.g., 'taking action' or 'to act') in the Australian Curriculum. We located 35 Content Descriptions that support the learning of *Agency*. The selection of content descriptions from the Australian Curriculum highlights the critical role of communication, engagement, and participation in fostering student *Agency*. *Agency*, in this context, refers to the capacity of students to act upon knowledge, influence society, and contribute to the common good.

The Cross-curriculum Priorities and General Capabilities of the Australian curriculum were also analysed. All teachers across all Learning Areas are responsible for integrating these aspects of the Australian

[1] 'Learning areas' is the overarching collective term for disciplines of study in the Australian Curriculum A Learning Area is akin to a 'subject' in other jurisdictions.

Table 3.1 Agency content descriptions in the Australian curriculum

Learning area	*Year level*	*Number of content descriptions and examples*
Languages	5–10	26 e.g., Take action in collaborative tasks, activities and experiences that involve negotiation, making arrangements, problem-solving and shared transactions (Chinese Year 7 & 8); Take responsibility by initiating interactions, solving problems and encouraging others to act (Indonesian Year 9 & 10); Plan, participate in and evaluate events and activities that involve taking shared action, negotiating and problem-solving and that draw upon personal and community knowledge (Framework for Aboriginal Languages and Torres Strait Islander Languages Year 7 to 10)
Humanities	3–10	7 e.g., Reflect on learning to propose actions in response to an issue or challenge and consider possible effects of proposed actions (HASS Year 3); Reflect on and evaluate findings of an inquiry to propose individual and collective action in response to a contemporary geographical challenge, taking account of environmental, economic, political and social considerations; and explain the predicted outcomes and consequences of their proposal (Geography Year 9)
Health and Physical Education	Foundation	2 e.g., Identify people and demonstrate protective behaviours and other actions that help keep themselves safe and healthy
English	10	1 e.g., Plan, rehearse and deliver presentations, selecting and sequencing appropriate content and multimodal elements to influence a course of action (English Year 10)

Curriculum into their Learning Area content. Two of the three Cross-curriculum Priority areas were found to advocate for *Agency* development. In *Asia and Australia's Engagement with Asia* Cross-curriculum Priority, we identified the following statement: "An understanding of Asia underpins the capacity of Australian students to be active and informed citizens working together to build harmonious local, regional and global communities..." (ACARA, n.d. -a). Another statement within the Humanities and Social Sciences calls for "students [to] reflect on how Australians can participate in the Asia region as active and informed citizens" (ACARA, n.d. -a). The *Sustainability* Cross-curriculum Priority (ACARA, n.d. -b) is quite explicit, advocating for:

- Students to participate critically and act creatively in determining more sustainable ways of living
- Students to undertake actions to improve sustainability, including developing the knowledge, skills, values and world views necessary for people to act in ways that contribute to living sustainably
- Actions that support more sustainable living
- Building capacities for thinking and acting in ways to create a more sustainable future, promoting reflective thinking processes to empower actions that will lead to a more equitable and sustainable future

There is an additional emphasis on a future-oriented education that focuses on protecting environments and creating a more ecologically and socially just world through informed action:

- The sustainability of ecological, social, and economic systems is achieved through informed individual and community action that values local and global equity and fairness across generations into the future.
- Actions for a more sustainable future reflect the values of care, respect, and responsibility and require us to explore and understand our environments.
- Designing action for sustainability requires evaluating past practices, assessing scientific and technological developments, and making balanced judgements based on projected economic, social, and environmental impacts.

- Sustainable futures result from actions designed to preserve and/or restore the quality and uniqueness of environments.

The *General Capabilities* "play a significant role in the Australian Curriculum, equipping students to live and work successfully in the twenty-first century. A capability encompasses knowledge, skills, behaviours, and dispositions" (ACARA, n.d. -c). The *Personal and Social Capability* is interesting as it advocates *social awareness* as an organising element. Students explore the role of advocacy and how to critique societal constructs. Students learn to appreciate diverse perspectives and contribute to civil society by developing and acting with personal and social capability (ACARA, n.d. -c).

The analysis of the 35 Content Descriptions, Cross-curricula Priorities, and General Capabilities revealed that the Australian Curriculum is structured to empower students to become active participants in their learning and broader societal discourses. Four themes emerged from the analysis.

3.4.1 Communication as a Form of Agency

A significant portion of the Content Descriptions are drawn from Language subjects, demonstrating the essential role of language in shaping identity, articulating viewpoints, and engaging in social discourse. The curriculum emphasises students expressing opinions and ideas in multiple languages, initiating and sustaining interactions, and engaging in discussions on social and cultural issues. These aspects highlight the recognition of language as a tool for *A*gency, enabling students to participate in diverse conversations crucial for civic engagement and social responsibility (Mouboua et al., 2024).

Effective communication is foundational to action in both personal and societal contexts. The ability to express ideas persuasively and engage in meaningful dialogue is critical for participation in democratic processes, advocacy, and social change (Mouboua et al., 2024). Within a Plexus Model of STEA^{+}M Education, the emphasis on communication aligns with the need for scientists, engineers, and technologists to articulate their findings, collaborate across disciplines, and engage with the public. Integrating strong communication skills within a Plexus Model of STEA^{+}M Education reinforces the broader objective of preparing students to be active and informed citizens.

3.4.2 *Citizenship, Ethical Reasoning, and Decision-Making*

Several Content Descriptions originate from Civics, Economics, and Humanities, which inherently focus on social systems, governance, and ethical considerations. These Content Descriptions encourage students to explore how actions impact individuals and communities, evaluate social and economic issues from multiple perspectives, and understand civic responsibilities and rights. Such elements are essential for fostering informed, responsible citizens who can critically assess societal structures and advocate for positive change (Yilmaz & Ayaz, 2021).

Incorporating these elements into a plexus STE*A*⁺M framework ensures that STEM education is not isolated from broader societal issues. Ethical reasoning and decision-making are crucial in scientific and technological fields where innovations can have far-reaching consequences. By embedding ethical and civic reasoning within STEM education, students develop the capacity to approach technological and scientific challenges with a balanced perspective that prioritises social justice, equity, and sustainability (Yilmaz & Ayaz, 2021).

3.4.3 *Agency: The Role of Participation and Change*

A notable characteristic of the selected Content Descriptions is the frequent use of physical action-oriented verb clauses such as participate, initiate, explore, present, engage, and prepare. The emphasis on these verb clauses signals a pedagogical approach that extends beyond knowledge acquisition to active engagement and real-world application. This aligns with transformative learning perspectives, where education is not simply about transferring knowledge but about empowering students to contribute meaningfully to their communities and the world.

Integrating action-oriented learning within the Plexus Model of STE*A*⁺M Education reinforces the idea that scientific and technological expertise must be paired with practical application and societal engagement. For example, students working on engineering projects should be encouraged to engage with communities to understand their needs and co-develop solutions. Similarly, STEM-related research should include participatory methodologies that involve stakeholders in decision-making processes (Goldenkoff, 2020). The presence of these verb clauses across language and humanities subjects suggests a shared pedagogical emphasis on active learning leading to *A*gency, which can be effectively harnessed within a Plexus Model of STE*A*$^{+}$M Education to cultivate a generation of students equipped to address complex global challenges.

3.4.4 The Role of Multimodal and Cross-Disciplinary Learning

Multiple disciplines include Content Descriptions calling for action, particularly Languages and Humanities, highlighting the importance of an interdisciplinary approach to fostering *A*gency. Students are encouraged to interpret and present information across different cultural contexts, critically engage with societal issues and make informed economic and civic decisions. Despite the transformative potential of STEM education, research indicates that students in STEM fields tend to be less engaged in political and social issues than their humanities peers (Goldenkoff, 2020). This disconnect may stem from traditional STEM curricula, which often prioritise technical mastery and problem-solving over broader societal considerations. While STEM subjects equip students with analytical and logical skills, they may not always provide explicit opportunities to examine scientific and technological advancements and ethical, political, and social dimensions (Kidman et al., 2024). As a result, STEM learners may struggle to see how their knowledge connects to real-world challenges that require civic participation, ethical reasoning, and an awareness of social justice issues. Garibay (2015) also observed that many STEM students perceive themselves as lacking the capacity to influence social change, a less common belief among students in the Humanities. This perception may be reinforced by the way STEM subjects are traditionally taught, often in isolated disciplinary silos that focus on technical expertise rather than interdisciplinary connections. Without exposure to discussions on the societal impact of their work, STEM students may view their knowledge as detached from broader human concerns, limiting their ability to see themselves as active contributors to social and political discourse.

To tackle this gap, modern STEM education should intentionally collaborate with the humanities to promote a more comprehensive understanding that crosses disciplinary lines (Bevan et al., 2019). This integration should not merely be an add-on but an essential component of STEM learning, where ethical inquiry, civic engagement, and interdisciplinary problem-solving are embedded into the curriculum. By situating STEM education within broader social and moral contexts, students can be encouraged to see their technical expertise as a tool for societal transformation. Through interdisciplinary projects, case studies on real-world challenges, and engagement with culturally relevant societal issues, students can develop a sense of *A*gency and responsibility, recognising their role in shaping the world.

A STEM curriculum that includes historical perspectives, philosophical debates, and ethical dilemmas related to scientific and technological progress allows students to evaluate their work's consequences critically. Thus, it becomes a Plexus Model of STE*A*+M Education. When students are given opportunities to examine how technological innovations impact communities, influence policy, and shape cultural narratives, they develop a more nuanced understanding of their potential to effect change. Encouraging STEM students to engage in civic participation, advocacy, and interdisciplinary collaboration will produce graduates skilled in their technical fields and prepared to address complex, real-world problems with social consciousness and ethical responsibility (Goldenkoff, 2020). This suggests that *A*gency is not confined to any single discipline but instead emerges from multiple fields of knowledge. The 2016 Charter (International Geographical Union Commission on Geographical Education [CGE], 2016) claims the role of students is to be active citizens who participate in community-driven inquiry and social action. We align with Banks and Barlex (2020), who propose the concept of *looking sideways* in education. Banks and Barlex suggest that students' learning experiences are enriched when teachers draw connections across disciplinary boundaries and build upon prior knowledge gained in other subject areas.

From a Plexus Model of STE*A*+M Education perspective, there is a need to embed social engagement within STEM education. Traditionally, STEM education has prioritised technical and scientific literacy, often without explicit connections to broader social and ethical considerations. However, expanding STEM to a Plexus Model of STE*A*+M Education focusing on *A*gency necessitates emphasising how scientific knowledge is communicated, applied, and used for societal improvement. STEM activities should involve technical problem-solving and incorporate social impact, ethical decision-making, and public engagement discussions. A plexus STE*A*+M framework fosters interdisciplinary learning and can produce well-rounded, technically proficient, and socially conscious students.

3.5 Implications for a STEA^{+}M Framework Where "A^{+}" Stands for Agency and Innovation

The insights derived from the Australian Curriculum Content Descriptions, Cross-curricula Priorities and General Capabilities provide a strong foundation for redefining STEM education as a Plexus Model of STEA^{+}M Education, with innovative Agency at its core. Three key strategies should be considered to integrate these insights into a Plexus Model of STEA^{+}M Education that emphasises the value of interdisciplinary STEM concepts, including culturally relevant societal scenarios to complement traditional school subjects.

First, ethical and civic reasoning should be integral to STEM learning in schools, ensuring that students connect their scientific and technological knowledge to real-world societal challenges (Wysocki et al., 2013; Yilmaz & Ayaz, 2021). Young learners should be encouraged to analyse the ethical implications of innovations, considering their impact on communities and the environment. By integrating case studies, classroom debates, and interdisciplinary projects that reflect culturally relevant scenarios, students can develop the ability to critically assess how STEM solutions align with social responsibility and contribute to the common good.

Second, effective communication and active participation in societal issues are essential for students to see the relevance of STEM beyond the classroom. Schools should provide opportunities for students to engage in science communication, advocacy, and community-based problem-solving (Soomro et al., 2023). This can be achieved through school projects that involve explaining scientific concepts to non-experts, collaborating with local organisations on STEM-related issues, and participating in culturally relevant initiatives that require students to translate their learning into practical, community-focused solutions.

Third, encouraging students to integrate knowledge from the Humanities and Social Sciences into their STEM studies strengthens their ability to apply technical skills in socially responsible ways (Garibay, 2015; Goldenkoff, 2020). School-based interdisciplinary projects, such as using engineering to design culturally sustainable housing, combining environmental science with Indigenous land management practices, or linking technology with ethical considerations in artificial intelligence, help students see how STEM knowledge is interconnected with broader societal concerns (Garibay, 2015; Goldenkoff, 2020). By fostering these cross-disciplinary connections, schools can empower students to become thoughtful, innovative problem-solvers who engage meaningfully with their communities.

3.6 Conclusion

The shift from traditional STEM to a Plexus Model of STE*A*⁺M Education represents a fundamental transformation in how students engage with knowledge and apply their learning. In contrast to disciplinary boundaries, the plexus model frames STE*A*⁺M as an interconnected, dynamic, and socially engaged educational framework that fosters critical thinking, ethical reasoning, and societal agency. A core argument presented in this chapter is that STE*A*⁺M education should not merely teach students technical skills but empower them to become changemakers. This means that students must understand their work's ethical, civic, and societal dimensions and ensure that they apply their knowledge in ways that contribute to the common good. By embedding interdisciplinary approaches within the curriculum, students gain a more holistic understanding of the world and can confidently navigate complexity (Bevan et al., 2019).

This chapter highlighted that ethical and civic reasoning must be integral to STE*A*⁺M education. Traditional STEM education has often overlooked scientific and technological advancements' social and ethical implications (Garibay, 2015; Goldenkoff, 2020). However, as global challenges become increasingly complex, students must develop the skills to engage with societal dilemmas. This requires technical expertise and the ability to critically evaluate their work's consequences (Moses et al., 2020).

Another critical component of STE*A*⁺M education is societal engagement. We have argued that effective communication, advocacy, and participatory problem-solving should be central to STEM learning. Schools must provide opportunities for students to engage in public discourse, collaborate with local communities, and develop solutions to real-world challenges (Karahan & Roehrig, 2016). These experiences enhance students' practical and technical skills and ensure their learning is rooted in meaningful, culturally relevant contexts.

Interdisciplinary, *A*gency-driven education is essential for preparing students for the future. However, further research is needed to explore how best to implement these strategies within diverse educational contexts. We need to know how teachers can be supported in adopting interdisciplinary pedagogies (Johnson et al., 2020), how curricula can be redesigned to facilitate cross-disciplinary connections, and how policies can be developed to prioritise student agency within national and global education frameworks. The remainder of this book aims to explore these further.

STEA^+M education, framed through a plexus framework, offers a transformative approach to learning. By emphasising ethical responsibility, interdisciplinary problem-solving, and student *A*gency, the Plexus Model of STEA^+M Education ensures that students are not just learners but active participants in shaping the world. Through this lens, STEA^+M education is about knowledge acquisition and empowering students to become innovators, decision-makers, and ethical leaders of the future.

References

Australian Curriculum, Assessment and Reporting Authority (ACARA). (2020). *The shape of the Australian Curriculum* (Version 5.0). https://www.acara.edu.au/docs/default-source/curriculum/the_shape_of_the_australian_curriculum_version5_for-website.pdf

Australian Curriculum, Assessment and Reporting Authority (ACARA). (n.d. -a). *Asia and Australia's engagement with Asia (Version 8.4)*. Australian Curriculum, Cross-curriculum Priorities. https://www.australiancurriculum.edu.au/f-10-curriculum/cross-curriculum-priorities/asia-and-australia-s-engagement-with-asia/

Australian Curriculum, Assessment and Reporting Authority (ACARA). (n.d. -b). *Sustainability (Version 8.4)*. Australian Curriculum, Cross-curriculum Priorities. https://www.australiancurriculum.edu.au/f-10-curriculum/cross-curriculum-priorities/sustainability/

Australian Curriculum, Assessment and Reporting Authority (ACARA). (n.d. -c). *Personal and social capability (Version 8.4)*. Australian Curriculum, General Capabilities. Accessed from https://www.australiancurriculum.edu.au/f 10 curriculum/general-capabilities/personal-and-social-capability/

Australian Education Council. (2018). *Optimising STEM industry school partnerships: Inspiring Australia's next generation*. Education Services Australia. https://www.chiefscientist.gov.au/sites/default/files/2019-11/optimising_stem_industry-school_partnerships_-_final_report.pdf

Banks, F., & Barlex, D. (2020). Teaching STEM in the secondary school: Helping teachers meet the challenge. *Routledge*. https://doi.org/10.4324/9780429317736

Bevan, B., Peppler, K., Rosin, M., Scarff, L., Soep, E., Wong, J. (2019). Purposeful pursuits: Leveraging the epistemic practices of the arts and sciences. In A. J. Stewart, M. P. Mueller, & D. J. Tippins (Eds.), *Converting STEM into* STE*A*$^+$M *programs* (pp. 21–38). Springer. https://doi.org/10.1007/978-3-030-25101-7_3

Education Council. (2015). *National STEM school education strategy, 2016–2026.* https://www.education.gov.au/download/12532/national-stem-school-education-strategy/23618/national-stem-school-education-strategy/pdf

English, L. (2016, August). Targeting all of STEM in the primary school: Engineering design as a foundational process [Paper presentation]. *ACER Research Conference 2016—Improving STEM Learning: What will it take?* https://research.acer.edu.au/cgi/viewcontent.cgi?article=1298&context=research_conference

Forbes, A., Chandra, V., Pfeiffer, L., & Sheffield, R. (2021). STEM education in the primary school: A teacher's toolkit. *Cambridge*. https://doi.org/10.1017/9781108868488

Garibay, J. (2015). STEM students' social agency and views on working for social change: Are STEM disciplines developing socially and civically responsible students? *Journal of Research in Science Teaching, 52*(5), 610–632. https://doi.org/10.1002/tea.21203

Goldenkoff, E. (2020, September 18). Civic engagement in the STEM classroom. *PLOS Blogs: ECR Community.* https://ecrcommunity.plos.org/2020/09/18/civic-engagement-in-the-stem-classroom/

Gouëdard, P., Pont, B., Hyttinen, S., & Huang, P. (2020). *Curriculum reform: A literature review to support effective implementation.* (OECD Working Papers No. 239). OECD Publishing. https://doi.org/10.1787/efe8a48c-en

Guyotte, K. W. (2019). Toward a philosophy of STE*A*$^+$M in the Anthropocene. *Educational Philosophy and Theory, 52*(7), 769–779. https://doi.org/10.1080/00131857.2019.1690989

Herranen, J., Fooladi, E. C., & Milner-Bolotin, M. (2021). Promoting STE*A*$^+$M in Education [Editorial]. *LUMAT: International Journal on Math, Science and Technology Education, 9*(2), 1–8. https://doi.org/10.31129/LUMAT.9.2.1559

Horn, I. S., & Gresalfi, M. (2021). Broadening participation in mathematical inquiry: A problem of instructional design. In R. G. Duncan & C. Chinn (Eds.), *International Handbook of Inquiry and Learning* (pp. 311–324). Routledge. https://doi.org/10.4324/9781315685779-21

International Geographical Union Commission on Geographical Education. (2016). *2016 international charter on geographical education.* International Geographical Union Commission on Geographical Education. http://www.igu-cge.org/2016-charter/

Johnson, J., Macalalag, A., & Dunphy, J. (2020). Incorporating socioscientific issues into a STEM education course: Exploring teacher use of argumentation in SSI and plans for classroom implementation. *Disciplinary and Interdisciplinary Science Education Research, 2*(1), Article 9. https://doi.org/10.1186/s43031-020-00026-3

Karahan, E., & Roehrig, G. (2016). Use of socioscientific contexts for promoting student agency in environmental science classrooms. *Bartin University Journal of Faculty of Education, 5*(2), 425–442. https://dergipark.org.tr/en/pub/buefad/issue/24349/258097

Ke, L., Sadler, T. D., Zangori, L., & Friedrichsen, P. J. (2020). Students' perceptions of socio-scientific issue-based learning and their appropriation of epistemic tools for systems thinking. *International Journal of Science Education, 42*(8), 1339–1361. https://doi.org/10.1080/09500693.2020.1759843

Kidman, G. & Tan, H. (2025). Exploring cognitive principles of instructional models and problem-solving processes in the STEM disciplines. In T. Barkatsas & W. Goff (Eds.), *Advances in mathematics and STEM education: Incorporating developments in neuroscience, brain science and neurodiversity* (pp. 1–37). Brill. https://doi.org/10.1163/9789004722651_001

Kidman, G., Tan, H., & Macdonald, S. (2024). Reimagining STEM education: Developing agency through STEA$^+$M and design thinking. In T. Barkatsas, P. McLaughlin, & W. Goff (Eds.), *Reimagining education for the second quarter of the 21st century and beyond: Dilemmas, challenges, advancements and innovations* (pp. 1–23). Brill. https://doi.org/10.1163/9789004688490_001

Liao, C. (2019). Creating a STEA$^+$M map: A content analysis of visual art practices in STEA$^+$M education. In M. Khine & S. Areepattamannil (Eds). *STEA$^+$M education: Theory and practice* (pp. 37–55). Springer. https://doi.org/10.1007/978-3-030-04003-1.pdf

Lingard, B. (2018). The Australian Curriculum: A critical interrogation of why, what and where to? *Curriculum Perspectives, 38*(1), 55–65. https://doi.org/10.1007/s41297-017-0033-7

Liu, S., & Li, C. (2023). Promoting design thinking and creativity by making: A quasi-experiment in the information technology course. *Thinking Skills and Creativity, 49*, Article 101335. https://doi.org/10.1016/j.tsc.2023.101335

Lombardi, D., Matewos, A. M., Jaffe, J., Zohery, V., Mohan, S., Bock, K., & Jamani, S. (2022). Discourse and agency during scaffolded middle school science instruction. *Discourse Processes, 59*(5–6), 379–400. https://doi.org/10.1080/0163853X.2022.2068317

Lyons, T. (2020). Seeing through the acronym to the nature of STEM. *Curriculum Perspectives, 40*(2), 225–231. https://doi.org/10.1007/s41297-020-00108-2

Marginson, S., Tytler, R., Freeman, B., & Roberts, K. (2013). *STEM: Country comparisons: International comparisons of science, technology, engineering and mathematics (STEM) education*. Australian Council of Learned Academies. https://acola.org.au/wp/project-2/

Mathers, B., Johnson, J., Kaufmann, A., Sinni, N., et al. (2024). Developing student agency through authentic application of socioscientific issues in STEM classrooms. *Turkish Journal of Education, 13*(5), 508–534. https://doi.org/10.19128/turje.1507933

Micheli, P., Wilner, S. J. S., Hussain Bhatti, S., Mura, M., & Beverland, M. B. (2019). Doing design thinking: Conceptual review, synthesis, and research agenda. *Journal of Product Innovation Management, 36*(2), 124–148. https://doi.org/10.1111/jpim.12466

Morales-Doyle, D. (2017). Justice-centered science pedagogy: A catalyst for academic achievement and social transformation. *Science Education, 101*(6), 1034–1060. https://doi.org/10.1002/sce.21305

Moses, L., Rylak, D., Reader, T., Hertz, C., & Ogden, M. (2020). Educators' perspectives on supporting student agency. *Theory into Practice, 59*(2), 213–222. https://doi.org/10.1080/00405841.2019.1705106

Mouboua, P. D., Atobatele, F. A., & Akintayo, O. T. (2024). Language as a tool for intercultural understanding: Multilingual approaches in global citizenship education. *Magna Scientia Advanced Research and Reviews, 11*(1), 19–30. https://doi.org/10.30574/msarr.2024.11.1.0071

O'Donoghue, R., van Staden, W., Bhurekeni, J., Snow-Macleod, J., & Ndlamlenze, L. (2024). A formative study towards the inclusion of Indigenous technologies and knowledge practices in Science, Technology, Engineering, Arts, and Mathematics (STE*A*+M) curriculum settings. *Educational Research for Social Change, 13*(1), 34–47. https://doi.org/10.17159/2221-4070/2024/v13i1a3

Pike, G. & Selby, D. (1988). *Global teacher, global learner*. Hodder and Stoughton.

Prinsley, R., & Baranyai, K. (2015). *STEM skills in the workforce: What do employers want?* Occasional Papers Series (9). Office of the Chief Scientist. https://www.chiefscientist.gov.au/wpcontent/uploads/OPS09_02Mar2015_Web.pdf

Roberts, D. A., & Bybee, R. W. (2014). Scientific literacy, science literacy, and science education. In N. G. Lederman & S. K. Abell (Eds.), *Handbook of research on science education, Volume II* (pp. 559–572). Routledge. https://doi.org/10.4324/9780203097267-38

Roos, M. (2024). Principles of complexity economics: Concepts, methods and applications. *Springer*. https://doi.org/10.1007/978-3-031-51436-4_7

Sarid, A., & Goldman, D. (2021). A value-based framework connecting environmental citizenship and change agents for sustainability—Implications for education for environmental citizenship. *Sustainability, 13*(8), 4338. https://doi.org/10.3390/su13084338

Soomro, S. A., Casakin, H., Nanjappan, V., & Georgiev, G. V. (2023). Makerspaces fostering creativity: A systematic literature review. *Journal of Science Education and Technology, 32*(4), 530–548. https://doi.org/10.1007/s10956-023-10041-4

Sutaphan, S., & Yuenyong, C. (2019). STEM education teaching approach: Inquiry from the context based. *Journal of Physics: Conference Series, 1340*, Article 012003. https://doi.org/10.1088/1742-6596/1340/1/012003

von Braun, J. (2017). Children as agents of change for sustainable development. In A. M. Battro, P. Léna, S. Sorondo, & von Braun (Eds.), *Children and sustainable development: Ecological education in a globalized world* (pp. 17–30). Springer. https://doi.org/10.1007/978-3-319-47130-3_2

White, D. W. (2014). What is STEM education and why is it important? *Florida Association of Teacher Educators Journal, 1*(14), 1–9. http://www.fate1.org/journals/2014/white.pdf

Wood, P., & Puttick, S. (2019). A plexus curriculum in school geography—A holistic approach to school geography for an endangered planet. In W. L. Filho & S. L. Hemstock (Eds.), *Climate change and the role of education* (pp. 385–399). Springer. https://doi.org/10.1007/978-3-030-32898-6_22

Wysocki, B. T., McDonald, N., Fanto, M., & McEwen, T. (2013). Designing STEM activities to complement neural development in children. In *Proceedings of the 2013 IEEE Integrated STEM Education Conference (ISEC)*, 1–5. https://doi.org/10.1109/ISECON.2013.6525234

Yilmaz, F., & Ayaz, E. (2021). STEM education practices and moral character education: McSTEM? *Research in Pedagogy, 11*(1), 45–62. https://doi.org/10.5937/IstrPed2101045Y

CHAPTER 4

Interdisciplinary Thinking in STE*A*$^+$M: Cognitive Integration, Makerspaces, and Knowledge Creation

4.1 Introduction

Interdisciplinary education is increasingly recognised as a crucial pedagogical approach for addressing complex, real-world challenges (Kidman, 2025). Within the context of STE*A*$^+$M education, where science, technology, engineering, innovative *A*gency (*A*$^+$), and mathematics), interdisciplinary learning encourages students to integrate diverse disciplinary perspectives, fostering creativity, problem-solving, and critical thinking toward developing *A*gency in societal contexts (Burnard & Colucci-Gray, 2019). This chapter examines the role of disciplinary integration in STE*A*$^+$M education, emphasising its significance in contemporary educational practices and the cognitive processes that underpin effective interdisciplinary learning.

The integration of disciplines has long been discussed in educational theory. Scholars differentiate between cross-disciplinary, multidisciplinary, interdisciplinary, and transdisciplinary approaches, each varying in the level of integration and synthesis of knowledge (Klaassen, 2018). However, it is common to see the terms used interchangeably. Cross-disciplinarity is an umbrella term encompassing the three other terms. Multidisciplinarity involves parallel engagement with multiple disciplines without substantial integration. Interdisciplinarity actively synthesises insights across disciplines to develop new understandings. Transdisciplinarity transcends traditional disciplinary boundaries, promoting a

G. Kidman et al., *Knowledge Generation in STEM and STEAM Education*, https://doi.org/10.1007/978-3-032-03247-8_4

holistic approach that incorporates non-academic knowledge and real-world problem-solving. These distinctions are critical for understanding how disciplinary integration shapes learning experiences (Bardecki, 2019) in STE*A*⁺M education.

The main focus of interdisciplinary thinking fosters cognitive flexibility, enabling students to draw connections between diverse domains. This cognitive flexibility is essential for STE*A*⁺M education, where students must navigate and synthesise knowledge from science, technology, engineering, the arts, and mathematics, or in our case, STE*A*⁺M education, where we advocate a plexus approach and *A*gency (Kidman et al., 2024) (see Chapter 3). This chapter introduces the *Cognitive Integration Framework* to provide a model for understanding cognitive integration. The Cognitive Integration Framework informs us that cognitive integration is an emergent process involving creating common ground and using integration techniques like analogical reasoning, conceptual mapping, and metacognitive reflection (Repko & Szostak, 2021). Students develop more profound learning experiences beyond rote memorisation and isolated disciplinary knowledge by engaging in these processes.

One of the most effective ways to facilitate interdisciplinary thinking in STE*A*⁺M education is through makerspaces (Marín-Marín et al., 2021). Makerspaces are dynamic learning environments where students engage in hands-on experimentation, prototyping, and iterative design. Integrating hands-on activities is critical to experiential learning in STE*A*⁺M education (Mater et al., 2023). These spaces encourage experiential learning, where cognitive and tactile interactions merge to produce tangible artefacts. Students often work in teams or small groups (Ortiz-Revilla et al., 2020), sharing ideas, discussing strategies, and providing feedback to one another (Rodier et al., 2021). The *Crafting Knowledge Framework* introduced in this chapter complements the Cognitive Integration Framework as it provides a model for understanding the cognitive processes involved in making. It emphasises the iterative cycle of conceptualisation, material engagement, and reflective refinement, illustrating how creators develop confidence and expertise through active making.

Many schools still operate within rigid disciplinary silos, making it challenging to implement holistic approaches to STE*A*⁺M education (Kurup et al., 2021). However, by incorporating makerspaces and emphasising cognitive integration, educators can create meaningful learning experiences that prepare students for plexus real-world problem-solving.

The chapter explores strategies for fostering interdisciplinary collaboration, including structured integration techniques, peer teaching, and collaborative problem-solving frameworks.

4.2 Cognates of 'Disciplinary'

The cognate terms of 'disciplinary' include intradisciplinary, multidisciplinary, interdisciplinary, transdisciplinary, cross-disciplinary, intradisciplinary, sub-disciplinary, post-disciplinary, uni-disciplinary, pluri-disciplinary, extra-disciplinary, and meta-disciplinary (Bardecki, 2019). The terms are interrelated and refer to the nature of integrating the disciplines in question to achieve common goals in research and education—the context of this chapter. The most common of these terms are the initial five in Bardecki's list, which is our current focus. While 'disciplinary' is rarely problematic, cross-, multi-, inter-, and transdisciplinary are often used interchangeably, resulting in ambiguous descriptions and misunderstandings. Bardecki informs us of the confusion and inconsistency where the "terms are often undefined and/or used with dramatically different meanings. In some circumstances, they are used as synonyms, whereas elsewhere, significant efforts are made to differentiate their meanings" (p. 1179).

To achieve clarity, we explored Philosophy literature, particularly focusing on the works of Eigenbrode et al. (2007), Newell (2007), and O'Rourke and Crowley (2013). Within Philosophy, cross-disciplinarity refers to research encompassing multiple disciplines, while the terms multi-, inter-, and transdisciplinary specify types of research that integrate insights from these disciplines. Below, we briefly describe each term based on literature from beyond education but interpreted through a STEM education lens.

Intradisciplinary—working within a single discipline. The discipline remains independent of other disciplines, their insights, and their methods. Intradisciplinary research "is any study or group of studies from two or more distinct subdisciplines within the same parent discipline. The research is a synthesis (or derivative) of concepts, models, and/or theoretical frameworks from those subdisciplines" (Schary & Cardinal, 2015, p. 177). An example is integrating botany (plants), physics (light), and chemistry (freshwater) in the study of life sciences in the early years or primary school settings.

Multidisciplinary is where several disciplinary insights from sole disciplines are considered individually, one at a time, to answer a question, solve a problem, or address a broad or complex topic that a sole discipline cannot deal with adequately. Each discipline makes a separate contribution "to address a question or problem pertaining to a single system" (Eigenbrode et al., 2007, p. 56), but the disciplinary insights are not integrated. The synthesis of insight does not occur. Separate voices are evident in multidisciplinary conversations, allowing disciplinary identity and perspectives to be maintained, with "the perspective of one discipline that may emerge as dominant within the project (Eigenbrode et al., 2007, p. 56). The methodology and guiding theory used in multidisciplinary studies are the methods and theories of each base discipline. Concerning STE*A*⁺M education, an example of a multidisciplinary experience is where the mathematics teacher focuses on graphing skills. The science teacher has students undertaking plant growth experiments where seedling heights are represented in graphical form.

Interdisciplinary—is a "mode of research, education, and problem-solving that integrates information, methods, tools, concepts, and theories from two or more disciplines" (National Academy of Sciences et al., 2005, 2). Interdisciplinary research is when two or more sole disciplines come together, and an integration process synthesises the ideas and methods (National Academy of Sciences et al., 2005). The integration process is where "participants must work together from problem formulation and hypothesis development to data analysis, interpretation, presentation, and application" (Eigenbrode et al., 2007, p. 55). Where a problem or issue cannot be dealt with adequately by a sole discipline, separate disciplines offer disciplinary insights drawn together to create a deeper understanding and solution (Newell, 2007). Disciplinary insights (concepts, data, perspectives, methods of investigation, and theories) are relied upon in the integration process. By combining and restructuring disciplinary insights through integration, an interdisciplinary insight is created and used to produce a solution (Newell, 2007). Interdisciplinary studies do not privilege an individual disciplinary method or theory. Pohl et al. (2021) view interdisciplinary collaboration as a spectrum, ranging from adopting ideas and techniques from other fields to creating novel frameworks that are applicable across multiple disciplines. A STE*A*⁺M education example is the study of plants that includes a) the historical consideration of botanical medicines, as held by women herbalists, b) the generation of scientific illustrations made from direct observations, and c) the utility of plants for food.

Transdisciplinary—The idea originated at the OECD's International Conference on Interdisciplinary Research and Education in 1970. What exactly is transdisciplinary thinking? Since then, transdisciplinary research has developed to incorporate knowledge, techniques, and skills from various fields to create innovative frameworks that drive scientific discovery and foster innovation. This term has a meaning that is quite different from multidisciplinary and interdisciplinary. According to Repko, Szostak, and Buchberger (2020), transdisciplinary research needs to involve problems in context as a case study. Being in context, the question, problem, or complex topic is addressed *between* the disciplines, *across* the disciplines, and *beyond* the disciplines (Repko, 2012). This trisect method of using disciplinary insight promotes a systems approach that unifies disciplinary knowledge that considers sectors of society. In transdisciplinary research, researchers from different disciplines co-produce new knowledge and understanding and include contextual stakeholders to solve societal problems (Pohl et al., 2021). We provide a case study example of this in Chapter 8. The interdisciplinary process is described as linking, blending, fusing, and synthesising. However, transdisciplinary research transcends, transgresses, and transforms (Klein, 2014). Transdisciplinary research challenges disciplinary boundaries (Pohl et al., 2021). To continue our STEA^+M education example around botany, a transdisciplinary example would see botanical science as historically grounded, where marginalised contributors lead to the reimagining of botany as an inclusive human enterprise, intersecting with women's history and Indigenous studies.

Cross-disciplinary—Heitzmann et al. (2021) conceptualise three forms of cross-disciplinary collaboration: multidisciplinary, interdisciplinary, and transdisciplinary, with O'Rourke et al. (2016) using cross-disciplinary as a cover term for both interdisciplinary and transdisciplinary activity. The disciplinary cognates are distinguished by the degree of integrativeness of the disciplinary insights (O'Rourke & Crowley, 2013). Disciplinary insights are the contributions of an expert member of a discipline relating to a question, solving a problem, or addressing a broad or complex topic. The type of cross-disciplinary collaboration affects the collaborative process and determines which problem-solving skills are employed. The discussion surrounding cross-disciplinary research differentiates between multidisciplinary, interdisciplinary, and transdisciplinary approaches based on their integrativeness, forming a continuum (see Fig. 4.1). Mennes (2020) describes the degree of integration as follows: multidisciplinarity is non-integrative, interdisciplinarity is integrative, transdisciplinarity is hyper-integrative, and it involves the integration of non-academic disciplines from society. However, Drake and Burns (2004) consider multidisciplinarity in education theme-based. Integration is necessary when considering curriculum standards from complementary disciplines introduced in the same lesson. The integration in multidisciplinary situations is weak — the distinct identity of each discipline is maintained.

The integration process is central to cross-disciplinary research, education, and problem-solving. Integration is mainly absent in both intra-and multidisciplinary problem-solving (Schary & Cardinal, 2015). Intradisciplinary insights and lines of thinking stay well within clearly defined disciplinary boundaries. Several disciplines offer their insights during multidisciplinary problem-solving, each contributing separately toward the solution (Klaassen, 2018). Disciplinary identity and perspectives remain distinctive. During interdisciplinary research, education, and problem-solving, disciplinary boundaries become less clear upon initiating the integration process, and a single discipline may offer more insights than others (Kurup et al., 2021). The disciplinary insights become linked. Perspectives are blended and fused, leading to a new synthesised understanding. However, research, education, and problem-solving integration are ineffective without context. The meaningful answering of a question, solving a problem, or addressing a broad or complex topic requires the

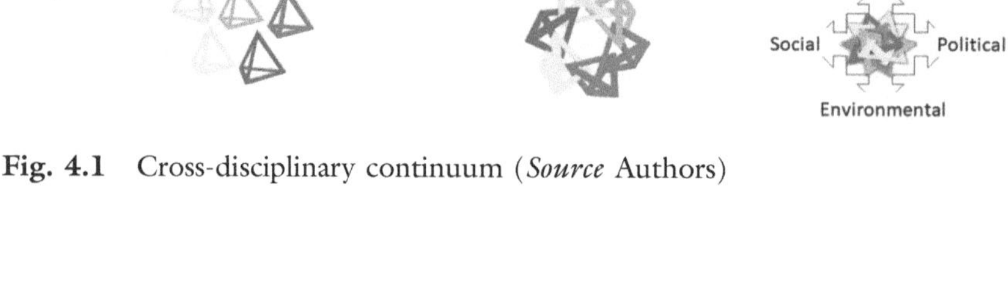

Fig. 4.1 Cross-disciplinary continuum (*Source* Authors)

social, economic, political, and environmental context to be considered. This results in transdisciplinary thinking that transcends (oversteps) disciplinary boundaries and transforms knowledge. Disciplinary boundaries are indistinct; insights and perspectives transform into widely accepted knowledge through newly developed protocols and innovations in practice and policy (Pohl et al., 2021). This indicates a paradigm shift (Drake & Burns, 2004).

Having reviewed the essentials of disciplinary cognates, we focus on understanding the integration process, which is central to the cognates of disciplinary research, education, and problem-solving.

4.3 Exploring the Integration Process

Central to Fig. 4.1 is the degree of integrativeness. The nature of the question, problem, or complex topic (the epistemic goal) determines the degree of integrativeness and type of integration required. Brigandt (2010) explains that integration is not the primary goal; instead, it is crucial for addressing the problem, as the problem's characteristics dictate the necessary intellectual integration required. This cognitive dimension to integration leverages the disciplinary perspectives, leading to integrative complexity and reflecting the different ways of knowing. When different knowledge communities cooperate, the kinds of integration will vary. Pohl et al. (2021) indicate that integration can vary as it weaves the insights of different knowledge systems, creates a balance between different knowledge systems and their ways of knowing, and accommodates the needs of the non-research participants in transdisciplinary research. Different kinds of integration, thus, are options. The key factors are determining the kind and the degree of integrativeness needed and whether it suits a particular type of societal problem at that stage of the research process.

For STEM education, effective approaches for integration continue to be pervasive and problematic (Roehrig et al., 2021). Unless we gain a better understanding of the integration of thinking skills in STEM and STEA^{+}M education contexts, STEM and STEA^{+}M models of education that purport to prepare students to meet the needs of society now and in the future may not be productive. Adding to this complexity, Giebels et al. (2020) identify a lack of in-depth understanding of how knowledge integration works as a human interaction process, while Tsai (2021) claims there is a need to disentangle how the cognitive dimensions influence knowledge sharing and decision-making between and within the disciplines during integration. To respond to the calls from Geibels et al. and Tsai to clarify the cognitive dimensions of knowledge integration during collaborative exchanges, we turn to research from outside the field of Education.

Table 4.1 provides a brief overview of a selection of papers that have explored discipline-combining research from outside of Education. We selected these papers as they provide a philosophical background for our STE*A*$^{+}$M Education theorising. Darden and Maull's (1977) research highlights the emergence of 'fields' in science and the nature of their central problems. Darden and Maull acknowledge the notion of 'discipline' yet choose to explore inter-field developments to avoid an analysis of evolutionary epistemology when making explicit the relations between fields.

Table 4.1 Discipline-combining research from outside education

What is integrated	*Key ideas*
Theories	Given that 'fields' are the basic units of scientific research, integration is conceptualised as an 'inter-field theory' (theories set out and explain the relationships between fields). Where relationships are already known, background knowledge is recognised, leading to inter-field questions (e.g., Darden & Maull, 1977)
Language and concepts	Conceptual change holds significance because concepts are organised by their similarities and differences regarding features and values. Individuals within the same community use a shared language. Consequently, interactions across disciplines yield far richer outcomes than those within a single discipline (e.g., Andersen, 2013)
Explanations	Integrating insights and explanations from different disciplines yields an overall explanation of a complex phenomenon. Explanatory integration gives rise to the origin of novelties (e.g., Brigandt, 2010)
General studies of integration	The nature of integration varies depending on the inter-multi- or transdisciplinary focal context. The variance also relates to how to achieve integration—the algorithmic (implementing fixed steps in an orderly fashion) versus heuristic (varying across contexts) divide. Definitions offered to describe integration are as a process/product with an input/output process structure (e.g., O'Rourke et al., 2016)

From Table 4.1 it is evident that the integration of disciplines in STE*A*$^+$M education must reflect four key ideas: theories, language and concepts, explanations, and general integration studies. Theories establish relationships between fields, creating an 'inter-field theory' that allows for recognising background knowledge and formulating new interdisciplinary questions (Darden & Maull, 1977). The background knowledge and formulation of new interdisciplinary questions are relevant to the Plexus Model of STE*A*$^+$M Education. Language and concepts facilitate interdisciplinary exchange by enabling conceptual change and fostering a shared language among disciplines, leading to richer interactions than those occurring within a single field (Andersen, 2013). The emphasis on language and communication in collaborative STE*A*$^+$M education experiences that lead to idea exchange and knowledge creation is of interest. Explanations explore knowledge growth and integrate insights from multiple disciplines to develop comprehensive understandings of complex phenomena, contributing to the emergence of novel ideas (Brigandt, 2010). Finally, general studies of integration highlight the differences between inter-, multi-, and transdisciplinary approaches, emphasising structured (algorithmic) versus adaptable (heuristic) integration methods (O'Rourke et al., 2016). In STE*A*$^+$M education, these ideas support interdisciplinary problem-solving by encouraging students to connect knowledge across fields, use shared conceptual frameworks, develop innovative explanations, and apply varied integration strategies, ultimately fostering deeper learning and creative agency.

Understanding the integration process in STE*A*$^+$M education provides a crucial foundation for examining the cognitive mechanisms that underpin interdisciplinary thinking. As disciplines merge to create new learning frameworks, the cognitive demands placed on students shift, requiring them to develop adaptive reasoning, metacognitive awareness, and systems thinking skills (Barile, & Saviano, 2021). The way knowledge is structured, synthesised, and applied across disciplines shapes not only how students approach problem-solving but also how they cultivate creativity and innovative *A*gency. By exploring how integration occurs, we can better understand why interdisciplinary learning enhances cognitive development. This transition from a process-driven perspective to a cognition-centred analysis allows us to investigate how learners construct, modify, and apply knowledge when engaging in STE*A*$^+$M learning experiences. In the following section, we delve deeper into the cognitive foundations of STE*A*⁺M thinking, illustrating how interdisciplinary education fosters critical and creative thinking, metacognition, and knowledge acquisition.

4.4 The Cognitive Foundations of STE*A*⁺M Thinking

At its core, STE*A*⁺M thinking involves the dynamic interplay between cognitive, metacognitive, and affective processes. As we saw in the Plexus Model of STE*A*⁺M Education, cognitive development in STE*A*⁺M learning is rooted in constructivist and inquiry-based pedagogies, where students actively engage with concepts through exploration, experimentation, and problem-solving.

Metacognition is a crucial element of STE*A*⁺M cognition. Students develop self-regulated learning habits when they reflect on their learning processes, evaluate their strategies, and adjust their approaches accordingly. This ability to monitor and control cognitive processes is particularly important in fostering *A*gency, as it enables students to take ownership of their learning, identify areas for growth, and seek innovative solutions to complex problems (Sarid & Goldman, 2021; von Braun, 2017). This ownership links closely to the affective factors, including motivation, curiosity, and resilience, which play a significant role in STE*A*⁺M thinking (Bandura, 1986; Mathers et al., 2024). According to Bandura's social cognitive theory, self-efficacy affects motivation and performance. In STE*A*⁺M learning, fostering self-efficacy through hands-on projects, collaborative problem-solving, and real-world applications encourages students to persist in facing challenges, enhancing their capacity for innovative *A*gency.

The Plexus Model of STE*A*⁺M Education for learning is inherently interdisciplinary, requiring students to draw on multiple fields to develop comprehensive solutions. This approach aligns with cognitive flexibility theory (Spiro et al., 1988), which posits that individuals who integrate knowledge across disciplines are better equipped to adapt to novel situations and generate innovative ideas. By engaging in such integrative thinking, students can approach problems from multiple perspectives, a skill essential for innovative *A*gency. Also involved is systems thinking theory, a cognitive approach that enables individuals to understand complex interdependencies within real-world challenges (Cabrera & Cabrera, 2023). Cabrera and Cabrera describe systems thinking as four cognitive skills: making distinctions, organising systems, recognising relationships, and taking multiple perspectives. These cognitive skills allow students to break down complex problems, reconstruct solutions, understand interactions, and view challenges from multiple viewpoints. The final cognitive skill, taking multiple perspectives, is particularly significant in fostering interdisciplinary STE*A*$^+$M thinking and encouraging the *A*gency to address global issues such as climate change, social equity, and technological advancements.

Recognising multiple perspectives implies collaboration is necessary in the Plexus Model of STE*A*$^+$M Education. Collaboration reflects real-world scientific and technological practices (Soomro et al., 2023). The cognitive process of social cognition plays a key role in collaborative learning environments where individuals must judge, reason, remember, and decide in conjunction with others (Sloman et al., 2021). Collaborative problem-solving enhances learning by promoting shared mental models, distributed cognition, and collective intelligence. Sloman et al. (2021) indicate that group intelligence and inventiveness can result in new levels of learning that are not possible when acting and thinking as an individual.

In STE*A*$^+$M classrooms, collaborative learning often involves project-based and inquiry-driven work, where students must negotiate roles, communicate ideas, and construct shared knowledge (Kidman et al., 2024). Through such experiences, students develop the social and emotional intelligence needed to work effectively in diverse teams, a critical component of an innovative *A*gency. By learning to value different perspectives, challenge assumptions, and engage in constructive dialogue, students refine their ability to lead and implement change in society.

Critical and creative thinking are two key cognitive processes underpinning STE*A*$^+$M thinking (Wised & Inthanon, 2024). Critical thinking requires the analysis, evaluation, and synthesis of information to make well-informed decisions. This skill is crucial for innovative *A*gency, as it enables students to assess the feasibility and impact of their ideas. The Socratic method, problem-based learning, and design thinking frameworks are commonly used in STE*A*$^+$M education to cultivate critical thinking skills and to develop more adaptable and well-rounded cognitive abilities (Wised & Inthanon, 2024). In contrast, creative thinking requires divergent thinking, which is the capacity to produce various solutions to a problem. Encouraging students to engage in brainstorming, experimentation, and iterative design processes nurtures their creative capacities.

When combined with critical thinking, creativity allows students to develop novel solutions that are both imaginative and practical (Liao, 2016), a hallmark of true innovation.

Having considered the cognitive foundations of STE*A*⁺M thinking through an exploration of how interdisciplinary education fosters metacognition and critical and creative thinking, the following section explores integrating cognitive factors involved in knowledge acquisition. Nevertheless, as highlighted in previous chapters, there is a lack of educational research in this field, and a working framework for cognitive integration in classroom environments is essential. We tackle this gap by examining the interconnectedness of knowledge and defining cognitive integration as a practical framework.

The *Cognitive Integration Framework*, presented in the following section, is founded on a comprehensive review of interdisciplinary research, drawing from sociology, anthropology, and psychology literature. It is significantly informed by the work of Allen Repko, William Newell, Julie Thompson Klein, Rick Szostak, and Michelle Philips Buchberger, whose contributions provide critical insights into interdisciplinary methodologies. Their work has been adapted through an educational lens to ensure its applicability in teaching and learning contexts. By situating the framework within an educational context, we aim to refine the understanding of cognitive integration in pedagogical applications. Additionally, the framework is enriched by principles from cognitive science and theories of conceptual change drawn from educational research. This integration ensures that the framework effectively captures the cognitive processes that underpin interdisciplinary learning, making it a robust tool for fostering innovation, critical thinking, and problem-solving in educational settings.

4.5 STE*A*⁺M Thinking and Knowledge Creation Through Cognitive Integration

Acquiring knowledge is inherently dynamic and interactive, relying on integrating cognitive factors that shape individual understanding. Knowledge acquisition or creation is not solitary in classroom settings but a collaborative process where individuals share, evaluate, and integrate new information into their existing cognitive frameworks (Sloman et al., 2021). This section explores how knowledge acquisition or creation is an emergent process of cognitive integration that occurs through communication and conceptual synthesis. Cognitive integration involves a series of interactions or knowledge sharing in which individuals negotiate meaning by incorporating external or shared knowledge into their pre-existing cognitive structures. Prior knowledge is at the core of this process, serving as the basis upon which newly shared information is built. Individuals engaging in collaborative learning encounter novel perspectives that challenge, reinforce, or extend their understanding (Ortiz-Revilla et al., 2020). The integration process, therefore, requires active cognitive engagement, where newly shared knowledge must be evaluated, restructured, and synthesised to create a cohesive understanding.

As Fig. 4.2 illustrates this integration process. The *Cognitive Integration Framework* is an input–output operational framework in which cognitive integration is conceptualised as an emergent process-driven framework in which structured integration techniques mediate knowledge inputs and outputs. Inputs (for example, abstract ideas, cognitive models, social perspectives, and artefacts) must be considered in terms of quality—do they make sense compared to prior knowledge? Similarly, the quantity of inputs (the number of ideas and degree of conceptual variation) must be considered. Outputs reflect the transformation of these inputs into new understandings, explanations, and conceptual models. We consider these outputs as ah ha moments, which are defined by Pilcher (2015) as:

> Ah ha is a moment of sudden understanding ("now I get it") when a learner seeks a new connection not previously seen. The understanding may lead to excitement, deeper comprehension, a change in perspective, and/or recognition of how it can be applied to real-life situations. (p. 62)

Pilcher further informs us that the most common pedagogies or strategies leading to or resulting in "ah ha" moments (in ranked order) are.

1. Real life examples
2. Questioning
3. Reflection/reflective activities
4. Analogies
5. Problem-solving in teams (2015, p. 62).

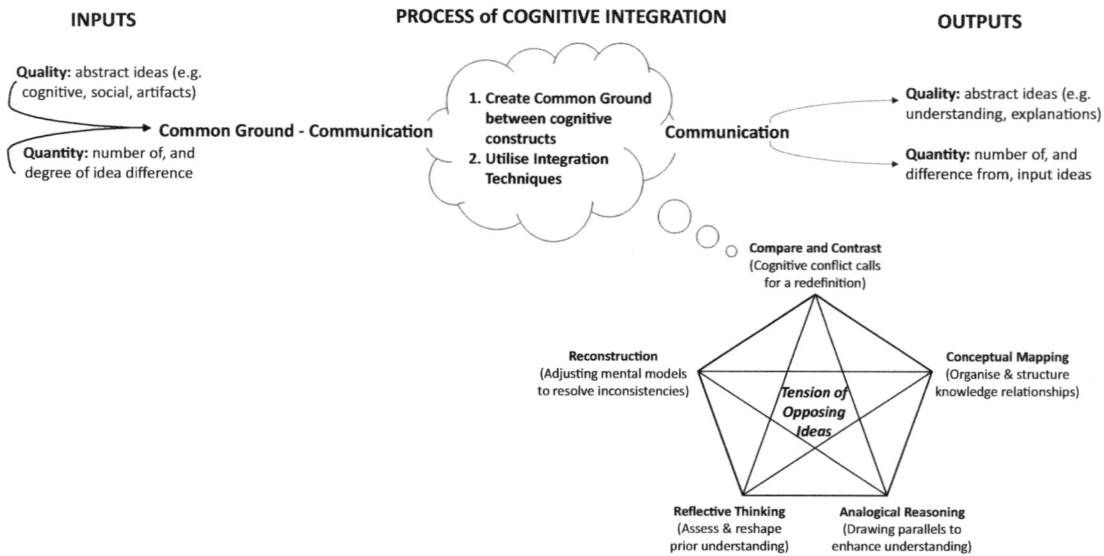

Fig. 4.2 Cognitive integration framework (*Source* Authors)

Outputs are also considered in terms of the quality of the understandings and explanations and the quantity of output ideas compared to the quantity of input ideas. The *Cognitive Integration Framework* is discussed further in the following sections.

4.5.1 Communication in Cognitive Integration

Communication plays a central role in the Cognitive Integration Framework by facilitating the exchange of ideas and the negotiation of meaning. In the discussion below, we highlight four key strategies required in classroom settings that facilitate communication and knowledge creation. These are broad strategies that promote the use of collaboration and knowledge sharing. Students must be provided with the opportunity to share ideas and knowledge so that they can update their mental models and create new knowledge. The Cognitive Integration Framework emphasises that cognitive integration is not a passive process but an active engagement where individuals co-construct knowledge through classroom discourse. This is critical in contemporary classrooms where integration practices are being supported.

An essential aspect of this process is dialogic learning, which harnesses the power of discourse to stimulate thinking, advance comprehension, and build arguments (Alexander, 2020). Through dialogic engagement, students verbalise their thought processes, reformulate their ideas, challenge assumptions, and critically evaluate concepts in collaboration with peers and teachers (Alrø & Skovsmose, 2004). Rejecting hierarchical structures in favour of joint meaning-making, dialogic learning fosters critical thinking, student agency, and democratic engagement (White, 2022). This emphasis on interactive dialogue aligns with the Plexus Model of STEA^{+}M Education, where interdisciplinary knowledge integration relies on shared reasoning and collaborative inquiry (Alexander, 2020; White, 2022).

Closely related to dialogic learning, collaborative problem-solving provides an essential mechanism for cognitive integration. Lin et al. (2021) explored how collaborative problem-solving skills intersect with decision-making processes and task outcomes, concluding that effective collaboration is strongly associated with improved cognitive engagement. Within collaborative learning environments, students adopt sociocognitive roles that shape their reflective and regulatory processes, reinforcing the intertwined nature of social and cognitive constructions in knowledge creation (Dowell et al., 2020). As ideas are negotiated and evaluated through discussion, collaborative knowledge building emerges, requiring students to compare perspectives critically, present evidence-based reasoning, and construct arguments supported by analytical rigour.

Argumentation further deepens cognitive engagement by encouraging students to question, evaluate, and integrate complex ideas. Nussbaum (2021) highlights the importance of critical integrative argumentation, which fosters higher-order thinking through formulating claims, data analysis, reasoning, and counterarguments. By incorporating interdisciplinary perspectives and structuring arguments with supporting evidence, warrants, qualifiers, and rebuttals (Erduran, 2018), students enhance their comprehension and refine their oral and written communication skills (Marthaliakirana et al., 2022). Encouraging argumentation strengthens students' ability to defend their perspectives and promotes critical reasoning and deep cognitive engagement, making it a fundamental component of knowledge integration in learning environments.

Another essential element of cognitive integration is peer teaching, which facilitates knowledge construction and communication development. Peer teaching strengthens students' ability to articulate complex concepts, reinforcing their technical understanding and enhancing communication skills (Govan, 2016). Through peer-led discussions and explanations, students gain access to the "hidden curriculum", fostering a more relaxed and collaborative learning atmosphere (Kunc et al., 2020). When students are given autonomy to select instructional materials and communication strategies, their engagement and motivation improve, resulting in more significant cross-curricular competency development, increased academic emotions, and enhanced participation in learning (Ribosa & Duran, 2022). Ultimately, peer teaching fosters active learning and strengthens students' cognitive abilities by reinforcing their capacity for independent knowledge construction.

4.5.2 *Common Ground and Integration Techniques in Cognitive Integration*

The Cognitive Integration Framework identifies two key mechanisms facilitating cognitive integration: creating common ground (Repko, 2012) between cognitive constructs and utilising integration techniques to reconcile differences and align conceptual frameworks. One of the fundamental prerequisites for cognitive integration is establishing interdisciplinary **common ground**, which enables meaningful communication between different knowledge bases. Common ground involves modifying shared concepts, assumptions, and language that allow individuals to exchange knowledge effectively (Repko & Szostak, 2021). This is especially relevant in interdisciplinary learning environments, where students bring diverse cognitive perspectives from various disciplines and aim to create a common understanding rather than focus on the differences. From the field of cognitive psychology, Bromme (2000) explained that successful communication where perspectives differ relies upon a common cognitive frame of reference between the communicators, known as common ground. Epistemology plays a role in creating common ground. Repko and Szostak (2021) describe narrow and wide interdisciplinarity in terms of the closeness of the disciplines. For example, physics and chemistry may have narrow interdisciplinarity, while art history and mathematics have wide interdisciplinarity. "The greater the epistemological distance between disciplines, the more challenging it is to create common ground between their insights" (Repko & Szostak, 2021, p. 273). When establishing common ground in interdisciplinary thinking, we need to be cognisant of not only connecting ideas but also leveraging perspectives.

A cognitive integration progression provides a structured approach for students to establish common ground and synthesise ideas into new knowledge. This process is facilitated through subconscious cognitive techniques and implicit cognitive adjustments, enabling students to navigate complex ideas and resolve conceptual tensions. While we describe the process linearly, we do not intend a linear conception to be formed. The cognitive techniques must be understood to function iteratively, allowing students to continuously refine their understanding and reconcile opposing ideas. Cognitive integration supports deep learning, innovation, and knowledge creation through this iterative cycle.

Figure 4.2 introduces five integration techniques as the essence of the cognitive integration process. The first integration technique is *compare and contrast*. This involves identifying similarities and differences between a new shared knowledge or idea and existing prior knowledge. This stage allows students to determine conceptual alignment and areas of divergence, forming the foundation for deeper cognitive engagement. Through this process, learners critically assess how new information relates to their existing cognitive structures, setting the stage for integration. *Conceptual mapping* is the second integration technique, providing a visual or cognitive framework for organising knowledge (Whittington et al., 2022). By mapping relationships between concepts, students develop a clearer understanding of how ideas interconnect, facilitating the restructuring of information in meaningful ways. This technique supports knowledge synthesis by making abstract relationships more tangible and accessible. The third integration technique is *analogical reasoning*, further enhancing integration by drawing parallels between familiar and new concepts (Gentner & Maravilla, 2017). Analogies enable learners to apply existing knowledge to unfamiliar domains, promoting cognitive flexibility and deepening comprehension. This form of reasoning allows for knowledge transfer across disciplines, fostering interdisciplinary thinking. The fourth integration technique is *reflective thinking*, which plays a crucial role in metacognition (Rodier et al., 2021). Students evaluate how new knowledge reshapes their prior understanding. Reflection fosters self-awareness, prompting learners to assess their learning progress and consider alternative perspectives. Finally, *reconstruction* involves adjusting mental models to accommodate novel insights and resolve inconsistencies. This final integration technique ensures that students integrate conflicting ideas into a coherent framework, solidifying their conceptual understanding and preparing them for future learning challenges. While each integration technique plays a distinct role, they work in tandem to help students synthesise knowledge, resolve conflicts between disciplines, and construct innovative solutions.

The integration techniques each serve a distinct function, as described above. They work together iteratively to make sense of opposing ideas and foster deep knowledge integration and interdisciplinary problem-solving. Our analysis indicates three iterative groupings, as described below.

Integration Techniques, Cognitive Flexibility, and Knowledge Transfer are essential cognitive abilities that enable students to navigate complex, interdisciplinary problems by shifting perspectives, recognising patterns, and applying insights across disciplines. These skills enable students to synthesise diverse knowledge frameworks, thus fostering creativity and innovation during problem-solving. Cognitive flexibility is the capacity to transition between different cognitive frameworks, problem-solving approaches, and disciplinary perspectives (Spiro et al., 1988). This ability allows students to explore and learn beyond disciplinary boundaries, facilitating the integration of knowledge from one discipline to another. In interdisciplinary contexts (for example, in STEA^{+}M education, as presented in Chapter 3), cognitive flexibility helps students reassess and restructure their understanding and prior knowledge when faced with new information or conflicting viewpoints. An effective integration technique that supports this process is the compare and contrast, where students evaluate similarities and differences between disciplinary approaches, identifying areas of convergence and divergence. By making comparisons, a dynamic integration of insights results in the student's cognitive ability to remain adaptable and open to new perspectives (Spiro et al., 1988).

Abstraction and generalisation are closely linked to cognitive flexibility, which involves recognising underlying principles across disciplines. Abstracting concepts from specific disciplinary contexts enables students to see patterns beyond subject boundaries. This skill is crucial for developing frameworks for integrating knowledge as part of a plexus STE*A*+M education. Conceptual mapping is beneficial in the abstraction and generalisation process, as it helps organise and structure knowledge, allowing students to visualise relationships between ideas and across disciplines (Whittington et al., 2022). Analogical reasoning further enhances knowledge transfer by enabling individuals to identify structural similarities between seemingly unrelated fields. Analogical reasoning supports interdisciplinary problem-solving by transferring conceptual models across disciplines and is an integration technique crucial for students drawing parallels between disciplines (Gentner & Maravilla, 2017).. Finally, reconstruction or adjusting mental models plays a significant role in cognitive flexibility and knowledge transfer. Students with conflicting disciplinary perspectives restructure their cognitive frameworks to resolve inconsistencies and build more integrated understandings. This iterative process ensures that knowledge is accumulated and reorganised to fit within a more cohesive interdisciplinary framework.

Integration Techniques, Reflective, and Evaluative Thinking is crucial in ensuring that interdisciplinary knowledge is critically examined, synthesised, and meaningfully integrated. Such thinking is necessary for students to engage in metacognition and critical reasoning to refine their understanding and resolve cognitive dissonance. By incorporating reflective thinking and reconstructing mental models, students can critically assess knowledge claims, identify biases, and create more coherent conceptual frameworks (Rodier et al., 2021). Inferential and critical thinking underpins the rigorous evaluation of interdisciplinary claims, emphasising the importance of recognising biases and questioning assumptions. Each discipline has its assumptions, methods, and standards of evidence (McElreavy et al., 2016), requiring students to adopt, compare, and contrast integration techniques to evaluate disciplinary differences and critically assess the credibility of knowledge claims and insights from diverse sources.

Metacognition and self-regulated learning further enhance reflective and evaluative thinking by promoting awareness of one's cognitive processes (Thibaut et al., 2018). Students engage in reflective thinking, assessing how they approach problems and whether disciplinary biases influence their reasoning. This metacognitive process supports conceptual mapping, allowing students to organise interdisciplinary knowledge systematically. Through self-regulation, students refine their ability to integrate insights from multiple disciplines, ensuring a balanced and well-informed approach to problem-solving (Zhan & Niu, 2023). Students refine their cognitive structures through reconstruction and adjusting mental models to accommodate new insights, ensuring that interdisciplinary synthesis is logically sound and methodologically robust.

Integration Techniques, Integrative, and Generative Thinking are essential cognitive abilities that foster interdisciplinary collaboration, drive innovation, and enable systemic problem-solving. They allow individuals to synthesise differing perspectives, generate novel insights, and construct holistic solutions to complex challenges. Students develop strategies that support interdisciplinary synthesis and creative problem-solving (Lin et al., 2021) by engaging in conceptual mapping, analogical reasoning, and reconstruction of mental models.

Perspective-taking and empathy play a foundational role in collaboration by promoting an understanding of different disciplinary viewpoints and their implications. This requires students to appreciate diverse values, priorities, and methodologies, ensuring that integration does not lead to hierarchical dominance but rather synthesis. Compare and contrast is essential in this process, enabling students to examine the similarities and differences between disciplinary perspectives critically. This fosters a deeper appreciation of alternative viewpoints. Problem-solving and systems thinking (Cabrera & Cabrera, 2023) facilitate the analysis of complex problems that do not fit neatly within a single discipline. Many real-world challenges require seeing interconnections (Karahan & Roehrig, 2016) between systems and their components, necessitating the ability to conceptually map out causal relationships, dependencies, and feedback loops. Conceptual mapping is particularly valuable in this context, as it helps students organise and structure interdisciplinary knowledge (Whittington et al., 2022), making it easier to identify gaps, overlaps, and emergent patterns.

Creativity and open-mindedness further enhance generative thinking by encouraging individuals to explore unconventional ideas and integrate diverse perspectives (Wised & Inthanon, 2024). Analogical reasoning supports this by enabling individuals to draw insights from one discipline and apply them innovatively to another, fostering knowledge transfer and breakthrough innovations. Finally, the reconstruction of mental models ensures that new interdisciplinary insights are meaningfully integrated, allowing learners to adapt their thinking frameworks to accommodate novel information and evolving perspectives. Through these integrative processes, interdisciplinary thinkers break down disciplinary silos (Liu & Li, 2023), cultivate innovation, and generate new knowledge that is both cohesive and transformative.

Interdisciplinary thinking is critical in education research, STEA^+M pedagogies, and environmental sciences—fields where integrating multiple perspectives leads to deeper understanding and more effective solutions to global challenges. The Cognitive Integration Framework (See Fig. 4.2) indicates that cognitive skills, supported by integration techniques such as compare and contrast, conceptual mapping, analogical reasoning, reflective thinking, and reconstructing mental models, form the foundation of interdisciplinary problem-solving, allowing students to navigate and integrate knowledge in a way that transcends traditional disciplinary silos, fostering intellectual adaptability and knowledge expansion.

The Cognitive Integration Framework emphasises that cognitive integration is not the passive absorption of information but an active, iterative engagement in which students construct knowledge through discourse, collaboration, and structured integration techniques. By fostering cognitive flexibility, reflective and evaluative thinking, and integrative and generative processes, students develop the ability to synthesise disciplinary perspectives, critically evaluate claims, and create novel insights that transcend traditional disciplinary boundaries. Through dialogic learning, collaborative problem-solving, argumentation, and peer teaching, learners enhance their cognitive integration skills, refining their ability to connect, critique, and construct knowledge meaningfully.

Cognitive integration is foundational in interdisciplinary learning; however, knowledge creation is not solely a conceptual endeavour. In

STEM education and in our approach to a plexus STEA^+M education, knowledge integration is only one aspect that needs consideration. A critical component of any STEM or STEA^+M education is the makerspace. Interdisciplinary learning in STEM education is also a materials and experiential process in a makerspace. In contemporary STEA^+M education, knowledge is increasingly constructed through discussion, reasoning, hands-on engagement, and artefact production. Makerspaces provide tangible learning environments where students can apply interdisciplinary knowledge, experiment with tools and materials, and bring abstract concepts to life through creative problem-solving (Liu & Li, 2023). The makerspace offers a practical learning space for developing critical and creative thinking, fine motor skills, idea development, and the ability to produce creative solutions (Soomro et al., 2023). In most school-based educational makerspace settings, it is what Soomro and colleagues call a *basic makerspace*. Teachers provide "simple tools such as pencils, erasers, rulers, tape, computers, and various reusable materials" (p. 539), fostering a learning-by-doing approach. Learning-by-doing is where artefacts are produced, enhancing cognitive "skills such as lateral thinking, problem-solving, creativity and innovation while working in groups" (Beyers, 2010, p. 453) and individual learning. Thus, the social aspects of makerspaces provide opportunities "to be expressive, enhances the chances of producing creative outcomes, and motivates their involvement in creative activities" (Soomro et al., 2023, p. 538). In the context of STEM or STEA^+M education, makerspaces serve as vibrant hubs that break through conventional disciplinary limits, promoting collaborative learning and innovation. By integrating tools and technologies, makerspaces facilitate interdisciplinary problem-solving, reflecting the interconnected nature of modern STEM fields (Soomro et al., 2023).

The next section explores how cognitive processes manifest in creating makerspace artefacts. Makerspaces serve as living laboratories for knowledge application, where students integrate scientific, technological, engineering, artistic, and mathematical principles through iterative design and prototyping. By examining the cognitive processes of making, we extend the discussion of knowledge integration beyond conceptual synthesis to include embodied, practical learning experiences that further enhance interdisciplinary thinking, innovation, and collaborative inquiry.

4.6 STEA$^+$M and Crafting Knowledge

Makerspaces inherently serve as interdisciplinary hubs where students from various backgrounds can collaborate, innovate, and create (Marín-Marín et al., 2021). Throughout a plexus STEA$^+$M experience (refer to Chapter 3), students acquire the ability to blend knowledge and skills from multiple disciplines, which aids in generating solutions to complex challenges. This hands-on learning method enhances the application of theoretical knowledge across disciplines, resulting in a richer comprehension of concepts and their practical uses (Mater et al., 2023). Ideas flow freely as learners investigate links between seemingly unrelated disciplines. This cognitive integration often sparks innovation as individuals draw inspiration from diverse sources, merging concepts from different areas to produce something novel and distinctive. Learning occurs in ways that surpass conventional disciplinary borders. The following sections explore cognitive integration and conceptual change within makerspace settings. We base our discussion on Fig. 4.3, the *Crafting Knowledge Framework*.

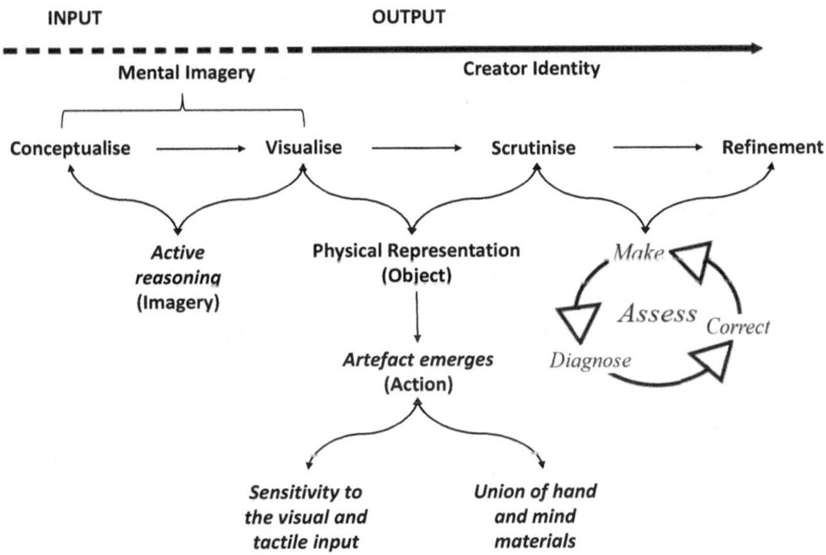

Fig. 4.3 Crafting knowledge framework (*Source* Authors)

The Crafting Knowledge Framework is an input–output model depicting the dynamic interplay between integrating cognitive reasoning and hands-on engagement. The Crafting Knowledge Framework is founded on over 200 interviews, conversations, and observations of students and teachers participating in makerspace activities. We also draw from literature in education, visual arts, anthropology, and psychology. The work of artisans significantly informs the model. We conducted 25 interviews with artisans at a Lost Trades Fair who demonstrated and shared their trade skills and gave insight into what it means to be a maker, what inspires them, and how they do what they do. An important factor communicated by the artisans was a maker's identity—the contention that making and being creative was an essential part of their identity (Culpepper & Gauntlett, 2021). Our artisans described their identities as fluid and malleable with each creation.

Conceptualisation is the starting point for a maker, where mental imagery and active reasoning shape an initial artefact vision. Doubts may arise, prompting problem-solving and refinement. As the creator transitions to physical making, their identity strengthens, and confidence grows. Sensitivity to visual and tactile inputs guides the shaping of the artefact. The emerging object undergoes assessment, where problems are recognised, causes are diagnosed, and corrections are applied. This iterative cycle highlights the unity of mind and hand, reinforcing both the creator's growth and the quality of the final artefact.

The process unfolds through a series of cognitive and physical stages, each influencing the next, ultimately leading to the emergence of a tangible artefact. The creation journey begins in the mind, where conceptualisation takes place. At this stage, the creator engages in active reasoning and mental imagery, constructing a vision of what will be made. Prior knowledge, experiences, and creative intuition shape this visualisation. Again, doubts may arise regarding the feasibility of bringing the envisioned object into existence. The creator identity is thus just starting to form. Uncertainties form an integral part of the creative process, as they drive problem-solving and the further refinement of ideas. As the creator moves from ideation to physical engagement, they interact with materials, translating their mental image into a tangible form. This transition significantly shifts the creator's identity (Culpepper & Gauntlett, 2021). Their confidence in their creative abilities strengthens as they actively manipulate materials and witness their ideas taking shape. The process reinforces the notion that creation is not merely about producing an object but also about constructing and reaffirming one's identity as a creator.

The creator is highly sensitive to visual and tactile inputs during the hands-on phase (Keller & Keller, 1996). The interplay between what is seen and what is felt guides the shaping of the object. The physical engagement of the hands with the materials allows for real-time adjustments, ensuring that the artefact aligns with the mental vision. This continuous interaction between the hand and the mind exemplifies the deeply embodied nature of creation. The tactile experience informs decision-making, refining the object's form, texture, and structure. The hands become an extension of the mind, translating abstract ideas into concrete reality.

As the artefact emerges, the creator enters a phase of assessment and refinement. At this stage, critical evaluation plays a crucial role. The creator scrutinises the object, identifying flaws, inconsistencies, or areas requiring improvement. Recognising problems is a cognitive process that involves comparing the physical representation with the original conceptualisation. If discrepancies arise, the creator must diagnose the underlying causes. This analytical diagnostic process requires understanding causality—why a specific element did not turn out as expected and what adjustments are necessary to rectify it (Keller & Keller, 1996). Once a problem is identified, the creator applies corrective procedures. These adjustments may involve reconfiguring components, refining shapes, altering materials, or even reconsidering foundational aspects of the design. This iterative cycle of making, assessing, diagnosing, and correcting is fundamental to creative work. It highlights the fluidity of the creative process, where errors are not failures but opportunities for learning and refinement (Culpepper & Gauntlett, 2021). Each correction brings the artefact closer to the intended vision while expanding the creator's knowledge and skills.

The entire process exemplifies a deep integration of cognitive and physical faculties. The mind conceptualises, reasons, and evaluates, while the hands bring forth the envisioned form and refine it through tactile interaction (Keller & Keller, 1996). The hands become the window into the creator's imagination. This union of hand and mind defines making as an embodied practice (Culpepper & Gauntlett, 2021). The process is not linear but cyclical, where each stage feeds backwards and forwards into the next, allowing for continuous evolution and improvement. The creator's evolving identity is essential to this process. With each creation, confidence grows, and self-perception as a capable maker solidifies (Culpepper & Gauntlett, 2021). This identity transformation is about skill acquisition and internalising the ability to navigate uncertainties, solve problems, and persist through challenges. The act of making, therefore, extends beyond the object itself; it shapes the creator as much as the artefact being produced. Ultimately, learning in a makerspace and creating an artefact is not merely about producing an artefact but the transformative experience of making itself.

4.7 Conclusion

This chapter explores interdisciplinary thinking in STE*A*⁺M education, emphasising cognitive integration and the importance of makerspaces as environments that support cognitive and creative development. By examining the theoretical and cognitive foundations of disciplinary integration and its practical applications in makerspaces, the chapter advocates for a transformative approach to education that encourages students to think beyond disciplinary boundaries, engage in hands-on creation, and develop the cognitive skills necessary for innovation and adaptability. This is presented as the Cognitive Integration Framework illustrates the input/output model of interdisciplinary cognitive integration where the mind creates common ground while utilising integration techniques.

Integrating interdisciplinary thinking in STE*A*⁺M education has far-reaching implications for pedagogy and educational policy. By fostering cognitive flexibility (Spiro et al., 1988), problem-solving skills, and creative engagement, interdisciplinary approaches prepare students for the complexities of the modern world (Education Council, 2015). This chapter has explored the role of disciplinary integration in STE*A*⁺M education, emphasising how cognitive processes such as analogical reasoning, conceptual mapping, and metacognitive reflection contribute to deeper learning through societal relevance. It has also highlighted the significance of makerspaces as physical environments where students can engage in hands-on experimentation and iterative design (Soomro et al., 2023).

One of the key implications of this discussion is the need for educational institutions to adopt more flexible and integrative pedagogical approaches. Traditional education systems often compartmentalise knowledge into distinct subjects, limiting students' ability to make meaningful connections across disciplines (Garibay, 2015). By implementing interdisciplinary STEA^{+}M curricula incorporating makerspaces, schools can provide students with opportunities to engage in holistic problem-solving and creative exploration.

Makerspaces are powerful environments for fostering integrative thinking (Marin-Marin et al., 2021; Soomro et al., 2023). Students can explore materials, use their theoretical knowledge in real-world situations, and participate in collaborative learning. This making process helps students cultivate vital skills, including adaptability, resilience, and iterative problem-solving (Soomro et al, 2023). The Crafting Knowledge Framework illustrates how makerspaces facilitate the union of cognitive and tactile engagement, reinforcing the development of creator identity and confidence. This model underscores the importance of allowing students to engage in hands-on creation to deepen their understanding of interdisciplinary concepts and develop a maker and creator identity.

Another critical implication is the need for professional development opportunities for educators. Teaching interdisciplinary STEA^{+}M curricula requires instructors to be comfortable navigating multiple disciplinary perspectives and facilitating integrative learning experiences (Forbes et al., 2021). Schools and policymakers should invest in teacher training and professional learning programs that emphasise interdisciplinary pedagogies, makerspace facilitation, and collaborative problem-solving techniques.

In conclusion, interdisciplinary thinking in STEA^{+}M education offers a robust framework for preparing students to engage with real-world challenges. By integrating makerspaces into educational practices and emphasising the cognitive synthesis, educators can cultivate environments that support creativity, problem-solving, and innovation. The insights from this chapter reinforce the importance of breaking down disciplinary silos and embracing a holistic approach to education. As schools evolve, interdisciplinary STEA^{+}M education will equip students with the skills to navigate an increasingly complex and interconnected world.

References

Alexander, R. (2020). A dialogic teaching companion. *Routledge*. https://doi.org/10.4324/9781351040143

Alrø, H., & Skovsmose, O. (2004). Dialogic learning in collaborative investigation. *Nordic Studies in Mathematics Education, 9*(2), 39–59. https://doi.org/10.7146/nomad.v9i2.147119

Andersen, H. (2013). Bridging disciplines: Conceptual development in interdisciplinary groups. In H. Frangerau, H. Geisler, T. Halling, & W. Martin (Eds.), *Classification and evolution in biology, linguistics and the history of science* (pp. 33–44). Franz Steiner Verlag. https://doi.org/10.25162/9783515105897

Bandura, A. (1986). *Social foundations of thought & action: A social cognitive theory*. Prentice-Hall.

Bardecki, M. J. (2019). Multi-disciplinarity. In W. Leal Filho (Ed.), *Encyclopedia of Sustainability in Higher Education* (pp. 1179–1184). Springer. https://doi.org/10.1007/978-3-030-11352-0_90

Barile, S., & Saviano, M. (2021). Interdisciplinary systems thinking for a new scientific paradigm: Toward a re-founding of human values. In G. Minati (Ed.), *Multiplicity and interdisciplinarity: Essays in honor of Eliano Pessa* (pp. 17–39). Springer. https://doi.org/10.1007/978-3-030-71877-0_3

Beyers, R. N. (2010). Nurturing creativity and innovation through FabKids: A case study. *Journal of Science Education and Technology, 19*(5), 447–455. https://doi.org/10.1007/s10956-010-9212-0

Brigandt, I. (2010). Beyond reduction and pluralism: Toward an epistemology of explanatory integration in biology. *Erkenntnis, 73*(3), 295–311. https://doi.org/10.1007/s10670-010-9233-3

Bromme, R. (2000). Beyond one's own perspective: The psychology of cognitive interdisciplinarity. In N. Stehr & P. Weingart (Eds.), *Practicing interdisciplinarity* (pp. 115–133). University of Toronto Press. https://doi.org/10.3138/9781442678729-008

Burnard, P., & Colucci-Gray, L. (Eds.) (2019). *Why science and art creativities matter. (Re-) Configuring* STEA⁺M *for future-making education*. Brill. https://doi.org/10.1163/9789004421585

Cabrera, D., & Cabrera, L. (2023). What is systems thinking? In J. M. Spector, B. B. Lockee, & M. D. Childress (Eds.), *Learning, design, and technology: An international compendium of theory, research, practice, and policy* (pp. 1495–1522). Springer. https://doi.org/10.1007/978-3-319-17461-7_100

Culpepper M. K., Gauntlett D. (2021). Inviting everyday creators to make, think, and talk. *Thinking Skills and Creativity, 42*, Article 100933. https://doi.org/10.1016/j.tsc.2021.100933

Darden, L., & Maull, N. (1977). Interfield theories. *Philosophy of Science, 44*(1), 43–64. https://doi.org/10.1086/288723

Dowell, N. M., Lin, Y., Godfrey, A., & Brooks, C. (2020). Exploring the relationship between emergent sociocognitive roles, collaborative problem-solving skills, and outcomes: A group communication analysis. *Journal of Learning Analytics, 7*(1), 38–57. https://doi.org/10.18608/jla.2020.71.4

Drake, S. M. & Burns, R. C. (2004). *Meeting standards through integrated curriculum*. Association for Supervision and Curriculum Development (ASCD), Virginia, USA. https://repository.bbg.ac.id/bitstream/606/1/Meeting_Standards_Through_Integrated_Curriculum.pdf

Education Council. (2015). *National STEM school education strategy, 2016–2026.* https://www.education.gov.au/download/12532/national-stem-school-education-strategy/23618/national-stem-school-education-strategy/pdf

Eigenbrode, S. D., O'Rourke, M., Wulfhorst, J. D., Althoff, D. M., Goldberg, C. S., Merrill, K., Morse, W., Nielsen-Pincus, M., Stephens, J., Winowiecki, L., & Bosque-Pérez, N. A. (2007). Employing philosophical dialogue in collaborative science. *BioScience, 57*(1), 55–64. https://doi.org/10.1641/B570109

Erduran, S. (2018). Toulmin's argument pattern as a "horizon of possibilities" in the study of argumentation in science education. *Cultural Studies of Science Education, 13*(4), 1091–1099. https://doi.org/10.1007/s11422-017-9847-8

Forbes, A., Chandra, V., Pfeiffer, L., & Sheffield, R. (2021). STEM education in the primary school: A teacher's toolkit. *Cambridge*. https://doi.org/10.1017/9781108868488

Garibay, J. (2015). STEM students' social agency and views on working for social change: Are STEM disciplines developing socially and civically responsible students? *Journal of Research in Science Teaching, 52*(5), 610–632. https://www.doi.org/10.1002/tea.21203

Gentner, D., & Maravilla, F. (2017). Analogical reasoning. In L. J. Ball & V. A. Thompson (Eds.), *International handbook of thinking and reasoning* (pp. 186–203). Routledge. https://doi.org/10.4324/9781315725697

Giebels, D., Carus, J., Paul, M., Kleyer, M., Siebenhüner, B., Arns, A., Bartholomä, A., Carlow, V., Jensen, J., Tietjen, B., Wehrmann, A. & Schroder, B. (2020). Transdisciplinary knowledge management: A key but underdeveloped skill in EBM decision-making. *Marine Policy, 119*, Article 104020. https://doi.org/10.1016/j.marpol.2020.104020

Govan, M. (2016). The application of peer teaching in digital forensics education. *Higher Education Pedagogies, 1*(1), 57–63. https://doi.org/10.1080/23752696.2015.1134198

Heitzmann, N., Opitz, A., Stadler, M., Sommerhoff, D., Fink, M. C., Obersteiner, A., Schmidmaier, R., Neuhaus, B. J., Ufer, S., Seidel, T., Fischer, M. R., & Fischer, F. (2021). Cross-disciplinary research on learning and instruction—Coming to terms. *Frontiers in Psychology, 11*(11), Article 562658. https://doi.org/10.3389/fpsyg.2021.562658

Karahan, E., & Roehrig, G. (2016). Use of socioscientific contexts for promoting student agency in environmental science classrooms. *Bartin University Journal of Faculty of Education, 5*(2), 425–442. https://dergipark.org.tr/en/pub/buefad/issue/24349/258097

Keller, C. M., & Keller, J. D. (1996). *Cognition and tool use: The blacksmith at work*. Cambridge University Press.

Kidman, G. (2025). STE*A*⁺M Theoretical and Conceptual Frameworks: Unpacking the Conflation of Terms. In S. J. Farenga, S. G. Garofalo, & D. Ness (Eds.), *International Handbook of Research on Steam Curriculum and Practice*. Routledge.

Kidman, G., Tan, H., & Macdonald, S. (2024). Reimagining STEM education: Developing agency through STE*A*⁺M and design thinking. In T. Barkatsas, P. McLaughlin, & W. Goff (Eds.), *Reimagining education for the second quarter of the 21st century and beyond: Dilemmas, challenges, advancements and innovations* (pp. 1–23). Brill. https://doi.org/10.1163/9789004688490_001

Klaassen, R. G. (2018). Interdisciplinary education: A case study. *European Journal of Engineering Education, 43*(6), 842–859. https://doi.org/10.1080/03043797.2018.1442417

Klein, J. T. (2014). Discourses of transdisciplinarity: Looking back to the future. *Futures, 63*, 68–74. https://doi.org/10.1016/j.futures.2014.08.008

Klein, J. T. (2017). Typologies of interdisciplinarity: The boundary work of definition. In R. Frodeman (Ed.), *The Oxford Handbook of Interdisciplinarity* (2nd ed., pp. 21–34). Oxford University Press. https://doi.org/10.1093/oxfordhb/9780198733522.013.3

Kunc, M., Groom, O., Barker S., & Nachiappan, N. (2020). Qualitative analysis of cognitive and social congruence in peer-assisted learning from the medical students perspective, *Medical Education Online, 25*(1), Article 1823619. https://doi.org/10.1080/10872981.2020.1823619

Kurup, P. M., Yang, Y., Li, X., & Dong, Y. (2021). Interdisciplinary and integrated STEM. *Encyclopedia of Social Sciences, 1*(4), 1192–1199. https://doi.org/10.3390/encyclopedia1040090

Liao, C. (2016). From interdisciplinary to transdisciplinary: An Arts-integrated approach to STE*A*⁺M education. *Art Education, 69*(6), 44–49. https://doi.org/10.1080/00043125.2016.1224873

Lin, Y., Dowell, N., & Godfrey, A. (2021). Skills matter: Modeling the relationship between decision making processes and collaborative problem-solving skills during hidden profile tasks. In *Proceedings of the LAK21: 11th International Learning Analytics and Knowledge Conference* (pp. 428–437). https://doi.org/10.1145/3448139.3448180

Liu, S., & Li, C. (2023). Promoting design thinking and creativity by making: A quasi-experiment in the information technology course. *Thinking Skills and Creativity, 49*, Article 101335. https://doi.org/10.1016/j.tsc.2023.101335

Marín-Marín, J. -A., Moreno-Guerrero, A. -J., Dúo-Terrón, P., & López-Belmonte, J. (2021). STE*A*+M in education: A bibliometric analysis of performance and co-words in Web of Science, *International Journal of STEM Education, 8*, Article 41. https://doi.org/10.1186/s40594-021-00296-x

Marthaliakirana, A. D., Suwono, H., Saefi, M., & Gofur, A. (2022). Problem-based learning with metacognitive prompts for enhancing argumentation and critical thinking of secondary school students. *Eurasia Journal of Mathematics, Science and Technology Education, 18*(9), Article em2148. https://doi.org/10.29333/ejmste/12304

Mater, N., Daher, W., & Mahamid, F. (2023). The effect of STE*A*+M activities based on experiential learning on ninth graders' mental motivation. *European Journal of Investigation in Health, Psychology and Education, 13*, 1229–1244. https://doi.org/10.3390/ejihpe13070091

Mathers, B., Johnson, J., Kaufmann, A., Sinni, N., et al. (2024). Developing student agency through authentic application of socioscientific issues in STEM classrooms. *Turkish Journal of Education, 13*(5), 508–534. https://doi.org/10.19128/turje.1507933

McElreavy, C., Tobin, V., Martin, T., Damon, M. B., Crate, N., Godinez, A., & Bennett, K. (2016). The history of the academy and the disciplines. In R. DeRosa (Ed.), *Interdisciplinary studies: A connected learning approach*. Rebus Community. https://press.rebus.community/idsconnect/chapter/the-history-of-the-academy-and-the-disciplines/

Mennes, J. (2020). Putting multidisciplinarity (back) on the map. *European Journal of Philosophical Science, 10*(2), Article 18. https://doi.org/10.1007/s13194-020-00283-z

Nussbaum, E. M. (2021). Critical integrative argumentation: Toward complexity in students' thinking. *Educational Psychologist, 56*(1), 1–17. https://doi.org/10.1080/00461520.2020.1845173

National Academy of Sciences, National Academy of Engineering, & Institute of Medicine. (2005). *Facilitating interdisciplinary research*. National Academies Press. https://doi.org/10.17226/11153

Newell, W. H. (2007). Decision making in interdisciplinary studies. In G. Morçöl (Ed.), *Handbook of decision making* (pp. 245–264). Routledge. https://doi.org/10.1201/9781420016918

O'Rourke, M., & Crowley, S. J. (2013). Philosophical intervention and cross-disciplinary science: The story of the Toolbox Project. *Synthese, 190*(11), 1937–1954. https://doi.org/10.1007/s11229-012-0175-y

O'Rourke, M., Crowley, S. J., & Gonnerman, C. (2016). On the nature of cross-disciplinary integration: A philosophical framework. *Studies in History and Philosophy of Science Part C: Studies in History and Philosophy of Biological and Biomedical Sciences, 56*, 62–70. https://doi.org/10.1016/j.shpsc.2015.10.003

Ortiz-Revilla, J., Adúriz-Bravo, A., & Greca, I. M. (2020). A framework for epistemological discussion on integrated STEM education. *Science & Education, 29*(4), 857–880. https://doi.org/10.1007/s11191-020-00131-9

Organisation for Economic Co-operation and Development. (2020). Addressing societal challenges using transdisciplinary research. *OECD Publishing*. https://doi.org/10.1787/0ca0ca45-en

Pilcher, J. (2015). A Modified Delphi Study to Define "Ah Ha" Moments in Education Settings. *Educational Research Quarterly, 38*(4), 51–67. https://www.proquest.com/scholarly-journals/modified-delphi-study-define-ah-ha-moments/docview/1692023351/se-2

Pohl, C. (2011). What is progress in transdisciplinary research? *Futures, 43*(6), 618–626. https://doi.org/10.1016/j.futures.2011.03.001

Pohl, C., Klein, J. T., Hoffmann, S., Mitchell, C., & Fam, D. (2021). Conceptualising transdisciplinary integration as a multidimensional interactive process. *Environmental Science and Policy, 118*, 18–26. https://doi.org/10.1016/j.envsci.2020.12.005

Repko, A. F. (2012). *Interdisciplinary research: Process and theory* (2nd ed.). SAGE.

Repko, A. F., Szostak, R. (2021). *Interdisciplinary research: Process and theory* (4th ed.). SAGE.

Repko, A. F., Szostak, R., & Buchberger, M. P. (2020). *Introduction to interdisciplinary studies* (3rd ed.). SAGE.

Ribosa, J., & Duran, D. (2022). Student-generated teaching materials: A scoping review mapping the research field. *Education in the Knowledge Society, 23*, Article 27443. https://doi.org/10.14201/eks.27443

Rodier, C., Galaleldin, M., Boudreau, J., Anis, H., & Peyton, L. (2021). STE*A*+M—Arts integration frameworks for transdisciplinarity. In *Proceedings of the Canadian Engineering Education Association (CEEA-ACEG) Conference June 20–23 PEI*. Paper 131. https://doi.org/10.24908/pceea.vi0.14918

Roehrig, G. H., Dare, E. A., Ellis, J. A., & Ring-Whalen, E. (2021). Beyond the basics: A detailed conceptual framework of integrated STEM. *Disciplinary and Interdisciplinary Science Education Research, 3*(1), 1–18. https://doi.org/10.1186/s43031-021-00041-y

Sarid, A., & Goldman, D. (2021). A value-based framework connecting environmental citizenship and change agents for sustainability—Implications for education for environmental citizenship. *Sustainability, 13*(8), Article 4338. https://doi.org/10.3390/su13084338

Schary, D. P., & Cardinal, B. J. (2015). Interdisciplinary and intradisciplinary research and teaching in kinesiology: Continuing the conversation. *Quest, 67*(2), 173–184. https://doi.org/10.1080/00336297.2015.1017586

Sloman, S. A., Patterson, R., & Barbey, A. K. (2021). Cognitive neuroscience meets the community of knowledge. *Frontiers in Systems Neuroscience, 15*, Article 675127. https://doi.org/10.3389/fnsys.2021.675127

Spiro, R. J., Coulson, R. L., Feltovich, P. J., & Anderson, D. (1988). *Cognitive flexibility theory: Advanced knowledge acquisition in ill-structured domains* [Technical Report 441]. Illinois University. https://files.eric.ed.gov/fulltext/ED302821.pdf

Soomro, S. A., Casakin, H., Nanjappan, V., & Georgiev, G. V. (2023). Makerspaces fostering creativity: A systematic literature review. *Journal of Science Education and Technology, 32*(4), 530–548. https://doi.org/10.1007/s10956-023-10041-4

Thibaut, L., Ceuppens, S., De Loof, H., De Meester, J., Goovaerts, L., Struyf, A., Boeve-de Pauw, J., Dehaene, W., Deprez, J., De Cock, M., Hellinckx, L., Knipprath, H., Langie, G., Struyven, K., Van de Velde, D., Van Petegem, P. & Depaepe, F. (2018). Integrated STEM Education: A systematic review of instructional practices in secondary education. *European Journal of STEM Education, 3*(1), Article 2. https://doi.org/10.20897/ejsteme/85525

Tsai, M. -F. (2021). Exploration of students' integrative skills developed in the design thinking of a Psychology course. *Thinking Skills and Creativity, 41*, Article 100893. https://doi.org/10.1016/j.tsc.2021.100893

von Braun, J. (2017). Children as agents of change for sustainable development. In A. M. Battro, P. Léna, S. Sorondo, & von Braun (Eds.), *Children and sustainable development: Ecological education in a globalized world* (pp. 17–30). Springer. https://doi.org/10.1007/978-3-319-47130-3_2

White, E. J. (2022). Dialogic pedagogy. *Oxford Bibliographies*. https://doi.org/10.1093/obo/9780199756810-0261

Whittington, J. C., McCaffary, D., Bakermans, J. J., & Behrens, T. E. (2022). How to build a cognitive map. *Nature Neuroscience, 25*(10), 1257–1272. https://doi.org/10.1038/s41593-022-01153-y

Wised, S., & Inthanon, W. (2024). The evolution of STEA^+M -Based programs: Fostering critical thinking, collaboration, and real-world application. *Journal of Education and Learning Reviews, 1*(4), 13–22. https://doi.org/10.60027/jelr.2024.780

Zhan, Z., & Niu, S. (2023). Subject integration and theme evolution of STEM education in K-12 and higher education research. *Humanities & Social Sciences Communications, 10*(1), Article 781. https://doi.org/10.1057/s41599-023-02303-8

PART II

Expanding the Boundaries of STEA^+M Education

This Part critically explores how STEA^+M education can transcend traditional boundaries to address the evolving needs of the 21st century and beyond. Through a progressive exploration of skills, thinking systems, and equitable practices, these chapters collectively argue for a reimagined STEA^+M education that is interdisciplinary, inclusive, and transformative. Each chapter builds on the previous one, creating a comprehensive narrative that expands the boundaries from 21st century foundational skills to thinking systems as meta-skills and finally to the hidden curriculum of STEM and STEAM education.

Chapter 5 serves as the foundation for Part II by examining the intersection of 21st century skills and STEM education. It traces the emergence of 21st century skills in educational discourse, defines their core components, and addresses critiques of the concept. Synthesis of the literature generated three notions of how STEM education is positioned relative to 21st century skills and vice versa. The first is the notion of *21st centurySTEM skills*. This conceptualisation is mainly concerned with determining the STEM skills needed so that we can prepare students for the demands of an evolving workforce for the remaining decades of the 21st century. The second notion concerns *STEM education as a learning approach* for developing 21st century skills. The third notion relates to *STEM literacy* as a foundational 21st century competency. Through reviewing the prevailing frameworks for 21st century skills and related STEM education literature, we call for more critical discussion and awareness of how the notions of 21st century skills are positioned within STEM education discourse.

Building upon the discussion of 21st century skills, Chapter 6 introduces the concept of *STEA⁺M thinking systems* as neo-21st century skills. The myriad of 21st century skills can be categorised into meta-skills such as cognitive skills, interpersonal skills, character qualities, and technological literacies. Neo-21st century skills extend existing conceptualisations of 21st century skills with more complexities, encompassing integrative thinking, meta-communicative and influence skills, social consciousness and *A*gency, and ethical technological literacies. These neo-21st century skills enable individuals to integrate knowledge and skills across disciplines, and synthesise them to form STE*A*⁺M thinking systems such as *Social Entrepreneurial Mindsets*, *Socio-environmental Sustainability Agency*, and *Resilience Engineering for Climate Adaptation and Human-centred Design*. Through hands-on, interdisciplinary learning, such as design thinking and makerspaces, students cultivate adaptability and problem-solving capabilities. By shifting the focus from isolated skills to interconnected thinking systems, this chapter positions STE*A*⁺M education as a transformative framework for developing adaptable, resilient, and innovative learners. It also bridges the gap between foundational skills and the need for equitable, inclusive practices, which are explored in depth in Chapter 7.

While STE*A*⁺M education explicitly promotes skill development, implicit biases and power structures shape learning experiences in ways that reinforce inequities. Chapter 7 examines the *hidden curriculum* of STEM and STEAM education from the perspectives of four curriculum ideologies: scholar-academic, social efficiency, learner-centred, and social reconstruction ideologies, as well as three contextual lenses: structural-functional, social phenomenological, and radical critical lenses. The chapter illustrates and reviews the hidden curriculum contexts and explores the tensions within each of these contexts, discussing unspoken norms, power structures, and disciplinary hierarchies that shape students' learning experiences and influence student engagement, agency, and inclusion. By making the hidden curriculum visible, the chapter advocates for the intentional design of STE*A*⁺M learning environments that prioritise inclusivity, agency, and meaningful participation. It proposes transformative practices, such as critically reflective and inclusive curriculum design and implementation that integrate diverse perspectives, and student-centred assessment strategies. The chapter also emphasises the importance of being attentive to: 1) the ***inclusion, roles, and representations*** of disciplines in STEM and STEAM; 2) ***non-content aspects*** of the curriculum such as neo-21st century skills; and 3) ***interconnected ways of knowing*** such as cultural self-knowing, relational knowing, critical knowing, visionary and ethical knowing, and knowing in action; in order for educators and curriculum designers to mitigate the adverse effects of STEM and STEAM's hidden curriculum. By challenging dominant narratives and valuing diverse knowledge systems, we envision a more equitable and empowering plexus model of STE*A*$^+$M education that prepares all learners to tackle global challenges. Chapter 7 not only concludes Part II but also serves as a call to action for systemic change, ensuring that the skills and thinking systems discussed earlier are accessible to all students, regardless of background or identity.

Part II builds on the theoretical foundation laid out in Part I, which calls for rethinking STE*A*$^+$M education as an integrative, interdisciplinary, and agency-driven endeavour. Part I employs a micro-view of STEM education and emphasises the importance of neuroeducation, epistemic flexibility, student *A*gency, and cognitive integration as the conceptual tools for reimagining STE*A*$^+$M education. Part II employs a macro-view and weaves a cohesive narrative relating to the practice and systemic realities that progressively expand the boundaries of STEM/STEAM/ STE*A*$^+$M education, moving from foundational skills to advanced thinking systems and, ultimately, to equitable and inclusive practices. Chapter 5 establishes the importance of 21st century skills and STEM education as a foundation for modern learning. Chapter 6 deepens this understanding by framing STE*A*$^+$M thinking systems as neo-21st century skills, emphasising the need for meta-skills and interconnected competencies to address complex challenges. Chapter 7 ties these ideas together by addressing the hidden curriculum and advocating for the design of an equitable and inclusive curriculum that ensures all students can benefit from the expanded vision of STE*A*$^+$M education.

Together, Parts I and II create a cohesive narrative that bridges theory and practice. Part I's emphasis on the interconnected nature of knowledge and the importance of fostering adaptable learners finds its practical counterpart in Part II's exploration of skills, thinking systems, and equitable practices. By expanding the boundaries of STE*A*$^+$M education, Part II not only reinforces the paradigm shift called for in Part I, but also provides actionable pathways for educators, researchers, and policymakers to create dynamic, interconnected, and future-oriented learning experiences. This integration of theory and practice underscores the transformative potential of STEA^+M education in preparing learners to address global challenges and contribute to a more just and equitable world.

Hazel Tan and Gillian Kidman

CHAPTER 5

Twenty-First Century Skills and STEM Education: A Review of Literature

5.1 Introduction

At the turn of the twenty-first century, there was widespread acknowledgement that the world is changing rapidly, as evidenced by globalisation, financial crises, the rise of the knowledge era, and technological advancement. This gave rise to an impending sense of urgency in educating our new generation to thrive in this new and complex world. A global movement started to debate and determine the foundation skills and competencies required in the education of all students (e.g., Reimers & Chung, 2016). An example of a foundation skill is literacy. Literacies evolved from the basic understanding and use of language in everyday life and work to encompass various forms such as media literacy, digital literacy, quantitative literacy, and multicultural literacy. Simultaneously, nations and international organisations have started to focus on and promote STEM education, aiming to equip the next generation with knowledge and skills for the future workplace (Freeman et al., 2014).

As we enter the next quarter of the twenty-first century, there is increasing global instability caused by divergent economic recovery from the aftermath of the COVID 19 pandemic, the food and energy crisis

G. Kidman et al., *Knowledge Generation in STEM and STEAM Education*, https://doi.org/10.1007/978-3-032-03247-8_5

arising from violent conflicts such as the Russia-Ukraine and the Israel-Palestine wars, multipolar geopolitical tensions with Trump's second presidency, and increased climate change impacts and disasters. The World Economic Forum's (WEF) Global Risk Report identifies perceived risks in 2024–2026: misinformation and disinformation, extreme weather events, social and political polarisation, cyber insecurity, disruption of food and energy supplies, and escalation or outbreak of armed conflicts (WEF, 2024). Insights from over 1200 experts across the WEF diverse network highlight the interconnections between risks to predict a potential "polycrisis" related to natural resource shortages, climate change, and socio-economic and environmental fall-out (WEF, 2023). In this new era, *what skills and competencies* are required to successfully navigate the complexities and positively impact our society? If 59% of the global workforce needs training or retraining by 2030 (WEF, 2025), *how do we prepare* ourselves and the next generation of learners for such a future? We contend that a plexus model of STEA^{+}M education (see Chapter 3) can promote the development of these neo-twenty-first century skills. In this chapter, a summary of major frameworks of twenty-first century skills is outlined, followed by a review and discussion of the literature on twenty-first century skills and their relations to STEM education.

5.2 The Rise of Twenty-First Century Skills in Education

This section traces the major milestones and frameworks in twenty-first century skills in education. A list of frameworks is displayed in chronological order in Table 5.1.

In reviewing international literature on 21st Century Skills, researchers point to the 1972 Faure Report *Learning to Be: The World of Education Today and Tomorrow*, published by the United Nations Educational, Scientific and Cultural Organization (UNESCO), as an important milestone document. Emerging during significant global change in the late 1960s and 1970s, including rapid technological advancements, globalisation, and social transformations, UNESCO sought to reimagine education's role to prepare for the changes. The Faure Report resulted from the International Commission on Education Development, chaired by Edgar Faure, a former French politician and Minister of Education of France. The vision of education is "to nurture the development of well-rounded individuals equipped to learn at all stages of life and committed

Table 5.1 Timeline of major frameworks for Twenty-First Century Skills

Year	*Framework*	*Key focus areas*
1972	Faure Report (UNESCO) (Faure et al., 1972)	Lifelong learning and education as a means for individual and societal development The report plants the seed for a global movement toward twenty-first century skills
1996	Delors Report (UNESCO) (Delors et al., 1996)	Four pillars of learning: learning to know, to do, to be, and to live together
2002	Partnership for 21st Century Skills (P21) (P21, 2002)	The comprehensive framework proposes fully integrating core subjects with learning skills, learning tools, and twenty-first century content as taught in twenty-first century contexts Learning and innovation, information and media literacy, and life and career skills are promoted (Trilling & Fadel, 2009)
2004	Key Competences for Lifelong Learning: A European Reference Framework (European Union, 2006)	Eight competences focus on communication, mathematical and digital competences, learning to learn, social and civic competences, initiative and entrepreneurship, and cultural awareness
2005	OECD Key Competencies Framework (DeSeCo) (OECD, 2005)	Three broad categories: use tools interactively (e.g., language, technology), interact in heterogeneous groups, and act autonomously
2010	Preparing 21st Century Students for a Global Society: An Educator's Guide to the "Four Cs" (US National Education Association [NEA], 2010)	Four Cs: critical thinking, communication, collaboration, and creativity
2012	Assessment and Teaching of 21st Century Skills (ATC21S) (Griffin et al., 2012)	Four groupings: ways of thinking, ways of working, tools for working and living in the world
2015	World Economic Forum's New Vision for Education (WEF, 2015)	Foundational literacies, competencies (e.g., critical thinking, collaboration), and character qualities
2021	UNESCO Futures of Education, Sahle-Work Report (Sahle-Work et al., 2021)	Broad competencies for global citizenship, sustainability, and adaptability in a complex world

to a profound aspiration for peace" (Faure et al., 1972, p. 348). In their report, Faure et al. (1972) also highlight that literacy is not an end in itself but enables individuals to participate in the functions of society, and that "learning content is no longer the usual educational subject-matter, but involves overall interdisciplinary responses to concrete problems" (p. 141). More than two decades later, in 1996, UNESCO also commissioned Jacques Delors, former president of the European Commission, to undertake consultation internationally and publish the Delors Report *Learning: The Treasure Within*. The document proposed four pillars of education for the new century: Learning to Know, to Do, to Be, and to Live Together (Delors et al., 1996).

In the 1990s, there were increasing discussions about the skills, often termed core skills, generic skills, key skills, employability skills, or transferable skills, that would be needed for the twenty-first century (Tight, 2021). The American Philosophical Association published the Delphi report (Facione, 1990) that defined critical thinking as a key tool of inquiry that is a "liberating force in education and a powerful resource in one's personal and civic life" (p. 2) and has both a skill dimension and a dispositional dimension. Core cognitive skills in critical thinking are: (1) interpretation, (2) analysis, (3) evaluation, (4) inference, (5) explanation, and (6) self-regulation (Facione, 1990, p. 4). In 1992, Reich reflected on the implications of information and communication technologies on the shifting demand for jobs from routine production to "symbolic-analytic services" that trade in the "manipulation of symbols – data, words, oral and visual representations" (p. 177). These services require problem-identifying, problem-solving, and strategic brokering of information. Educating symbolic-analysts, according to Reich (1992), entails refining four basic skills of "*abstraction*, *systems thinking*, *experimentation*, and *collaboration*" (p. 229, emphasis in original).

Bridges (1993) explored the concept of transferable skills, distinguishing between two types. The first includes skills that can be applied with minimal or no adaptation across various social settings, such as using a word processor. The second involves skills that enable individuals to adapt, modify, or extend knowledge, understanding, or abilities gained in one cognitive domain or social context to apply them in another, exemplified by negotiating skills (p. 50). Additionally, Bridges emphasised the significance of meta-skills or second-order skills, which allow individuals to “select, adapt, adjust, and apply one’s other skills to different situations, across various social contexts, and perhaps across different cognitive domains” (p. 50). These meta-skills require social awareness, responsiveness, and adaptability.

In 1999, various US departments (Commerce, Education, and Labor) and the National Institute for Literacy and Small Business Administration published a report on *Twenty-First Century Skills for Twenty-First Century Jobs*, which highlights the shifts in the workplace organisation and management (e.g., from hierarchical and rigid to flat and flexible) and job scope (from narrow, specialised and fixed operations, to broad, multi/cross-skilled, high autonomy, and constant changing conditions). Skills in technology, literacy, numeracy, and communication, as well as increased education and training, are required to meet the anticipated demands.

The turn of the millennium brought about a more intensive focus on twenty-first century skills in countries globally due to the interconnected combination of factors:

- **Shift in Economic Focus.** The shift from manufacturing to information and knowledge services in advanced economies contributed to the need for new skills. This shift highlighted the need for individuals who can communicate, share, and use information to solve complex problems (Blinkey et al., 2012).
- **Impact of Technology.** The rapid development of information and communication technologies (ICT) transformed work and social interactions. The increasing potential of ICT to take over routine tasks has led to a growing demand for skills in communication, problem-solving, and information mediation (Voogt & Roblin, 2012).
- **Response to Globalisation and Geopolitical Changes.** The world’s increasing interconnectedness due to globalisation and modernisation also contributed to the need for individuals to adapt to changing technologies and make sense of vast amounts of information. These changes also emphasised the importance of interacting and working in diverse groups (OECD, 2005). The collapse of the Soviet Union heralded the birth or rebirth of new countries grappling with developing new education systems, and the opening up of China also hastened the sense of urgency for countries to be

responsive to ensure innovation, creativity, economic growth, and global competitiveness (Power, 2015).

- **Influence of International Educational Reports and Initiatives.** The Faure et al. (1972), Delors et al. (1996), and Sahle-Work et al. (2021) Reports, commissioned by UNESCO, that sought to determine the kind of education needed for the future, are collectively a response towards the changing times and a call for more responsive action. Various international organisations and frameworks have played a role in shaping the concept of twenty-first century skills, which also has a snowball effect. These include the OECD's DeSeCo project, the European Union's Key Competencies for Lifelong Learning framework, and the Partnership for 21st Century Skills (P21) in the United States, which are highlighted next. These frameworks aimed to identify and define the skills needed for individuals to succeed in the modern world.

In 2002, the US government, together with private organisations such as AOL Time Warner Foundation, Apple Computer Inc., CISCO Systems Inc., Dell Computer Corporation, and Microsoft Corporation, formed the Partnership for 21st Century Skills (P21) to envision a model of learning that embeds twenty-first century skills into the US education system (Partnership for 21st Century Skills (P21), 2002). The document *Learning for the 21st Century: A Report and MILE Guide for 21st Century Skills* stipulates a comprehensive framework that includes six key elements of emphasising core subjects and learning skills, using twenty-first century tools to develop learning skills, teaching and learning twenty-first century content in a twenty-first century context, and using twenty-first century assessments that measure twenty-first century skills (P21, 2002). A guide to support teachers and school leaders is provided, with nine steps to build momentum.

1. Embrace a powerful vision of public education that includes twenty-first century skills
2. Align leadership, management and resources with educational goals
3. Use this tool to assess where schools are now
4. Develop priorities for twenty-first century skills
5. Develop a professional development plan for twenty-first century skills

6. Make sure students have equitable access to a twenty-first century education
7. Begin developing assessments to measure student progress in twenty-first century skills
8. Collaborate with outside partners
9. Plan collectively and strategically for the future (P21, 2002).

In 2004, the European Union released the *Key Competences for Lifelong Learning* report, which sets out eight key competences combining knowledge, skills, and attitudes (Blinkey et al., 2012). These competences are updated and extended over the years to reflect shifts in thinking, as seen in Table 5.2.

The following international milestones are two related projects undertaken by the Organisation for Economic Co-operation and Development (OECD)—the *Definition and Selection of Key Competencies* (DeSeCo) in education through expert consultations (OECD, 2005), and the Program for International Student Assessment (PISA), a three-yearly program to measure "15-year-olds' ability to use their reading, mathematics and science knowledge and skills to meet real-life challenges" (OECD, n.d., para. 1). Fifteen-year-olds were selected because, in many countries,

Table 5.2 Key competencies for lifelong learning—A European reference framework (EU, 2006, 2019)

2006	*2019*
1. Communication in the mother tongue	1. Literacy competence
2. Communication in foreign languages	2. Multilingual competence
3. Mathematical competence and basic competences in science and technology	3. Mathematical competence and competence in science, technology and engineering
4. Digital competence	4. Digital competence
5. Learning to learn	5. Personal, social and learning to learn competence
6. Social and civic competences	6. Citizenship competence
7. Sense of initiative and entrepreneurship	7. Entrepreneurship competence
8. Cultural awareness and expression	8. Cultural awareness and expression competence

this is the age at which compulsory education in science and mathematics often ends and the young adults are being prepared to meet the challenges of our knowledge societies (Adams & Wu, 2003). Unlike traditional assessments, which focus retrospectively on curriculum-based learning to evaluate readiness for further education, PISA sets out to assess students' preparedness for life. This involved evaluating their ability to apply knowledge and skills to real-world contexts (OECD, 2005).

After the turn of the twenty-first century, two subsequent seminal works were *21st Century Skills: Learning for Life in Our Times* by Trilling and Fadel (2009), based on the framework developed by P21 in which skills were grouped into three categories: learning and innovations skills; digital literacy skills; and life and career skills. The other was a major international research project supported by large multi-national companies and major institutions such as Cisco, Intel, Microsoft, the World Bank, UNESCO and OECD, under the umbrella of *Assessment and Teaching of 21st Century Skills* (ATC21S) (Griffin et al., 2012). Based on the analysis of twelve relevant frameworks from various countries and organisations, considering the knowledge, skills, attitudes, values, and ethics (KSAVE) aspects, the researchers outlined ten twenty-first century skills, grouped into four themes: *Ways of Thinking*, *Ways of Working*, *Tools for Working*, and *Living in the World.*

In 2015, the World Economic Forum (WEF) published a report on *New Vision for Education* examining the skills gap and proposing sixteen twenty-first century skills, grouped into three categories: *Foundational literacies* (how students apply core skills to everyday tasks), *Competencies* (how students approach complex challenges), and *Character qualities* (how students approach their changing environment) (WEF, 2015). The WEF updated their top 10 skills for jobs in 2015, 2020, and 2025 (see Table 5.3). The WEF Future of Jobs 2025 report (p. 41) noted a shift in skill demands, predicting that emerging skills such as environmental stewardship, AI and big data, and networks and cybersecurity will become critical in 2030.

The recent UNESCO report, published in 2021, is the Sahle-Work Commission's *Reimagining Our Futures Together: A New Social Contract for Education,* by the International Commission on the Futures of Education chaired by Sahle-Work Zewde, the President of Ethiopia (Sahle-Work et al., 2021). The report highlights the challenges to the equity of education and the ecological crisis, and poses questions about the nature and form of education amidst disruptions in the ways we

Table 5.3 Top 10 skills (retrieved from WEF future of jobs reports, 2015, 2020, 2025 https://www.weforum.org)

2015	*2020*	*2025*
1. Complex problem-solving	1. Complex problem-solving	1. Analytical thinking
2. Coordinating with others	2. Critical thinking	2. Resilience, flexibility and agility
3. People management	3. Creativity	3. Leadership and social influence
4. Critical thinking	4. People management	4. Creative thinking
5. Negotiation	5. Coordinating with others	5. Motivation and self-awareness
6. Quality control	6. Emotional intelligence	6. Technological literacy
7. Service orientation	7. Judgment and decision making	7. Empathy and active listening
8. Judgment and decision making	8. Service orientation	8. Curiosity and lifelong learning
9. Active listening	9. Negotiation	9. Talent management
10. Creativity	10. Cognitive flexibility	10. Service orientation and customer service

work, how we live and interact with technology, and our governance systems. It then proposes that "curricula should emphasize ecological, intercultural and interdisciplinary learning that supports students to access and produce knowledge while developing their capacity to critique and apply it" (p. 77). Four guiding principles for dialogue and action were recommended.

- **Knowledge Commons.** Curricula should empower learners to access and contribute to humanity's collective knowledge, fostering creative and critical engagement. Students should be taught to question assumptions, address omissions, and ensure the knowledge commons remains open, diverse, and inclusive for all.
- **Ecological Responsibility**. Education must reframe humanity's relationship with the planet by prioritising climate change education and teaching respectful, sustainable living to address ecological crises caused by human activity.
- **Combating Misinformation.** Curricula should promote scientific, digital, and humanistic literacies to counter misinformation. Students must learn to distinguish rigorous research from falsehoods, develop

digital skills, and critically engage with science and technology to influence their ethical applications.

- **Human Rights and Inclusion.** Learning should prioritise human rights, agency, and democratic participation. Curricula must address gender equality, challenge oppressive stereotypes, and teach students to confront racism and discrimination, fostering a moral commitment to equity and inclusion.

Besides the frameworks highlighted, researchers cited others. These are US-centric and include the more technology-specific frameworks such as the *enGauge 21st Century Skills: Digital Literacies for a Digital Age* by the North Central Regional Educational Laboratory and the Metiri Group (Lemke, 2002), *National Educational Technology Standards* by the International Society for Technology in Education (ISTE), and the US National Assessment of Educational Progress (NAEP) *Technology and Engineering Framework* (Voogt & Roblin, 2012). The National Research Council (NRC) (2012) also developed the *Education for Life and Work: Developing Transferable Knowledge and Skills in the twenty-first Century.* These frameworks are commonly used in U.S. schools and research. The P21's framework is popular in educational practice as it also considers the curriculum and instruction support systems, standards and assessments, professional development, and learning environment. The NRC framework is based on a systematic review of the research on 21st Century Skills and is used as the theoretical framework in studies published in a later book, *Teaching and Learning for the Twenty-first Century: Educational Goals, Policies and Curricula from Six Nations* (Reimers & Chung, 2016).

In summary, it can be seen that there is a plethora of frameworks relating to twenty-first century skills. The essence of twenty-first century skills within the diverse frameworks is discussed next.

5.3 Capturing the Essence of Twenty-First Century Skills

The concept of "twenty-first century skills" has become a cornerstone in discussions around education reform and workforce preparedness. As economies and societies grapple with rapid technological advancements, globalisation, and shifting cultural norms, the demand for individuals equipped with more than traditional academic knowledge has surged. These skills are positioned as essential for navigating the complexities of modern life, ranging from critical thinking and creativity to collaboration and digital literacy.

Despite its prominence, the term "twenty-first century skills" remains a contested and multifaceted concept. It encompasses diverse competencies that vary depending on the framework or context. Educational organisations, policymakers, and industry leaders often use the term to describe overlapping but distinct sets of abilities aimed at fostering adaptability, innovation, and global citizenship.

The challenge of defining these skills lies in their breadth and evolving nature. While widespread agreement exists on their importance, differing priorities among stakeholders have led to various interpretations. Understanding these nuances is crucial for assessing the strengths and limitations of the term, as well as for identifying areas where it may fall short in addressing contemporary educational and societal needs. Concerning the definitions of terms, there seemed to be a range of words used to describe what people should be able to do to live and succeed in the twenty-first century, and to meet the demands of this day and age. The first mention of the term "21st Century Skills" was attributed to the early 1980s US report *A National at Risk: The Imperative for Educational Reform*, and its usage evolved through appropriations by various stakeholders in academia, education, industry, and policymakers (Mehta et al., 2020). The Oxford Learners' Dictionary defines *skill* as "the ability to do something well" and "a particular ability or type of ability" and competency as "the ability to do something well". Both terms are similar in meaning.

The Oxford English Dictionary provides a more nuanced definition:

> Skill—the capability of accomplishing something with precision and certainty; practical knowledge in combination with ability; cleverness,

expertness. Also, an ability to perform a function, acquired or learnt with practice. https://www.oed.com/view/Entry/180865

There is a notion of successfully translating practical knowledge into actions within a context or situation to accomplish a purpose. The prerequisite knowledge and know-how are an important pre-condition in the actioning process. The ability to transfer the content knowledge and procedural knowledge (knowing when, where, why and how to apply the knowledge) in new situations and solve problems is the essence of twenty-first century skills (NRC, 2012). The composite of transferable knowledge and associated skills can be considered as *competency* (NRC, 2012). Besides knowledge and skills, some twenty-first century frameworks include attitudes and dispositions and personal character qualities such as resilience, initiative, and conscientiousness (see Table 5.1). The EU Key Competences for Lifelong Learning (Gordon et al., 2009) defined competence as "a combination of skills, knowledge, aptitudes and attitudes crucial for personal fulfilment, active citizenship and inclusion and employability" (p. 3). This is also illustrated in the OECD's definition below.

> A competency is more than just knowledge and skills. It involves the ability to meet complex demands, by drawing on and mobilizing psychosocial resources (including skills and attitudes) in a particular context. For example, the ability to communicate effectively is a competency that may draw on an individual's knowledge of language, practical IT skills and attitudes towards those with whom he or she is communicating. (OECD, 2005, p. 4)

Attitudes are included in addition to the common elements of knowledge and skills to achieve a particular purpose. For example, Critical Thinking is conceptualised as having cognitive skills (e.g., interpretation, analysis, self-regulation) and affective dispositions (personal traits and habits of mind associated with good critical thinkers) (Facione, 1990). As researchers, educators, and the industry continue to determine, redefine, and predict what the set of "twenty-first century skills" or "twenty-first century competencies" constitute, the notion of "skills" has expanded to encompass not only the attitude and character qualities but include the judgment, values and ethics, such as the KSAVE ATC21C framework by Griffin et al. (2012).

In summary, there is no commonly agreed-upon set of twenty-first century skills, and even the terminologies are somewhat loose (e.g., competencies versus skills). However, most of the internationally recognised frameworks highlighted above seemed to follow three or four main domains, roughly corresponding to (a) cognitive/ thinking/problem-solving, (b) social/ collaboration & communication/interpersonal, and (c) character qualities/life-skill/intrapersonal domains, with particular focus on information and digital literacy as a fourth domain (see Table 5.4). Statistically significant, moderate positive correlations between various cognitive, interpersonal, and intrapersonal competencies, and success in desirable adult outcomes had been found in the review of research from fields such as industrial-organisational psychology, developmental psychology, human resource development, and economics (NRC, 2012). However, there are other critiques of the notion of twenty-first century skills, which are discussed next.

5.4 Critiques of the Term "Twenty-First Century Skills"

Though widely embraced, the concept of "twenty-first century skills" has not escaped scrutiny. Scholars, educators, and policymakers have raised critical concerns about its clarity, cultural relevance, and practical implementation. These critiques overlap and highlight the need for a deeper examination of the term's implications and its role in shaping education and society.

5.4.1 Lack of Conceptual Clarity

The term "twenty-first century skills" has been criticised for its ambiguity and lack of a universally agreed-upon definition (Tight, 2021). While many frameworks share common elements, such as critical thinking and problem-solving, their scope, and priorities differ significantly. Many educators and researchers use the term as a placeholder, idealising abstract future skills (Mehta et al., 2020). The twenty-first century skills seemed to be "an overwhelmingly universally-accepted and all-encompassing term that appears to include virtually everything that contemporary students could potentially need" (Mehta et al., 2020, p. 362). Critics argue that this lack of conceptual clarity complicates efforts to implement and measure these skills effectively (Silva, 2009), with little differentiation of

Table 5.4 Major frameworks of twenty-first century skills, with their elements aligned into four domains

ATC21S (Griffin et al., 2012)	*World Economic Forum's Top 10 Core Skills in 2025 (WEF, 2025)*	*21C skills grouped into 3 clusters of competence (NRC, 2012)*	*Framework for 21st Century Learning (P21) (Trilling & Fadel, 2009)*
Ways of Thinking – Creativity and innovation – Critical thinking, problem-solving, decision making – Learning to learn, Metacognition	*Cognitive skills* – Analytical thinking – Creative thinking	*Cognitive Competencies* – Cognitive processes and strategies (critical thinking) – Creativity (innovation)	*Learning and Innovation Skills—4 Cs* – Critical thinking – Communication – Collaboration – Creativity
Tools for Working – Information literacy – ICT literacy	*Technology Skills* – Technological literacy	*Cognitive Competencies* – Knowledge (information literacy, reasoning and argumentation)	*Information, Media, & Technology Skills* – Information literacy – Media literacy – ICT literacy

ATC21S (Griffin et al., 2012)	*World Economic Forum's Top 10 Core Skills in 2025 (WEF, 2025)*	*21C skills grouped into 3 clusters of competence (NRC, 2012)*	*Framework for 21st Century Learning (P21) (Trilling & Fadel, 2009)*
Ways of Working – Communication – Collaboration (teamwork)	*Working with Others* – Leadership and social influence – Empathy and active listening *Management Skills* – Talent management *Engagement Skills* – Service orientation and customer service	*Interpersonal Competencies* – Teamwork and collaboration (communication and collaboration) – Leadership (responsibility and conflict resolution)	*Life and Career Skills* – Social and cross-cultural skills – Leadership and responsibility
Living in the World – Citizenship—local and global – Life and career – Personal and social responsibility—including cultural awareness and competence	*Self-efficacy* – Resilience, flexibility and agility – Motivation and self-awareness – Curiosity and lifelong learning	*Intrapersonal Competencies* – Intellectual Openness (flexibility) – Work ethic/ Conscientiousness (initiative, appreciation for diversity) – Positive Core Self-Evaluation (metacognition)	*Life and Career Skills* – Flexibility and adaptability – Initiative and self-direction – Productivity and accountability ara>

the types of skills needed for particular contexts (Mehta et al., 2020). A creative performance in the arts might look very different from a creative mathematical solution or a creative science experiment. Some researchers highlight the dangers of focusing heavily on twenty-first century skills at the expense of content-specific knowledge and competencies, which, although they need updating with time, are still crucial (Allen & van der Velden, 2012). As mentioned in the previous section, the essence of twenty-first century skills encompasses more than just "skills" and knowledge per se, and also includes various overlapping social and personal attributes such as ethics, values, dispositions, attitudes, and interpersonal capabilities (see Table 5.4).

5.4.2 *Overemphasis on Skills and Technology*

A recurring critique is that the emphasis on skills, particularly in isolation from content knowledge, undermines the importance of a strong educational foundation (Allen & van der Velden, 2012; Silva, 2009). Scholars have argued that skills such as critical and creative thinking are deeply dependent on subject-specific knowledge (Ananiadou & Claro, 2009). Without a robust disciplinary knowledge base, learners cannot apply critical and creative thinking and may lack the depth needed to address complex challenges. A core body of knowledge is essential for the development and use of skills since learners need a foundation of knowledge to engage in higher-order thinking and problem-solving (Silva, 2009; Tight, 2021; also see Chapter 2). Additionally, the focus on adaptability and future uncertainty has been criticised for marginalising the role of values and the past in education (Dishon & Gilead, 2021).

Other critics argue that an excessive focus on technology can distract from teaching core content (Ananiadou & Claro, 2009). Greenlaw (2015) asserts that while the twenty-first century skills movement offers many practical benefits, its overarching narrative of salvation through technology lacks a balanced perspective on what should be considered meaningful knowledge and pedagogy in schools. In short, the movement places excessive emphasis on the acquisition and management of information while undervaluing the pursuit of wisdom (Greenlaw, 2015). The emphasis on technology is a consequence of the perspective of employers and policy-makers looking to equip future workers with skills for the workforce. Scholars call for a more balanced and critical approach to technological skills, ensuring that technology is used purposefully and ethically (Allen & van der Velden, 2012), and "embracing technological fluidity that is directed toward social justice and not just corporate interests" (Mehta et al., 2020, p. 375).

5.4.3 *Western-Centric Perspective*

Many frameworks for twenty-first century skills are developed in Western contexts and reflect values such as individualism, innovation, and market-oriented thinking. Critics point out that these frameworks may not be universally applicable or culturally neutral (Allen & van der Velden, 2012; Mehta et al., 2020). For example, foreign language skills, multilingualism, and intercultural understanding are absent in the major frameworks (see Table 5.4), despite globalisation. This suggests that the frameworks may be developed with a specific cultural and linguistic perspective, potentially marginalising other viewpoints. Additionally, the global digital environment may reinforce deficit perspectives on non-Western or under-resourced communities' cultural and literacy practices (Mehta et al., 2020). Technology giants like Google, Facebook, Amazon, and Microsoft risk perpetuating new forms of cultural imperialism by promoting their products as essential educational tools for thriving in a globalised world. These companies often reflect a Western, English-dominated worldview, potentially marginalising minority languages and cultures with less influence or representation (Mehta et al., 2020). Dishon and Gilead (2021) propound that twenty-first century skills frameworks sidestep the notion of values in education due to two political reasons. The frameworks were created by multi-national organisations or countries where their members or communities have varying or conflicting views, so these frameworks avoid committing to specific substantive values to enhance their broad applicability and prevent alienating certain groups. A second reason is that the frameworks claim authority by presenting themselves as evidence-based and grounded in objective data. Embracing explicit values, often seen as subjective, could weaken this perceived objectivity and authority (Dishon & Gilead, 2021). Overall, the criticisms surrounding the Western-centric perspective on twenty-first century skills imply that it is important to recognise that the conceptions of twenty-first century skills are not culturally neutral nor universally applicable. A more critical approach is needed, one that recognises the value of diverse perspectives, local knowledge, and equity in educational practices.

5.4.4 *Commodification of Education*

Related to the previous critiques, the alignment of twenty-first century skills with economic and workforce demands has led some to view the concept as a form of education commodification (Allen & van der Velden, 2012). This critique argues that education becomes narrowly focused on producing "employable" individuals, often at the expense of cultivating well-rounded, critically engaged citizens (Biesta, 2021; Dishon & Gilead, 2021). By framing education around specific skills, especially linked to technology, education is at risk of increased neoliberal influences and practices, such as focusing excessively on standardised testing and school rankings of measurable skills, detracting from the deeper, less quantifiable and more humanistic aspects of learning such as creativity, life-long learning, and personal development (Biesta, 2021; Elfert, 2015). In the narrative of twenty-first century skills, creativity becomes a tool for innovation, problem-solving, and economic productivity, rather than a valuable end in itself (Mehta et al., 2020). This market-driven approach may also contribute to inequalities by prioritising skills that align with dominant economic systems and may not be equally accessible or relevant to all students (Mehta et al., 2020). Overall, the critiques suggest that the focus on twenty-first century skills risks reducing education to a commodity, shaped by economic and corporate interests. This commodification can lead to the neglect of humanistic values, ethical considerations, and the importance of context and equity.

5.4.5 *Challenges in Implementation and Assessment*

The practical challenges of embedding twenty-first century skills into educational systems are significant. Educators often lack the training, resources, or institutional support to teach these skills effectively (Anainadou & Claro, 2009). Moreover, the complexity of skills like collaboration, creativity, and ethical reasoning makes them difficult to assess using traditional standardised tests (Child & Shaw, 2016). There is also a lack of clarity surrounding how to effectively integrate twenty-first century skills into subject-based curricula (Tight, 2021). Many education systems are still structured around discipline-based curricula and subjects taught in a silo-approach, adding to the challenge of integrating whichever set of twenty-first century skills that were being emphasised. There is a need for teaching and learning models that explain how various types of skills can best be developed in relation to each other (Allen & van der Velden, 2012), as well as how teacher education programs might support the teaching of twenty-first century skills (Anainadou & Claro, 2009).

Besides the above critiques, there is also a noticeable lack of hands-on skills, such as those in embodied learning and makerspace education (see Chapter 4). Theories of embodied cognition and neuroscience studies found that the mind is not an isolated, abstract entity but is deeply integrated with the body's sensorimotor systems (Macedonia, 2019). Neuroscientific research shows that processing concepts like objects, music, or flavours activate brain areas tied to sensory and motor systems, forming neural networks based on past experiences (Macedonia, 2019). Embodied cognition research suggests that life experiences shape cognitive strategies for problem-solving, with individuals adapting their thinking and acting patterns based on their physical capabilities (Allen et al., 2024). The development of skills through physical activities involves both the incorporation of embodied knowledge or kinesthetic thinking and the ability to imagine oneself in someone's position or empathy (Henriksen, 2018). Hands-on physical interactions play a crucial role in this process, allowing for exploratory learning and constructing understanding through touch and manipulation, such as makerspaces (Adler-Beléndez et al., 2021). In Chapter 4, we proposed the *Crafting Knowledge Framework* illustrating the dynamic interplay between integrating cognitive reasoning and hands-on engagement, based on data from participants in our makerspace activities. In the WEF Future of Jobs Report 2025, Physical Abilities such as sensory-processing abilities and manual dexterity, endurance and precision are part of the global skills taxonomy (WEF, 2025). Thus, embodied cognition or hands-on learning and thinking skills, are implicated in mathematics and language learning (Macedonia, 2019), and might be another component lacking in the current twenty-first century skill frameworks.

5.5 Addressing the Critiques

The critiques of "twenty-first century skills" highlight important gaps and challenges in the concept's application, but they also pave the way for thoughtful refinements and adaptations. By understanding these tensions, educators, policymakers, and researchers can ensure that the term remains relevant and effective across diverse contexts. Below are suggested key approaches to reconciling definitions and addressing critiques.

5.5.1 *Balancing Skills and Knowledge with Values, Dispositions, and Agency*

Critics argue that "twenty-first century skills" often overemphasise skills at the expense of foundational knowledge. A more integrated approach, combining deep content mastery with skill development and growth mindsets, can resolve this tension. For instance:

- Emphasising domain-specific critical thinking in subjects like science, history, and literature.
- Encouraging interdisciplinary learning where students apply knowledge from multiple areas to solve real-world problems.

Frameworks such as the OECD Learning Compass (https://www.oecd.org/en/data/tools-compass-2030.html) have begun incorporating this balance, showcasing how skills and knowledge intersect to foster lifelong learning. It is also the reason why we prefer STE*A*⁺M education rather than STEM or STEAM education in this book, and discuss the *Plexus Model of STEA⁺M Education* with *A*⁺ for *A*gency and innovation (Chapter 3), and the disciplinary (Chapter 2) and interdisciplinary (Chapter 4) cognitive, affective, and conative aspects in integrated STEM and STEAM education.

5.5.2 *Promoting Cultural Responsiveness*

To address concerns about Western-centrism, frameworks for twenty-first century skills must integrate culturally responsive and locally relevant perspectives. This can include:

- Developing flexible definitions that allow regions to tailor skills to their unique cultural and socio-economic contexts.
- Encouraging collaboration with educators, policymakers, and community leaders from diverse backgrounds to co-create frameworks.

For example, UNESCO's Global Citizenship Education (GCED) framework (https://www.unesco.org/en/global-citizenship-peace-education/need-know) strongly emphasises cultural diversity and global–local interconnectedness. In Chapter 9, we also highlighted our Planning and Design Learning (PaDL) model, which we developed with Southeast Asian educators.

5.5.3 Innovating Assessment Practices

The challenges of assessing complex skills like creativity and ethical reasoning call for innovative, formative assessment methods (Henriksen, 2018). Promising approaches include:

Project-Based Assessments: Students demonstrate skills through real-world projects, such as designing a sustainable solution for a community issue.

Digital Portfolios: Students curate a collection of their work over time, showcasing their development in areas like collaboration and problem-solving.

Simulation-Based Assessments: Using virtual environments to evaluate how students apply skills in dynamic, scenario-based contexts.

Emerging tools and frameworks, such as those developed by the Assessment and Teaching of 21st Century Skills (ATC21S) initiative, and interdisciplinary STEM and STEAM makerspaces (Gesthuizen et al., 2025), provide practical models for these methods.

5.5.4 Addressing the Commodification of Education

To counter the view that twenty-first century skills reduce education to workforce preparation, there is a growing movement to emphasise humanistic and civic dimensions. This includes:

- Framing education as a means of fostering engaged, ethical citizens capable of contributing to society beyond the marketplace.
- Highlighting values such as empathy, sustainability, and global awareness alongside technical and cognitive competencies.

Efforts like the United Nations' Sustainable Development Goal 4 (SDG 4) and UNESCO's Sahle-Works report emphasise the broader societal role of education in promoting equity, inclusion, and sustainability. Our Plexus Model of STE*A*⁺M Education (Chapter 3) positions the learner as the agent of change within a trialectic framework featuring interconnecting elements: (1) using STE*A*⁺M as a problem-solving framework to engage with societal issues; (2) *A*gency as the foundational core of student-driven learning; and (3) Design Thinking as the process of inquiry and innovation. In Chapter 6, we propose a model for neo-twenty-first century skills in STE*A*⁺M education that extends existing competencies into thinking systems.

5.5.5 *Redefining the Term for Lifelong Learning*

Finally, the concept of "twenty-first century skills" may benefit from a shift toward lifelong learning and adaptability. As the pace of change continues to accelerate, the focus should move beyond preparing students for a single career trajectory toward equipping them to continuously learn, unlearn, and adapt. This broader vision aligns with frameworks like the World Economic Forum's Future of Jobs Report (WEF, 2025), highlighting the importance of adaptability, resilience, and emotional intelligence in the future workforce. This calls for a transformative education that integrates various disciplines and promotes learner *A*gency in learning and doing (see Chapter 3).

As the discussion on twenty-first century skills has illustrated, these competencies are indispensable for navigating the complexities of a rapidly changing global landscape. Building on this foundation, the following section examines the relationship between twenty-first century skills and STEM education. While the previous section explored the conceptual evolution and critiques of twenty-first century skills, here we examine how these skills are both developed within and transformed by STEM learning environments. This exploration highlights three key streams of thought in the literature: the alignment of twenty-first century skills with workforce demands in STEM fields, the use of STEM learning as a pedagogical approach for fostering these skills, and the emergence of STEM literacy as a holistic framework. Together, these perspectives demonstrate how STEM education equips individuals with essential skills and addresses broader societal challenges, bridging the theoretical and practical dimensions of twenty-first century education.

5.6 Twenty-First Century Skills and STEM Education

Reviewing the literature on twenty-first century skills and science, technology, engineering, and mathematics (STEM) education, we found that the papers broadly fall into three categories. There is a body of work that considers what twenty-first century skills relate to the STEM careers we want to develop in students and incorporate into STEM education (e.g., Jang, 2016). This perspective of ***twenty-first century skills needed in STEM fields*** may be considered the neoliberal or economic perspective, with its starting point being the needs of the STEM workforce and employers (e.g., WEF, 2025). The narrative, with this skills perspective, is that twenty-first century skills are essential for navigating complex global challenges and driving innovation in science, technology, engineering, and mathematics fields. STEM education, traditionally focused on discipline-specific knowledge and technical expertise, is now evolving to incorporate interdisciplinary and holistic approaches. The inclusion of the arts—transforming STEM into STEAM—has further expanded the scope of these educational frameworks, with the narrative of fostering creativity and innovation alongside technical proficiency. However, despite the growing recognition of the importance of twenty-first century skills, there remains a need to systematically explore how these skills are being taught, assessed, and integrated into STEM and STEAM curricula. This body of literature involves the identification, teaching, learning, and assessing of twenty-first century skills valued in STEM education. The literature surrounding twenty-first century skills and STEAM education is emerging and, to a certain extent, follows the STEM literature. Overall, the skills and competencies emphasised are critical thinking, creativity, collaboration, and digital literacy. The main question asked in this body of work relates to:

- What twenty-first century skills are needed in STEM fields, from the views of various stakeholders such as employers, STEM workers, and STEM graduates?

Another set of literature considers STEM activities (in particular, integrated STEM learning) to be a highly appropriate and recommended pedagogical approach for fostering students' twenty-first century skills (e.g., Fajrina et al., 2020). These researchers' starting point is from the pedagogical perspective of twenty-first century skills and how STEM learning can strengthen or enhance the development of some of these skills—***STEM as a learning approach for developing twenty-first century skills***. Unlike traditional, siloed subject instruction, integrated STEM activities emphasise interdisciplinary problem-solving, hands-on learning, and real-world applications (see Chapters 3 and 4). These approaches naturally align with the development of skills such as creativity, collaboration, and critical thinking, as they require students to engage in complex, open-ended tasks that mirror the challenges of the modern workforce. For example, makerspace learning in STEM encourages students to work in teams, design innovative solutions, and communicate their ideas effectively—all while applying knowledge from multiple disciplines (Gesthuizen et al., 2025). The main question asked in this body of work related to:

- What are the defining characteristics of integrated STEM learning, and how do they align with the development of twenty-first century skills?

A recent turn is the idea of ***STEM as a twenty-first century skill*—STEM literacy**. This literacy perspective views STEM as more than just a collection of disciplines or a teaching method. It frames the set of STEM as a form of literacy—a necessary competency that people today (and in the future) need to succeed in navigating and addressing the complex, interconnected issues in the twenty-first century. STEM (or STEA^{+}M) literacy encompasses the integrative knowledge, skills, and dispositions needed to understand and solve societal problems, such as climate change, public health crises, and technological disruptions. It emphasises the ability to think critically across disciplines, apply systems thinking, and engage in ethical decision-making, positioning STEM as a foundational literacy for lifelong learning, civic engagement, and global citizenry. The ***integrative STEA⁺M thinking systems***, such as the socio-entrepreneurial mindset and socio-environmental sustainability agency (which we will discuss in the next chapter), are illustrations of STEM literacy. The body of work in this literacy perspective seeks to conceptualise STEM as a form of twenty-first century literacy, ascertaining individuals' STEM literacy, and understanding the pedagogical knowledge on equipping individuals with the integrative knowledge, skills, and dispositions needed to navigate and solve societal problems. The main question asked in this body of papers concerns:

- How is STEM literacy defined, and what are its core components (e.g., knowledge, skills, dispositions, interdisciplinary thinking)?

Next, we discuss each of the three bodies of literature, starting from the skills perspective of specific twenty-first century skills within STEM, then moving to the pedagogical perspective of developing twenty-first century skills using STEM as a learning approach, and finally to the literacy perspective of STEM/STE*A*⁺M literacy as an amalgamation of integrative twenty-first century meta-skills.

5.6.1 *Twenty-First Century Skills Needed in STEM fields: STEM Employability Skills*

This skills perspective relating to twenty-first Century Skills and STEM education is concerned with the twenty-first century skills for STEM graduates (the term "STE*A*⁺M graduates" is yet to be popularised, as with "STE*A*⁺M fields", https://en.wikipedia.org/wiki/STEAM_fields), also known as twenty-first century STEM skills or competencies. In their report on *Challenges in STEM Learning in Australian Schools: Literature and Policy Review*, Timms et al. (2018, p. 2) stated that.

> There is an understanding that STEM education should develop a set of personal attributes that are variously known in the education sector as twenty-first century skills, soft skills or general capabilities. To employers they are also known as employability skills, and include competencies such as problem solving, collaboration, creativity and innovation. This is often associated with an inquiry-based pedagogy.

The consensus is that everyone needs foundational literacies and skills, as highlighted above. However, the critique is on the philosophy of the education of these skills. It has been problematised that the development of twenty-first century skills has, over time, taken a neoliberal turn, from the perspective that we are developing the child holistically for their self-actualisation (a humanistic agenda), to that of preparing the child "to keep themselves employable in fast-changing labour markets" (Biesta, 2021, p. 656), becoming their duty to learn for an economic agenda. Countries promote STEM education in order to (re)gain competitiveness in global socioeconomic status (Freeman et al., 2014). We want to advance research and development in science and technologies, as well as high-tech industries, and create a powerhouse for the STEM workforce. This has also led the industry to have a say and impose their voice in terms of employability skills and STEM skills into the educational discourse and policies as twenty-first century skills such as the 4Cs (critical thinking, creativity, collaboration, communication) (Fajrina et al., 2020; Mehta et al., 2020). So, while the philosophy of STEM education is not the focus of our chapter, we caution against the over-emphasis on grounding twenty-first century skills in STEM as employability skills, which is indicative of "a view of education which is deterministic about the future and the supposed skills needed to deal with changes spurred by global digitization, the transforming economy and the shifting landscape of education" (Mehta et al., 2020, p. 362). We should be critical about how we define the twenty-first century skills for STEM education, as not just valuing the voice of the private sector but considering the holistic development of the whole person to be stewards of the community and the world. In the next chapter, we propose meta-skills, which are the syntheses of foundational skills, disciplinary and interdisciplinary knowledge, and dispositions (see Chapter 6). Those skills are informed by the general literature on twenty-first century skills and specific literature relating to STEM skills desired for the twenty-first century workplace, presented in this section.

5.6.2 *Types of Twenty-First Century STEM Competencies and Skills*

UNESCO published a document on *Exploring STEM Competences for the 21st Century* (Ng, 2019), which defines STEM competence as "an individual's ability to apply STEM knowledge, skills and attitude appropriately in his or her everyday life, workplace or educational context. It should not be confined and developed within the traditional boundaries of discrete bodies of knowledge (e.g., physics competence or computing competence)" (p. 11). It comprises the "know-what": knowledge, attitudes, and values associated with disciplines, and the "know-how": skills to apply the knowledge, considering ethical values and attitudes to act appropriately and effectively in a given context. These components are interconnected, contextualised, and holistic in nature (Ng, 2019). Exploring educational programs, curriculum and policy, a list of STEM knowledge, skills, practices, values, and ethics are summarised in Table 5.5.

Jang (2016) analysed responses to three different surveys about the knowledge, skills, and work activities of 50,527 STEM and non-STEM employees through the O*NET, the US National Occupational Information database. She compared the responses of the two groups of individuals and categorised important STEM skills, knowledge, and work activities into five categories of STEM competency. There were 18 STEM skills, seven types of knowledges, and 27 work activities, of which the top three (if more than three) are found in Table 5.6. An interesting gap in the results highlighted was that STEM "competencies relevant to working with an organization or a system [system skills] were not reported as important in frameworks of 21st century skills" (Jang, 2016, p. 290). System skills are not found in education-based twenty-first century frameworks such as Ng (2019) and P21. We suggest that this competency might be seen as more related to humanities discipline (e.g., business management, leadership) than STEM. Hence, this may justify the need for using the phrase "21st century STEAM competencies or skills" rather than just "STEM skills".

Besides Jang's framework of five categories of STEM competencies, there are similar frameworks, from the perspectives of STEM students (e.g., Marbach-Ad et al., 2019; Mark et al., 2018; Sarkar et al., 2016), employers (e.g., McGunagle & Zizka, 2020), and universities (e.g.,

Table 5.5 List of STEM skills, practices, values and ethics outlined in Ng (2019)

STEM Knowledge	*STEM Skills*	*STEM Practices*	*STEM Values and Ethics*
Big ideas within disciplines	Cognitive skills	Asking questions and defining problems	Curiosity
Big ideas across STEM subjects	Information processing—data interpretation and data analysis	Developing and using models	Integrity
Technical knowledge associated with STEM careers	Problem-solving and engineering thinking	Planning and carrying out investigations	Objectivity
	Scientific investigation	Analysing and interpreting data	Open-minded
	Computational thinking and ICT	Using mathematics and computational thinking	Diligent and perseverance
	Design thinking, creativity and innovation	Constructing explanations and designing solutions	Systematic
	Manipulative and technological skills	Engaging in argument from evidence	Cooperative
	Collaboration and communication skills	Obtaining, evaluating, and communicating information	Responsible
			Precision
			Appropriate risk-taking
			Ethical decision-making
			Appreciate the contribution of STEM to our everyday living

Table 5.6. Twenty-First Century STEM competencies proposed by Jang (2016), with examples of skills, knowledge, and work activities

21st century STEM competencies	*Skills*	*Knowledge*	*Work activities (Skills in practice)*
(Ill-defined) Problem-solving skills	Critical thinking Complex problem-solving Mathematics Science	Mathematics English language	Getting information Making decisions and solving problems Analysing data or information
Social communication skills	Active listening Speaking Writing Social perceptiveness	Customer and personal service Education and training	Communicating with supervisors, peers, or subordinates Establishing and maintaining interpersonal relationships Interpreting the meaning of information for others
Technology and engineering skills	BLANK*	Computers and electronics Engineering and technology	Interacting with computers Processing information Inspecting equipment, structures, or material

(continued)

Table 5.6. (continued)

21st century STEM competencies	*Skills*	*Knowledge*	*Work activities (Skills in practice)*
System skills	Monitoring Systems analysis Systems evaluation	Administration and management	Monitor processes, materials, or surroundings Judging the qualities of things, services, or people Developing objectives and strategies
Time, resource, and knowledge management skills	Time management Learning strategies Active learning	BLANK*	Organising, planning, and prioritising work Scheduling work and activities Updating and using relevant knowledge

*There were gaps found, which Jang (2016) suggests are deficiencies in the twenty-first century frameworks and instruments used

Rayner & Papakonstantinou, 2015). McGunagle and Zizka (2020) examined twenty-first century STEM employability skills from a survey of 250 US human resource managers in five manufacturing industries (Aerospace and defence, Automotive, Consumer products, Electronics, and Industrial manufacturing). They synthesised recent literature on STEM skills/competencies/attributes for graduates, to a list of 16 skills (ranked in order of importance by participants):

1. Team player
2. Self-motivation
3. Verbal communication
4. Problem-solving
5. Proactive
6. Decision making
7. Adaptability
8. Leadership
9. Written communication
10. Ability to gather data
11. High self-confidence
12. Customer focus
13. Assertiveness
14. Oral presentations
15. Ability to synthesise
16. Negotiation skills

Abina et al. (2024) surveyed companies in Slovenia and Norway that employed ecotechnologies and civil and environmental engineering program graduates, and found that there are significant differences in what different countries' organisations value as important STEM skills. They proposed a framework of ten employability skills for STEM students, with the first four skills being identified as important in both countries:

1. Managing and designing processes, technologies, and people's work
2. Adapting to changes
3. Ability to manage uncertainty
4. Innovation
5. Systemic, analytical, and critical thinking
6. Creativity
7. Self-initiative and self-management
8. Business and entrepreneurial skills
9. Advanced computer/IT skills (e.g., artificial intelligence)
10. Ensuring safety (data, privacy, human health, environmental protection, working safely with devices)

It is interesting to note that all STEM skills and competencies found in Jang (2016), Ng (2019), McGunagle and Zizka (2020), and Abina et al. (2024) are quite different in terms of structure and organisation (detailed frameworks with knowledge, skills, values, and attitudes, versus lists of top skills) even though they share some commonalities. Some of the skills do not seem to overlap with the conventional twenty-first century skills frameworks (e.g., systems skills, customer focus, assertiveness, ability to synthesise, business and entrepreneurial skills). At the same time, they are clearly grounded in neoliberal ideologies—some frameworks are lacking in the humanistic elements of ethics, creativity, inquiry, and engagement with the ways in which STEM disciplines shape our perspectives of the material, intellectual, social, and cultural world (Akerson & Buck, 2020; Mehta et al., 2020). Nonetheless, even from an economic perspective, there were gaps raised in the literature relating to equipping future STEM graduates with twenty-first century skills that have implications for STEM education. They are briefly described in the following.

Mismatch between STEM/STEAM education and workplace needs.

Employers often note that while STEM/STEAM graduates possess strong technical knowledge, they often lack essential soft skills such as communication, collaboration, adaptability, and problem-solving (McGunagle & Zizka, 2020; Teshome & Oumer, 2024). These skills are often underemphasised in STEM curricula, which tend to focus more on technical knowledge (Börner et al., 2018; Rayner & Papakonstantinou, 2015). The gap between academic training and industry expectations is evident across various countries and STEM fields (Abina et al., 2024). In a world where interdisciplinary teamwork is the norm, the ability to articulate complex ideas, work effectively in diverse teams, and adapt to new challenges is critical. For example, an engineer may excel in designing a product but struggle to explain its benefits to non-technical stakeholders or collaborate with designers to enhance its aesthetic appeal. As Börner et al. (2018) found in their analysis of a large volume of research publications, course syllabi and job advertisements published in 2010-2016, "even with the rise of data analytic, machine learning, and AI-related skills, people will continue to be needed to communicate complex ideas, negotiate, and lead" (p. 12637). This disconnect underscores the importance of integrating arts and humanities into STEM education, fostering well-rounded professionals who can bridge the gap between technical expertise and real-world application.

Growing demand for transferable skills in STEM.

In an era defined by rapid technological change and global interconnectedness, transferable skills are more important than ever. Skills such as critical thinking, adaptability, independent learning, taking initiative, and emotional intelligence enable STEM/STEAM professionals to navigate complex challenges and transition seamlessly between roles and industries (McGunagle & Zizka, 2020; Sarkar et al., 2016). For instance, a data scientist with strong analytical skills can apply their expertise to fields as diverse as healthcare, finance, or environmental science. By emphasising these competencies and shifting from "knowing" to "doing" with experiential learning, educational institutions can prepare graduates to excel in a dynamic job market where versatility is key (Abina et al., 2024).

Need for business and entrepreneurial skills in STEM careers.

Many STEM graduates enter the workforce with a limited understanding of business fundamentals, such as project management, financial literacy, and organisational dynamics (Abina et al., 2024; Jang, 2016). This gap can hinder their ability to contribute strategically or understand the broader context of their work. For example, a software developer who lacks knowledge of budgeting or client management may struggle to align their work with organisational goals. In fact, Sarkar et al. (2016) found that *commercial awareness* is the top skill that both STEM industry employers were dissatisfied with, and STEM graduates would have liked more opportunities to develop within their degree. Mark et al. (2018) also found that both STEM and non-STEM graduates perceive that entrepreneurship is needed to succeed in the gig economy and is a gap in their tertiary education. To address this, STEM programs should incorporate business and entrepreneurship education, equipping graduates with the skills needed to drive innovation and lead projects effectively.

To conclude, the exploration of twenty-first century STEM competencies reveals a critical need for a more holistic and integrated approach to education. While the various frameworks highlight the importance of skills such as critical thinking, creativity, collaboration, and adaptability, the literature lacks a consistent or overarching model that fully addresses the evolving demands of STEM industries. It is a moving goalpost and will further evolve as the world becomes more complex. A recurring theme is the growing emphasis on business and entrepreneurial skills and the balance of hard and soft skills, underscoring the need for STEM graduates who are not only technically proficient but also capable of navigating complex organisational and societal challenges. To prepare individuals for the STEM fields of the future, we must move beyond traditional silos and embrace interdisciplinary STEAM learning. Next, we review the literature on the pedagogical perspective of STEM education as a learning approach to developing twenty-first century skills.

5.7 STEM Education for Developing Twenty-First Century Skills: STEM as a Learning Approach

In the body of literature with the pedagogical perspective of STEM as a learning approach, the common rhetoric is that we need to develop and nurture twenty-first century skills in the new generation to meet the demands of the twenty-first century and the Fourth Industry Revolution. STEM (and STEAM) education—particularly interdisciplinary and transdisciplinary learning approaches—is widely regarded as a key strategy to equip students with these essential skills (Fajrina et al., 2020). However, despite the common rhetoric emphasising the need for twenty-first century skills, there is a lack of consensus among stakeholders—ranging from educators to industry leaders—on precisely what these skills entail (Akerson & Buck, 2020; Mehta et al., 2020). This ambiguity has not diminished the enthusiasm for STEM education as a pathway to fostering skills such as critical thinking, creativity, communication, and collaboration—often referred to as the "Four Cs" (Fajrina et al., 2020).

5.7.1 The Role of STEM in Developing Twenty-First Century Skills

STEM education has emerged as a powerful approach for cultivating twenty-first century skills, particularly through its emphasis on inquiry-based and problem-solving methodologies. Research indicates that STEM-based learning enhances not only the "Four Cs" but also STEM literacies, problem-solving, and innovation (Ärlebäck & Albarracín, 2019; Sen et al., 2018). For instance, integrating STEM with project-based and problem-based learning models has been shown to improve student outcomes by fostering innovation, problem-solving, and digital literacy (Leavy et al., 2023). These skills are critical for preparing students to navigate the challenges of the twenty-first century workplace, where adaptability and creativity are paramount (Kennedy & Sundberg, 2020). Inquiry-based, real-world, problem-based learning approaches in STEM classrooms are known to be effective in promoting social learning (teamwork, collaboration, communication), creativity, critical thinking, and complex problem-solving (Tan & Kidman, 2021).

The historical context of STEM education further underscores its relevance for the twenty-first century. The space race of the mid-twentieth century prioritised science, mathematics, and technology, laying the groundwork for today's emphasis on STEM. In the 2000s, this focus expanded to include twenty-first century skills, driven by the demands of a knowledge-based economy and the rise of AI and big data (Freeman et al., 2014). This evolution has led to a conflation of STEM education and twenty-first century skills, with schools, researchers and communities increasingly viewing STEM as a forward-looking and attractive educational approach (Newman et al., 2020). For example, Bybee (2010) highlighted that "in STEM programs, student investigations and projects present the time and opportunity for teachers to help students develop twenty-first century skills" (p. 31). At the ground level, there is also a need for schools to 'work smart' to incorporate these educational policies filtering down from the top. In analysing how STEM schools portrayed themselves to the public via their school websites and social media, Newman et al. (2020) found that a common theme is the portrayal of STEM as developing twenty-first century skills such as critical thinking, collaboration, innovation, and communication. It would seem that these twenty-first century skills are part of the STEM narrative, and others such as "ethical decision making" or "appreciate the contribution of STEM to our everyday living" (Ng, 2019) or "metacognition" or "living in the world" (see Table 5.4), are not as popular.

5.7.2 *Challenges and Nuances in Implementation*

Despite its potential, STEM education faces significant challenges in implementation. One key issue is the difficulty of developing "soft" skills, such as creativity and collaboration, within traditional classroom settings (Ng, 2019). Science and mathematics, in particular, are often taught in ways that prioritise content mastery over creative thinking, leading to a perceived lack of creativity among students (Mehta et al., 2020). This has prompted the inclusion of engineering and the arts in STEM, transforming it into STEAM, to better address affective and creative dimensions of learning (Leavy et al., 2023).

Another challenge lies in the assessment of twenty-first century skills. Despite the rhetoric, there are very few studies that investigate an entire suite of twenty-first century skills in students. There are also very few lessons that explicitly address and assess the suite of skills. Stehle and Peters-Burton (2019) used a set of learning design rubrics developed by Microsoft and SRI Education, that measure the extent to which twenty-first century skills are developed in teachers' lesson plans and students' work. Their rubrics had six categories: (a) knowledge construction, (b) real-world problem-solving, (c) skilled communication, (d) collaboration, (e) use of ICT for learning, and (f) self-regulation. The researchers acknowledged that skills are not developed in isolation and reinforce one other when used in context. For example, students develop communication skills as they learn to collaborate, and their self-regulation improves their knowledge construction and real-world problem-solving skills. Analysing 67 teacher lesson plans and 29 student work samples in seven STEM high schools, Stehle and Peters-Burton (2019) found that nearly 75% of the lessons addressed at least one twenty-first century skill, with a majority (67%) involving two or more skills. However, these skills were addressed at a low level, such as giving instructions to form a group (for collaboration), without evidence of defining group roles, ways to share responsibility, and the interdependence of group members.

5.7.3 *Moving Forward: STEA⁺M as a Catalyst for Reform*

The integration of STE*A*⁺M education offers a promising way forward. By incorporating the *A*gency into STEM for *A*ll (see Chapter 3), educators can create more holistic learning experiences that foster creative, meta-cognitive, humanistic, and emotional engagement alongside technical skills. Inquiry-based learning and authentic assessment, which emphasise real-world problem-solving and collaborative projects, provide a framework for developing twenty-first century skills in meaningful ways (Kidman et al., 2024; Tan & Kidman, 2021). These approaches not only align with the goals of STEM and STEAM education but also address the paradox of preparing students for the future while meeting the demands of accountability and assessment.

However, it is important to recognise that not all twenty-first century skills can be developed uniformly across all contexts or students. Individual differences in learning needs, interests, and prior experiences mean that the same activity may yield varying outcomes for different learners (Gesthuizen et al., 2025). By providing diverse and challenging inquiry-based learning experiences—ranging from group projects to individual tasks—educators can create opportunities for students to develop their unique strengths and skills. For example, makerspace activities, which emphasise hands-on, collaborative learning, have shown promise in inspiring both teachers and students by bridging the gap between theory and practice (Nugraha et al., 2024). In our STE*A*⁺M projects, we can see the difference in the teachers and students who were inspired by our makerspace activities (see Part III Chapters 8 and 9; Mangao et al., 2022).

To conclude, the plexus model of STE*A*+M education holds significant potential for developing twenty-first century skills, but its success depends on the awareness and prioritising of which twenty-first century skills to emphasise, thoughtful planning and implementation of learning experiences, and a willingness to address its inherent challenges such as assessment, teacher pedagogical content knowledge and efficacy, and school organisational constraints.

5.8 STEM as Twenty-First Century Skills: STEM Literacy

In this body of literature with a literacy perspective, STEM competencies are viewed as foundational skills and, therefore, a type of literacy that is required by every individual to function in society. Especially seen in the pandemic, the need for critical literacy to understand the science, technology, and mathematics relating to the coronavirus and its different strands, how it is spread, how fast it is spreading, social distancing needs, and evaluating sources of claims and their validity (Braun, 2021).

STEM literacy is a multifaceted concept encompassing the ability to apply knowledge and skills from science, technology, engineering, and mathematics to solve complex problems (Tang & Williams, 2019; Zollman, 2012). It involves cognitive, affective, and psychomotor domains (Zollman, 2012) and includes identifying, organising, and integrating information; creating representations; communicating for help; documenting processes; and sharing finished products (Tucker-Raymond & Gravel, 2019). While some argue for a unitary STEM literacy, others emphasise distinct science, technology, engineering, and mathematics literacies to reflect discipline-specific requirements (Tang & Williams, 2019). STEM literacy development can occur in formal and informal learning environments and is crucial for addressing societal, economic, and personal needs (Zollman, 2012). Recent research highlights the importance of critical STEM literacy in understanding complex issues like the COVID-19 pandemic (Braun, 2021) and the need to consider various learning theories in conceptualising STEM literacy (Tang & Williams, 2019).

The distinction between STEM and STEAM literacy remains unclear, with arguments for both unitary and separate literacy constructs, as mentioned earlier (Tang & Williams, 2019). STEAM education, a popular extension of STEM that includes the Arts, has gained attention for its potential to enhance creativity and critical thinking skills. However, recent research shows mixed results regarding its effectiveness in enhancing creativity compared to STEM alone (Aguilera & Ortiz-Revilla, 2021). While STEAM may offer advantages in developing what STEM education can offer, such as computational thinking skills (Syafe'i et al., 2023), its roll-out needs deliberate planning and implementation, including effective integration of arts and teacher preparedness (Kang, 2019).

In terms of the separate disciplinary literacies, digital literacy or information, media, and technology skills are a common theme in the various frameworks for twenty-first century skills (see Table 5.4). While less prominent, science literacy and mathematics literacy/numeracy are also seen as important competencies for the twenty-first century, as can be seen in their role in PISA assessment. Engineering is usually more conflated with technology than with science and mathematics. Technological and engineering literacy can include engineering design and systems thinking, and incorporate other disciplinary (scientific, mathematical, technological) knowledge and skills, and engineering habits of mind (Hoepfl, 2020).

Regarding the unitary construct of *STEM literacy*, the concepts were adapted and expanded from scientific literacy (Braun, 2021). Bybee (2010, p. 31) proposed that STEM literacy refers to the following set of competencies:

- Acquiring scientific, technological, engineering, and mathematical knowledge and using that knowledge to identify issues, acquire new knowledge, and apply the knowledge to STEM-related issues.
- Understanding the characteristic features of STEM disciplines as forms of human endeavours that include the processes of inquiry, design, and analysis.
- Recognising how STEM disciplines shape our material, intellectual, and cultural world.
- Engaging in STEM-related issues and with the ideas of science, technology, engineering, and mathematics as concerned, affective, and constructive citizens.

The concept of STEM literacy is still evolving, and needs more clarity in how it is used for policy, research, curriculum, and instruction (Hoepfl, 2020; Tang & Williams, 2019). Zollman (2012) stated that there is a general consensus that STEM literacy is for everyone, but distinguished between "STEM literacy" and "STEM literate" in that STEM literacy is not a content area but the synergising of skills, abilities, content and procedural knowledge, and metacognition, to gain further learning for personal, societal and economic needs. We agree with Zollman's concept of STEM literacy and take it further to propose a set of integrative STE*A*$^{+}$M thinking skills and meta-skills that are needed to educate our new generations in the next quarter of the twenty-first century (see next Chapter).

5.9 Conclusion

In this chapter, we have examined the prevailing twenty-first century skills frameworks and their connection to STEM education. As we move forward from Industry 4.0 to Industry 5.0 (Ziatdinov et al., 2024) and beyond (Chourasia et al., 2022), disciplinary expertise alone is no longer sufficient. Instead, interdisciplinary competencies—spanning cognitive (e.g., critical and creative thinking), affective (e.g., social-emotional intelligence, growth mindset, lifelong learning), and conative (e.g., values, ethics, agency) dimensions—are essential. More importantly, fostering metacognition about how these dimensions interact is crucial. Developing self-awareness and the ability to manage these competencies in both educators and students will be key to preparing for the complex challenges of the future.

References

Abina, A., Salaj, A. T., Cestnik, B., Karalič, A., Ogrinc, M., Lukman, R. K., & Zidanšek, A. (2024). Challenging 21st-century competencies for STEM students: Companies' vision in Slovenia and Norway in the light of global initiatives for competencies development. *Sustainability, 16*, Article 1295. https://doi.org/10.3390/su16031295

Adams, R., & Wu, M. (Eds.). (2003). *Programme for International Student Assessment (PISA): PISA 2000 technical report*, PISA, OECD Publishing. https://doi.org/10.1787/9789264199521-en

Adler-Beléndez, D., Hoppenstedt, E., Husain, M., Chng, E., & Schneider, B. (2021). How are 21st century skills captured in makerspaces? A review of the literature. In *Proceedings of the FabLearn 2020—9th Annual Conference on Maker Education, April 4–5, 2020, NY*, (pp. 40–45). Association for Computing Machinery. https://doi.org/10.1145/3386201.3386214

Aguilera, D., & Ortiz-Revilla, J. (2021). STEM vs. STE*A*⁺M education and student creativity: A systematic literature review. *Education Sciences, 11*, Article 331. https://doi.org/10.3390/educsci11070331

Akerson, V. L., & Buck, G. A. (Eds.). (2020). *Critical questions in STEM education*. Springer. https://doi.org/10.1007/978-3-030-57646-2Allen, K. R., Smith, K. A., Bird, L-A., Tenenbaum, J. B., Makin, T. R. & Cowie, D. (2024). Lifelong learning of cognitive styles for physical problem-solving: The effect of embodied experience. *Psychonomic Bulletin & Review, 31*, 1364–1375. https://doi.org/10.3758/s13423-023-02400-4

Allen, K. R., Smith, K. A., Bird, L-A., Tenenbaum, J. B., Makin, T. R. & Cowie, D. (2024). Lifelong learning of cognitive styles for physical problem-solving: The effect of embodied experience. *Psychonomic Bulletin & Review, 31*, 1364–1375. https://doi.org/10.3758/s13423-023-02400-4

Allen, J. P., & van der Velden, R. K. W. (2012). Skills for the 21st century: Implications for education. *Researchcentrum voor Onderwijs en Arbeidsmarkt, Faculteit der Economische Wetenschappen*. ROA Research Memoranda No. 11 https://doi.org/10.26481/umaror.2012011

Ananiadou, K., & Claro, M. (2009). *21st century skills and competences for new millenium learners in OECD countries*. (OECD Education Working Papers No. 41). Organisation for Economic Cooperation and Development Publishing. https://doi.org/10.1787/218525261154

Ärlebäck, J. B., & Albarracín, L. (2019). The use and potential of Fermi problems in the STEM disciplines to support the development of twenty-first century competencies. *ZDM—Mathematics Education, 51*, 979–990. https://doi.org/10.1007/s11858-019-01075-3

Biesta, G. (2021). Reclaiming a future that has not yet been: The Faure report, UNESCO's humanism and the need for the emancipation of education. *International Review of Education, 68*, 655–672. https://doi.org/10.1007/s11159-021-09921-x

Blinkey, M, Erstad, O., Herman, J., Raizen, S., Ripley, M., Miller-Ricci, M., & Rumble, M. (2012). Defining twenty-first century skills. In P. Griffin et al. (Eds.), *Assessment and teaching of 21st century skills* (pp. 17–66). Springer. https://doi.org/10.1007/978-94-007-2324-5_2

Börner, K., Scrivner, O., Gallant, M., Ma, S., Liu, X., Chewning, K., Wu, L., & Evans, J. A. (2018). Skill discrepancies between research, education, and jobs reveal the critical need to supply soft skills for the data economy. *Proceedings of the National Academy of Sciences (PNAS), 115*(50), 12630–12637. https://doi.org/10.1073/pnas.1804247115

Braun, M. (2021). Critical STEM literacy and the COVID-19 pandemic. *Canadian Journal of Science, Mathematics and Technology Education, 21*, 339–356. https://doi.org/10.1007/s42330-021-00150-w

Bridges, D. (1993). Transferable skills: A philosophical perspective. *Studies in Higher Education, 18*(1), 43–51. https://doi.org/10.1080/03075079312331382448

Bybee, R. W. (2010). Advancing STEM education: A 2020 vision. *Technology and Engineering Teacher, 70*, 30–35.

Child, S., & Shaw, S. (2016). Collaboration in the 21st century: Implications for assessment. *Research Matters, 22,* 17–22. https://doi.org/10.17863/CAM.100344

Chourasia, S., Tyagi, A., Pandey, S.M., Walia, R. S., & Qasim, M. (2022). Sustainability of Industry 6.0 in global perspective: Benefits and challenges. *MAPAN Journal of Metrology Society of India, 37*, 443–452. Springer. https://doi.org/10.1007/s12647-022-00541-w

Delors, J., Al Mufti, I., Amaği, I, Carneiro, R., Chung, F., Geremek, B., Gorham, W., Kornhauser, A, Manley, M., Padró Quero, M., Singh, K., Stavenhağen, R., Myong, W. S., & Zhou, N. (1996). *Learning: The treasure within. Report to UNESCO of the International Commission on Education for the Twenty-first Century.* United Nations Educational, Scientific and Cultural Organization (UNESCO). https://unesdoc.unesco.org/ark:/48223/pf0000109590

Dishon, G., & Gilead, T. (2021). Adaptability and its discontents: 21st-century skills and the preparation for an unpredictable future. *British Journal of Educational Studies, 69*(4), 393–413. https://doi.org/10.1080/00071005.2020.1829545

Elfert, M. (2015). UNESCO, the Faure report, the Delors report, and the political utopia of lifelong learning. *European Journal of Education, 50*(1), 88–100. https://doi.org/10.1111/ejed.12104

European Union. (2006). Recommendation of the European Parliament and the Council of 18 December 2006 on key competences for lifelong learning—A European reference framework. *Official Journal of the European Union*. http://data.europa.eu/eli/reco/2006/962/oj

European Union. (2019). Directorate-General for Education, Youth, Sport and Culture (European Commission). *Key competences for lifelong learning*. https://doi.org/10.2766/569540

Facione, P. A. (1990). *Critical thinking: A statement of expert consensus for purposes of educational assessment and instruction (The Delphi Report)*. California Academic Press.

Fajrina, S., Lufri, L., & Ahda, Y. (2020). Science, Technology, Engineering, and Mathematics (STEM) as a learning approach to improve 21st century skills: A review. *International Journal of Online and Biomedical Engineering (iJOE), 16*(7), 95–104. https://doi.org/10.3991/ijoe.v16i07.14101

Faure, E., Herrera, F., Kaddoura, A. R., Lopes, H., Petrovski, A. V., Rahnema, M., & Ward, F. C. (1972). *Learning to be: The world of education today and tomorrow*. UNESCO/Harrap. https://unesdoc.unesco.org/ark:/48223/pf0000001801

Freeman, B., Marginson, S., & Tytler, R. (Eds.). (2014). *The age of STEM: Educational policy and practice across the world in Science, Technology, Engineering and Mathematics*. Routledge. https://doi.org/10.4324/9781315767512

Gesthuizen, R., Tan, H., & Kidman, G. (2025). Inspired to build understanding: Learning about sustainability and environmental education from within a virtual world. *International Research in Geographical and Environmental Education, 34*(1), 25–42. https://doi.org/10.1080/10382046.2024.2348265

Gordon, J., Halasz, G., Krawczyk, M., Leney, T., Michel, A., Pepper, D., Putkiewicz, E., & Wisniewski, J. (2009). *Key competences in Europe: Opening doors for lifelong learners across the school curriculum and teacher education*. Center for Social and Economic Research.

Greenlaw, J. (2015). Deconstructing the metanarrative of the 21st century skills movement. *Educational Philosophy and Theory, 47*(9), 894–903. https://doi.org/10.1080/00131857.2015.1035156

Griffin, P., McGaw, B., Care, E. (Eds.). (2012). *Assessment and teaching of 21st century skills*. Springer. https://doi.org/10.1007/978-94-007-2324-5

Henriksen, D. (2018). The 7 transdisciplinary cognitive skills for creative education. *SpringerBriefs in Educational Communications and Technology*. https://doi.org/10.1007/978-3-319-59545-0_1

Hoepfl, M. (2020). Defining technological and engineering literacy. *Technology and Engineering Teacher, 80*(3), 1–6. https://par.nsf.gov/biblio/10210625

Jang, H. (2016). Identifying 21st century STEM competencies using workplace data. *Journal of Science Education and Technology, 25*, 284–301. https://doi.org/10.1007/s10956-015-9593-1

Kang, N. -H. (2019). A review of the effect of integrated STEM or STEA^+M (science, technology, engineering, arts, and mathematics) education in South Korea. *Asia-Pacific Science Education, 5*, Article 6. https://doi.org/10.1186/s41029-019-0034-y

Kennedy, T. J., & Sundberg, C. W. (2020). 21st century skills. In B. Akpan, & T. J. Kennedy (Eds.), *Science education in theory and practice* (pp. 479–496). Springer. https://doi.org/10.1007/978-3-030-43620-9_32

Kidman, G., Tan, H., & Macdonald, S. (2024). Reimagining STEM education: Developing agency through STEA^+M and design thinking. In T. Barkatsas, P. McLaughlin, & W. Goff (Eds.), *Reimaging education for the second quarter of the 21st century and beyond* (pp. 1–23). Brill. https://doi.org/10.1163/9789004688490_001

Leavy, A., Dick, L., Meletiou-Mavrotheris, M., Paparistodemou, E., & Stylianou, E. (2023). The prevalence and use of emerging technologies in STEA^+M education: A systematic review of the literature. *Journal of Computer Assisted Learning, 39*(4), 1061–1082. https://doi.org/10.1111/jcal.12806

Lemke, C. (2002). *enGauge 21st century skills: Digital literacies for a digital age*. North Central Regional Educational Laboratory, & the Metiri Group. https://eric.ed.gov/?id=ED463753

Macedonia, M. (2019). Embodied learning: Why at school the mind needs the body. *Frontiers in Psychology, 10*, Article 2098. https://doi.org/10.3389/fpsyg.2019.02098

Mangao, D. D., Kidman, G., Tan, H., Gesthuizen, R., Macdonald, S., & Sinniah, D. N. (2022). *SEAMEO STEM planning and design learning (PaDL) framework: Towards 21st century skills and design thinking*. SEAMEO RECSAM. http://recsam.edu.my/images/everything/pdf/2022_SEAMEO_STEM_PaDL.pdf

Marbach-Ad, G., Hunt, C., & Thompson, K. V. (2019). Exploring the values undergraduate students attribute to cross-disciplinary skills needed for the workplace: An analysis of five STEM disciplines. *Journal of Science Education and Technology, 28*, 452–469. https://doi.org/10.1007/s10956-019-09778-8

Mark, K. P., So, J. C. H., Chan, V. C. W., Luk, G. W. T., & Ho, W. T. (2018). Surviving in the gig economy: Change of STEM students' perceptions on the generic skills for the workplace. In *Proceedings of the 2018 IEEE International Conference on Teaching, Assessment, and Learning for Engineering (TALE)*, IEEE, Wollongong, NSW, Australia, 4–7 December 2018. https://doi.org/10.1109/TALE.2018.8615414

McGunagle, D., & Zizka, L. (2020). Employability skills for 21st-century STEM students: The employers' perspective. *Higher Education, Skills and Work-Based Learning, 10*(3), 591–606. https://doi.org/10.1108/HESWBL-10-2019-0148

Mehta, R., Creely, E., & Henriksen, D. (2020). A profitable education: Countering Neoliberalism in 21st century skills discourses. In J. Keengwe & G. Onchwari (Eds.), *Handbook of research on literacy and digital technology integration in teacher education* (pp. 359–381). IGI Global. https://doi.org/10.4018/978-1-7998-1461-0.ch020

National Education Association (NEA). (2010). *Preparing 21st century students for a global society: An educator's guide to the "Four Cs"*. https://dl.icdst.org/pdfs/files3/0d3e72e9b873e0ef2ed780bf53a347b4.pdf

National Research Council (NRC). (2012). *Education for life and work: Developing transferable knowledge and skills in the 21st century*. The National Academies Press. https://doi.org/10.17226/13398

Newman, S., Kahn, T., Guo, M., Gerber, A., Burgess, A., & Akerson, V. L. (2020). Public portrayals of Indiana STEM certified schools. In V. L. Akerson, & G. A. Buck (Eds.), *Critical questions in STEM education* (pp. 167–184). Springer. https://doi.org/10.1007/978-3-030-57646-2_10

Ng, S. B. (2019). Exploring STEM competences for the 21st century. *In-progress reflection No. 30. Current and Critical Issues in Curriculum, Learning and Assessment*. United Nations Educational, Scientific and Cultural Organization (UNESCO) and International Bureau of Education. https://unesdoc.unesco.org/ark:/48223/pf0000368485.locale=en

Nugraha, M. G., Kidman, G., & Tan, H. (2024). Interdisciplinary STEM education foundational concepts: Implementation for knowledge creation. *EURASIA Journal of Mathematics, Science and Technology Education, 20*(10), em2523. https://doi.org/10.29333/ejmste/15471

Organisation for Economic Cooperation and Development. (OECD). (n.d.). *What is PISA?* https://www.oecd.org/pisa/

Organisation for Economic Cooperation and Development. (OECD). (2005). *Definition and selection of key competencies: Executive summary*. https://www.deseco.ch/bfs/deseco/en/index/02.html

Organisation for Economic Cooperation and Development. (OECD). (2019). *OECD future of education and skills 2030: Conceptual learning framework: Student agency for 2030*. https://www.oecd.org/education/2030-project/teaching-and-learning/learning/student-agency/

Partnership for 21st Century Skills (P21). (2002). *Learning for the 21st century. A report and a mile guide for 21st century skills*. https://eric.ed.gov/?id=ED480035

Power, C. (2015). The power of education: Education for all, development, globalisation and UNESCO. *Springer*. https://doi.org/10.1007/978-981-287-221-0_6

Rayner, G., & Papakonstantinou, T. (2015). Employer perspectives of the current and future value of STEM graduate skills and attributes: An Australian study. *Journal of Teaching and Learning for Graduate Employability, 6*(1), 100–115.

Reimers, F. M., & Chung, C. K. (2016). *Teaching and learning for the twenty-first century: Educational goals, policies, and curricula from six nations*. Harvard Education Press. https://hep.gse.harvard.edu/9781612509228/teaching-and-learning-for-the-twenty-first-century/

Sahle-Work, Z., Nóvoa, A., Aoyagi, M., Appadurai, A., Awuah, P., Ben Hassen, A. B., Buarque, C., Guerra, E., Jafar, B., Kim, D.-Y., Lin, J. Y., Morozov, E., Mundy, K., Reimers, F. M., Rivera Zea, T., Thiam, S. M., Vike-Freiberga, V., & Yahya, M. (2021). *Reimagining our future together: A new social contract for education*. UNESCO. https://unesdoc.unesco.org/ark:/48223/pf0000379707

Sarkar, M., Overton, T., Thompson, C., & Rayner, G. (2016). Graduate employability: Views of recent science graduates and employers. *International Journal of Innovation in Science and Mathematics Education, 24*(3), 31–48.

Sen, C., Ay, Z. S., & Kiray, S. A. (2018). STEM skills in the 21st century education. In M. Shelley & M. Kiray, A. (Eds.), *Research highlights in STEM education* (pp. 81–101). ISRES Publishing. https://www.isres.org/stem-skills-in-the-21st-century-education-103-s.html

Silva, E. (2009). Measuring skills for 21st century learning. *Phi Delta Kappan, 90*(9), 630–634. https://doi.org/10.1177/003172170909000905

Stehle, S. M., & Peters-Burton, E. E. (2019). Developing student 21st century skills in selected exemplary inclusive STEM high schools. *International Journal of STEM Education, 6*. Article 39. https://doi.org/10.1186/s40594-019-0192-1

Syafe'i, S. S., Widarti, H. R., Dasna, W., Habiddin, Parlan, & Wonorahardjo, S. (2023). STEM and STEA^+M affects computational thinking skill: A systematic literature review. *Orbital: The Electronic Journal of Chemistry, 15*(4), 208–216. https://doi.org/10.17807/orbital.v15i4.18323

Tan, H., & Kidman, G. (2021). Authentic assessment in STEM education: An integrative review of research. In T. Barkatsas & T. McLaughlin (Eds.), *Authentic assessment and evaluation approaches and practices in a digital era* (pp. 24–52). Brill. https://doi.org/10.1163/9789004501577_002

Tang, K.-S., & Williams, P. J. (2019). STEM literacy or literacies? Examining the empirical basis of these constructs. *Review of Education, 7*(3), 675–697. https://doi.org/10.1002/rev3.3162

Teshome, A., & Oumer, J. (2024). STEM education and labor needs: Engineering graduates in Ethiopia. *International Journal of Research in STEM Education, 6*(1), 46–58. https://doi.org/10.33830/ijrse.v6i1.1664

Tight, M. (2021). Twenty-first century skills: Meaning, usage and value. *European Journal of Higher Education, 11*(2), 160–174. https://doi.org/10.1080/21568235.2020.1835517

Timms, M. J., Moyle, K., Weldon, P. R., & Mitchell, P. (2018). *Challenges in STEM learning in Australian schools: Literature and policy review.* Australian Council for Educational Research (ACER). https://research.acer.edu.au/policy_analysis_misc/28

Trilling, B. & Fadel, C. (2009). *21st century skills: Learning for life in our times.* Partnership for 21st Century Skills. Jossey-Bass, A Wiley Imprint.

Tucker-Raymond, E., & Gravel, B. E. (2019). STEM literacies in makerspaces: Implications for learning, teaching, and research. *Routledge.* https://doi.org/10.4324/9781351256728

Voogt, J., & Robin, N. P. (2012). A comparative analysis of international frameworks for 21st century competences: Implications for national curriculum policies. *Journal of Curriculum Studies, 44*(3), 299–321. https://doi.org/10.1080/00220272.2012.668938

World Economic Forum (WEF). (2015). *New vision for education: Unlocking the potential of technology.* World Economic Forum. https://widgets.weforum.org/nve-2015/index.html

World Economic Forum (WEF). (2023). *The global risks report 2023 18th edition: Insight report.* https://www.weforum.org/reports/global-risks-report-2023/

World Economic Forum (WEF). (2024). *The global risks report 2024, 19th edition: Insight report.* https://www.weforum.org/publications/global-risks-report-2024/

World Economic Forum (WEF). (2025). *The future of jobs report 2025: Insight report.* https://www.weforum.org/publications/the-future-of-jobs-report-2025/

Ziatdinov, R., Atteraya, M. S., Nabiyev, R. (2024). The fifth industrial revolution as a transformative step towards Society 5.0. *Societies*, *14*. Article 19. https://doi.org/10.3390/soc14020019

Zollman, A. (2012). Learning for STEM literacy: STEM literacy for learning. *School Science and Mathematics, 112*(1), 12–19. https://doi.org/10.1111/j.1949-8594.2012.00101.x

CHAPTER 6

STEA^{+}M Thinking Systems as Neo-21st Century Skills

6.1 Introduction

> A key aspect of teaching students in the twenty-first century is preparing them for an increasingly global community, global economy and global workplace... it is crucial that young people view themselves as "players in the world", participating in global events—not simply observing them; taking action "to improve conditions" in the world—not simply agonising over them; and reflecting on their actions and their participation, with a view to continue to improve and enhance this participation. (Reynolds et al., 2017)

The world has become more interconnected and complex. Global challenges such as climate change, artificial intelligence, the COVID-19 pandemic, and geopolitical instability are profoundly reshaping the landscape of education. The pandemic, for instance, emerged as a stark reminder of global interdependence, triggering widespread disruption. Despite the many issues and challenges that came in its wake, the pandemic also catalyses breakthroughs and reforms across healthcare, biomedical, economic, technology, infrastructure, and education (Anderson et al., 2021). In response to these global challenges, there has been a growing recognition of the critical nexus between digital

G. Kidman et al., *Knowledge Generation in STEM and STEAM Education*, https://doi.org/10.1007/978-3-032-03247-8_6

literacy and human-centred skills such as resilience, adaptability, collaboration, empathy, and ethical decision-making (Bozkurt & Sharma, 2021; Sahle-Work et al., 2021).

The rapid shift to online learning, remote work, and digital interactions not only exposed systemic inequities but also underscored the importance of socio-emotional competencies and technological proficiency (Sahle-Work et al., 2021). These developments call for a reimagined educational framework that equips students to navigate a dynamic and uncertain world while empowering them to tackle pressing global issues, from public health crises and climate change to social justice. To achieve this, education systems are shifting to embrace interdisciplinary and flexible approaches, such as those embodied by Science, Technology, Engineering, Arts, and Mathematics (STEAM) education. In Chapter 3, we introduced a plexus model of STE*A*$^{+}$M education that places emphasis on *A*$^{+}$ as *A*gency and innovation. Integrated STEM and STEAM education, as STE*A*$^{+}$M, offers a promising pathway to address the evolving demands of the 21st century (e.g., Kidman et al., 2023).

6.2 Twenty-First Century Skills and Meta-Skills

As we propel toward the end of the third decade of the twenty-first century, artificial intelligence (AI) and technological race will continue, global economic and geopolitical powers will shift, and climate change will exacerbate (PricewaterhouseCoopers, 2004). Experts predict that people will rely more on rapidly evolving digital tools for both good and ill (Anderson et al., 2021) and that we are facing a multitude of global challenges compounding into "polycrisis" (WEF, 2023). We will need to have not only the twenty-first century skills to deal with these challenges, but also enculture a strong sense of social justice, resilience, and *A*gency to chart a way forward to a better world.

In Chapter 5, we reviewed the literature surrounding twenty-first century skills frameworks and STEM education. While there is no consensus about what twenty-first century skills are precisely, they are broadly defined to be the general skills that can be applied across various contexts in order for an individual to function and contribute to the twenty-first century society. Generally, there are two ways that twenty-first century skills are conceptualised: as a discrete list of skills (e.g., World Economic Forum's Future of Jobs report, 2025) and as a set of skills categorised into broad categories (e.g., Partnership for 21st Century P21,

Trilling & Fadel, 2009). There is a struggle to define the skills or competencies because these skills include more than just knowing what to do, but when and how to effectively and successfully apply the skills in various contexts (Senova, 2020). Thus, twenty-first century skills expanded to include the set of knowledge, skills, attitudes, dispositions, values, and ethics in various frameworks (Griffin et al., 2012). Yet, others started using the term *meta-skills*.

The term meta-skill is used in a variety of ways across different disciplines and industries, and there is not yet a universal definition (Stephen et al., 2020). However, scholars reviewing the literature in the fields of psychology, education, work-based learning, and industries suggest that the definitions of meta-skills are generally grouped into two main categories:

- **Meta-skills as a label for metacognitive skills.** This definition refers to "the ability to use introspection to further develop learning, thinking, and understanding (cognitive) capacities" (Stephen et al., 2020, para. 7). It is about being aware of and regulating one's own cognitive processes, and includes skills such as monitoring one's own mental and emotional states through knowledge, life experiences, and abilities (Das et al., 2024), and managing one's own mental states to achieve adaptability, autonomy, and life satisfaction (Mitsea et al., 2023).
- **Meta-skills as higher-order skills**. This definition refers to skills that are applicable across domains and disciplines and that enable individuals to improve or accumulate "hard" or technical skills beyond the discipline or context where they were first developed (Stephen et al., 2020). These include skills such as communication, problem-solving, creativity, analysis, self-promotion (Yadollah & Yazdani, 2020), and the ability to learn new skills and engage others in learning (Tammeaid et al., 2022).

No matter which definition, meta-skills are seen as skills that "integrate meta-cognitive, social-emotional, and motivational attributes that enable individuals to be self-motivated, self-regulated, and adaptive in every context of human life" (Mitsea et al., 2023, p. 2). They are essential to lifelong learning (Das et al., 2024; Moore, 2002) and are necessary for adapting to the rapidly changing world (Stephen et al., 2020). Meta-skills are often described as advanced capabilities that help individuals gain other relevant skills for better outcomes and are considered essential for mastery (Das et al., 2024). Another key characteristic is the transferability of the meta-skills from situation to situation (Senova, 2020) and across different work environments (Finch et al., 2016).

Meta-skills might also be domain-specific, such as technology scanning, technology assessment, technology comprehension and debugging, which are proposed by Westfall (1999) as key meta-skills in information technology crucial to navigating the rapid changes in the field. One could argue that the first three of these meta-skills are essential to everyone in the current day and age—the skills of knowing how to access, glean, select, review, and integrate information and content into one's personal knowledge (Moore, 2002). Meta-skills that are discipline-specific are transferable across various contexts within the field, in contrast to context-specific skills (Finch et al., 2016).

In their *Skills 4.0—A skills model to drive Scoltand's future* report, Skills Development Scotland (2018) proposed 12 meta-skills that are organised into three themes.

- Self management:
  - o Focusing—sorting information, attention, filtering;
  - o Integrity—self awareness, ethics, self control;
  - o Adapting—openness, critical reflection, adaptability, self-learning, resilience; and
  - o Initiative—courage, independent thinking, risk-taking, decision making, self belief, self motivation, responsibility, enterprising.
- Social intelligence:
  - o Communicating—receiving information, listening, giving information, storytelling;
  - o Feeling—empathy, social conscience;
  - o Collaborating—relationship building, teamwork and collaboration, social perceptiveness, global and cross cultural competence;
  - o Leading—inspiring others, influencing, motivating others, developing others, change catalyst.

- Innovation:
  - o Curiosity—observation, questioning, information sourcing, problem recognition;
  - o Creativity—imagination, idea generation, visualising, maker mentality.
  - o Sensemaking—pattern recognition, holistic thinking, synthesis, opportunity recognition, analysis;
  - o Critical thinking—deconstruction, logical thinking, judgement, computational thinking.

Considering the body of literature about twenty-first century skills and meta-skills, it is clear that just the term "skills" is inadequate to convey the idea of skills we project as integral for the future. The use of the term "meta-skills", though complex and multifaceted, makes a case for the need to expand the concept of twenty-first century skills. The twenty-first century skills do not exist in isolation; instead, they are "developed and applied in combination with and in relation to other skills" (Allen & van der Velden, 2012, p. 24). A combination of basic skills such as reading, motor skills, oral and written communication, and numeracy are considered the foundation for developing twenty-first century and work-specific job skills (Allen & van der Velden, 2012). The OCED Programme for International Student Assessment (PISA) measures the basic skills of reading, mathematics, and science knowledge and skills. It includes other twenty-first century competencies, such as Collaborative Problem Solving in PISA 2015 (https://www.oecd.org/pisa/innovation/collaborative-problem-solving/) and Global Competence in PISA 2018 (https://www.oecd.org/pisa/innovation/global-competence/). On their website, Gobal Competence is defined to be "students' capacity to examine local, global and intercultural issues, to engage in open, appropriate and effective interactions with people from different cultures, and to act for collective well-being and sustainable development". This is a multi-dimensional construct that considers the application of Knowledge, Skills, Attitudes, Values, and Ethics (KSAVE) aspects (Griffin et al., 2012) to global and/or intercultural contexts (see Chapter 5). This phenomenon of extending beyond foundational competencies heralds a need for a more holistic (re)interpretation of twenty-first century skills.

As the world grows increasingly complex, individuals must possess specialised knowledge and adaptive abilities to navigate technological landscapes and organisational structures (Spencer & Lucas, 2021). The acquisition and application of twenty-first century skills—critical thinking, collaboration, creativity, communication, and more—function as a networked system where skills reinforce and enhance one another. For instance, critical thinking strengthens communication, as clear reasoning leads to well-articulated explanations. To cultivate this "complete package" of competencies, a holistic approach that integrates skills across disciplines is essential (Allen & van der Velden, 2012, p. 24).

We propose that the neo-twenty-first century skills will have themes similar to existing twenty-first century frameworks or meta-skills, such as cognitive skills, interpersonal skills, character qualities, and technological literacies, but with shifts of emphasis within each theme (see Table 6.1). This transformation is toward integrating basic or foundation twenty-first century skills into ***neo-twenty-first century skills*** in order to take action to solve authentic problems in a world that is full of volatility, uncertainty, complexity, and ambiguity (abbreviated as VUCA, see Babu & Hisrich, 2023). Table 6.1 shows a list of neo-twenty-first century skills that are emerging as a response to making sense of the VUCA world.

These neo-twenty-first century skills in Table 6.1 can be integrated into the next level of *meta-skills*, which are *Thinking Systems*—interrelated networks of cognitive processes and knowledge involved in human thought and decision-making. The sine qua non of a thinking system is *purposive learning* (Waldman, 2007). In systems research, a *system* includes individual components that interact, which can be for:

- a consistent predictable outcome (**a machine system** e.g., a bicycle moving when pedals are turned),
- instinctively adapting to survival (**a complex adaptive system**, e.g., a flock of geese on their migration flight), or
- "actions that require decisions where there is no established right answer" (Waldman, 2007, p. 272) (**a thinking system**, e.g., a sports team planning and carrying out their play strategy).

When applied to the individual, the context is important since it provides the background as well as influences the *self-generated goal/s* the individuals set for themselves. A teacher making decisions during

Table 6.1 A transformation From twenty-first century meta-skills to Neo-twenty-first century skills

Twenty-first century meta-skill	*Neo-twenty-first century skills*
Cognitive skills	**Integrative thinking**—the interconnection of foundational thinking skills, attitudes, knowledge, and action. It includes –*Generative thinking:* using creativity and abductive reasoning to generate problems for open-ended inquiry (Gao, 2024; Ragusa, n.d.). Abductive reasoning is a way to generate and test hypotheses in situations where there are enormous amounts of information, unknown interdependencies, and limited feedback (e.g., used in complex innovations such as drug discovery, Dunne & Dougherty, 2016) –*Transformative thinking:* exploring multiple paradigms (thinking without a box) with imagination and innovation to generate new sustainable change. Bukusi (2020) suggests a six-stage transformative thinking process of Reflection, Analysis, Deconstruction & Ideation, Change of Perspective, Synthesis & Modelling, and Creation & Integration –*Interdisciplinary thinking*: integrating knowledge and modes of thinking from different disciplines or established areas of expertise to advance knowledge (e.g., explaining a phenomenon or solving a problem) in ways that would be impossible or unlikely through a single discipline (Boix Mansilla et al., 2000, p. 17) –*Embodied thinking*: integrating physical interaction, sensory experiences, and kinesthetic learning to solve problems, develop insights, test hypotheses, and generate solutions. It emphasises the role of the body and movement in understanding and shaping ideas, often through hands-on activities like building, prototyping, or experimenting (Meyer, 2012). It is a key skill in maker cultures, where learning and creativity emerge through direct interaction with materials, tools, and environments (Iwata et al., 2020) –*Scenario thinking:* strategic analysis and synthesis of relevant past (hindsight), present (lateral thinking) and future (foresight) information into insights (Sarpong, 2011)

(continued)

Table 6.1 (continued)

Twenty-first century meta-skill	*Neo-twenty-first century skills*
	–*Systems thinking*: analysis of "how parts of a whole interact with each other to produce overall outcomes in complex systems" (P21, 2009, p. 4). It is important to recognise the parts and wholes, the relationships between parts, what is systematised, how the system operates, the purpose of the system, and what is inherent behind the system (Buchanan, 2019)
Interpersonal skills	**Meta-communicative and influence skills**—applying communication and influence skills to deal with the complexities of the world. This includes 'soft' skills of having an awareness and understanding of how to navigate contexts to influence others (e.g., teamwork, conflict resolution, advocacy), such as
	–*Inter-cultural communication, inclusive and influence skills*: communicate effectively one's ideas across diverse audiences; teamwork and leadership skills in collaborating in and leading teams that are linguistically, culturally, and socially diverse (Boix Mansilla & Jackson, 2011)
	–*Interdisciplinary communication and influence skills*: collaborative skill involving reflecting on disciplinary knowledge and attitudes and committing to working together, respecting one another's competence, and having the trust and assurance to creatively negotiate to solve problems (Wahid et al., 2024; Winowiecki et al., 2011)
	–*Transformative leadership skills:* incorporating basic communication and influence skills to motivate and inspire, be responsive to and empower others (Riggio & Tan, 2013)
Character qualities/Personal disposition	**Social consciousness and agency**—having the confidence, readiness, and will to positively influence one's own life and the world. This includes the (a) disposition to act and transform the environment, (b) the motivation to regulate one's actions (e.g., persistence) and make choices and decisions, and (c) the positionality to negotiate and interact within complex social contexts (Vaughn, 2020)
	–*Social consciousness and social justice*: having a sense of responsibility as one participates in society and global citizenry, and aims to influence people, events, and situations for the better (OECD, 2019)

Twenty-first century meta-skill	*Neo-twenty-first century skills*
	–*Sustainability and socio-environmental agency*: having an awareness of one's positionality in both individual and collective change-making for taking action to address socio-environmental challenges or work toward greater sustainability (Tayne, 2022)
Technological literacies	**Ethical technological literacies**— digital and multi-literacies guided by ethics as the compass to navigate the sea of a complex, rapidly changing world. New technologies develop faster than protocols governing their use and misuse, and are anticipated to exacerbate inequalities (WEF, 2023) –*Socio-technological literacy*: understanding the potential and risks of how technologies are developed and used within the ecological, social, institutional, economic, and political contexts (van Mele, 2008) –*Ethical Artificial Intelligence (AI) literacy* having the knowledge, skill, and disposition to (a) apply AI (intelligent machines and algorithms) ethically for social good; (b) evaluate and create AI; and (c) consider the potential risks and issues related to AI such as fairness, accountability, transparency, and ethics (Ng et al., 2021)

classroom teaching will be influenced by their classroom goals, understanding of their students and how they learn the subject, the classroom dynamics and culture, the intended curriculum, and accessible resources (e.g., books, digital devices, and connections). The teacher's thinking and decision-making processes will incorporate not just their cognitive aspects, but affective and volitional dimensions of cognition. Their thought system, in this context, is a complex network including their pedagogical content knowledge (PCK) and skills, beliefs about teaching, learning and the role of new technologies such as AI, and educational and developmental goals for their students and themselves (if the teacher is intentional about collecting classroom evidence for their professional development). As a twenty-first century learner, the teacher's creativity, critical thinking, knowledge, skills, attitudes, values, and ethics are also integrated into the thought system dynamically. Reflection and metacognition are also necessary for professional development. This makes teaching complex and dynamic, and the outcomes of teacher's thoughts and decisions highly nuanced and capricious. According to Waldman (2007), the 'thinking' in thinking systems refers to "the ability to create new ideas, to make things never seen before, to behave emotionally and/ or illogically and to act purposefully" (p. 273).

Applying the concept of thinking systems to our discussion, the neo-twenty-first century skills highlighted in Table 6.1 can be explained as the compounding of foundational twenty-first century skills and meta-skills into purposive thought and action pathways. These skills interact with one another in the context of STE*A*$^+$M learning experiences as *integrative thinking systems*. Such integrative thinking systems can be seen in examples such as social entrepreneurial mindsets, sustainability or green transformational leadership, global competence/mindset, and resilience engineering for climate adaptation. Note that the difference between *integrative thinking systems* and *systems thinking* is that a "thinking system" refers to the interrelated networks of cognitive, affective, and conative processes and knowledges involved in human thought and decision-making, whereas "systems thinking" refers to a way of thinking ("a perspective, a language, and a set of tools", Monat & Gannon, 2015, p. 17) that focuses on how the components of a system interact with one another within a whole (Buchanan, 2019). Thus, we are applying a systems thinking approach to the concepts of neo-twenty-first century skills, highlighting the interrelations and dynamics between skills and meta-skills, within a thinking system. The connections between neo-twenty-first century skills and STE*A*⁺M integrative thinking systems are discussed next, with vignettes from our past research as examples.

6.3 Developing Neo-Twenty-First Century Skills and STE*A*⁺M Integrative Thinking Systems

6.3.1 *Interconnections Between Neo-Twenty-First Century Skills and Integrative Thinking Systems*

In Chapter 4, we proposed the *Cognitive Integration Framework* for STE*A*⁺M integrative thinking, and described the cognitive processes and mechanisms involved in makerspace contexts using the *Crafting Knowledge Framework*. By extension, the principles behind these frameworks also apply to the *development of Thinking Systems* through the integration of various twenty-first century meta-skills (cognitive skills, interpersonal skills, character qualities, and technological literacies, see Table 6.1). To visualise this, we draw on the work of Repko and Szostak (2021). When encountering new contexts and tasks, one's existing knowledge and skills in different areas need to be *Redefined* to be tapped upon within the context, *Extended* to be reinterpreted into the new task, *Transformed* to be grounded within the lived experience, and *Organised* into a new set of actions and thinking processes to enact the neo-twenty-first century skills and meta-skills (RETO, see Fig. 6.1).

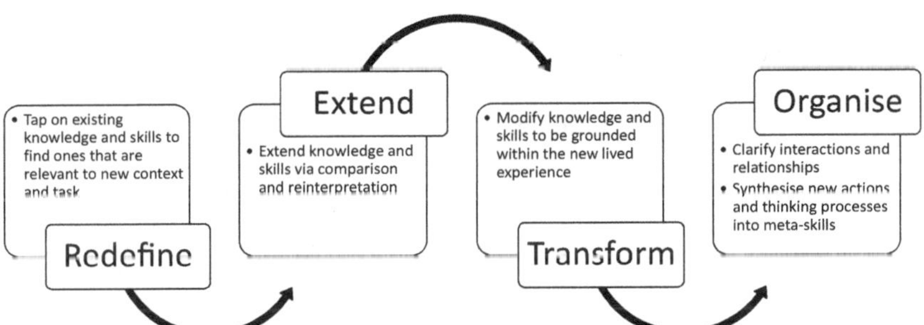

Fig. 6.1 Model of integration of skills and knowledge into neo-twenty-first century skills and meta-skills (*Source* Authors)

We envision that the neo-twenty-first century skills, highlighted in Table 6.1, required to meet the world's challenges, are interconnected and integrative, combining to form thinking systems that also have trans-disciplinary STEM/STEAM/STEA^+M connections. The following are some examples of the nexus.

6.4 Social Entrepreneurial Mindset

Advances in STEM and STEAM fields—such as Artificial Intelligence (AI), quantum computing, and biotechnology—affect many aspects of our lives and the world. The claim that "every job is becoming a STEAM job" (Dodge, 2017, para. 15) is evidenced by the success of STEAM integration into traditionally non-technical fields such as fashion and cosmetics, restaurant operations, environmental preservation, public policy, hospitality and professional sports (Dodge, 2017). The nature of work has changed since COVID, with greater mobility (working at any time, from anywhere), flexibility (more part-time and based on needs than full-time and ongoing), complexity (e.g., involving more sophisticated technologies), and sociality (working with different groups of people with greater diversity, from different parts of the world). Changes in society and culture in the neo-COVID world have resulted in a greater need for an entrepreneurial mindset with characteristics such as being risk-taking, innovative, proactive, resilient and adaptable (Ratten, 2020). Entrepreneurs, working in such a backdrop, need to "assemble all the essential resources and work with people who possess different capabilities" (Babu & Hisrich, 2023, p. 4) to solve social and economic problems faced by potential customers through innovation. In fact, business and entrepreneurial skills are found to be on the wishlist of STEM graduates as a top skill they would like to have more opportunity to develop within their degree, and a top skill that STEM industry employers were dissatisfied with for their graduate employees (Sarkar et al., 2016).

Entrepreneurship and social entrepreneurship are related yet distinct concepts. The former involves an individual's ability to innovate and create a new product or solution to solve a problem, for profit. In contrast, *social entrepreneurship* "aims at developing tools and technologies that produce positive social changes in the world" (Penna & Campelo, 2021, p. 160), prioritising social impact alongside, or sometimes above, financial gains. Social entrepreneurs aim to solve societal or environmental issues through innovative business models that are ethically and sustainably driven, as opposed to using the banner of social good merely as a marketing tactic. The positive impacts of social entrepreneurship are evident in practices such as microfinance, which empowers individuals in impoverished areas by providing them with small loans to start businesses, and in social enterprises that reinvest profits into environmental sustainability or community development. As science and technology are powerful drivers for change, STE*A*+M education with an entrepreneurship mindset can "ensure that the future is filled with scientists who are capable of being successful leaders of change, mobilising competencies, people, and resources, of improving the world" (Riberiro et al., 2023, p. 33). There are successful examples of social entrepreneurs in social media (e.g., Unstoppable, 2019), highlighting the tangible positive impacts their ventures can have on local communities and societies.

An entrepreneurial mindset, we contend based on our experiences and the literature (Kidman et al., 2023), consists of a continuum of eight dimensions that are each multifaceted and involve thinking, attitudes, and behaviours (p. 96):

1. Thinking that is visionary and creative, and based on curiosity
2. Ability to take risks when opportunities arise
3. Self-motivation, and self-regulation
4. Tenacity and resilience when faced with failure
5. Ability to be innovative and a user of novel approaches
6. Ability to identify goods and services with unexpected value
7. Understanding that resources, once transformed into goods or services, have a monetary value
8. An approach to resource allocation that involves sequencing, low commitment, and reallocation of resources.

Adding to this, a social entrepreneurial mindset includes the above dimensions, as well as the.

9. Ability to create social value and social empathy
10. Agency to search and advocate for a more inclusive, dignified, and just life for all (Penna & Campelo, 2021).

In the following section, we share a case study where we, as STEM educators, were involved in a 6-week design sprint during the COVID-19 pandemic. Through reflecting on the experience, we learned to cultivate a social entrepreneurial mindset (Redefine, Extend, Transform, and Organise process; see Fig. 6.1). We reinterpreted the above eight dimensions with respect to the neo-twenty-first century skills highlighted in Table 6.1.

6.4.1 Developing Social Entrepreneurial Mindset—A Case Study

During the COVID-19 pandemic, we had the opportunity to be involved in a design sprint with a commercial makerspace hub to mass-produce medical equipment (Kidman et al., 2023). A team of engineers, educators, and health workers was established to collaborate on the design and prototype of a novel face-shield and then registered it for use as "therapeutic goods" for mass production. Reflecting on the emerging twenty-first century skills illustrated in the project, we.

- engaged in *integrative thinking* (thinking outside the box, interdisciplinary thinking, generative thinking, systems thinking, creativity);
- utilised *meta-communication and influence skills* (soft skills of communication across diverse disciplinary fields and walks of life);
- with a sense of *social agency and sustainability orientation* to drive our tenacity to forge through issues and challenges (considering the issues of cost-effectiveness, easy-to-use personal protective equipment for community workers during the pandemic, and sustainable for the environment); and
- integrated the soft skills with the hard STEM skills to develop our *socio-technological literacy* (considering the prolonged pandemic lockdown constraints, opportunities, risks and threats toward how we work together using technologies, what we want to design, how to design collaboratively, and how to manufacture and deliver a quality product within the time and cost constraints).

We used an *appreciative inquiry* methodology to frame our process: (1) Discovery, (2) Dream, (3) Design, and (4) Delivery (Cooperrider et al., 2003). There is an emphasis on celebrating past successes and small wins, in order to drive innovation, imagination, and agency to Dream, Design and Deliver. These involved the affective (emotions, feelings and attitudes) and conative (values, volition and drive) dimensions, not just the cognitive knowledge and skills.

This example illustrates the social entrepreneurial mindset as an example of the synergising of various neo-twenty-first century skills into a *STEA+M Integrative Thinking System*. Through the iterative RETO process (see Fig. 6.1) of.

1. **Redefining** one's existing knowledge and skills to select (consciously or intuitively) elements relevant to the new contextual base,
2. **Extending** the skills through reinterpretation and combination of selected elements to adapt to the new context,
3. **Transforming** the various combinations of extended skill elements into actions to be grounded and "tested" in lived experiences, and
4. **Organising and consolidating** perceived effective or successful combinations of extended skill elements into a higher-order skill and eventually into a system of thinking and action.

The steps in the process are dynamic and fluid. One may only go through some steps, backtrack to previous steps, but not skip a step in the process unless they have already been through the previous step in another context (e.g., have "wired together" and applied some combination of skills in a previous experience). As we apply our existing skills or combination of skills in new contexts, more elements of the newly learnt knowledge and skills can be integrated, and thus developing our skills and meta-skills into a thinking system.

6.5 Socio-Environmental Sustainability Agency

Another example of an integrative thinking system that combines STEA+M generative thinking with social consciousness and *A*gency (see neo-twenty-first century skills in Table 6.1) is socio-environmental sustainability agency—understanding socio-environmental sustainability issues and developing the agency to take action, investigate, and design solutions to address them.

The United Nations' seminal Brundtland Report by the World Commission on Environment and Development (WCED, 1987) called for a new economic development model that is socially and environmentally sustainable. *Sustainable* development aims to "meet the needs of the present without compromising the ability of future generations to meet their own needs" (WCED, 1987, p. 16). The environmental issues and ecosystems in the world we live in cannot be seen as separate from human social development. This makes sustainable development an interdisciplinary field and leverages STE*A*⁺M education as the key to creating socio-environmental sustainability agency in our future generation. The Sustainable Development Goals (SDGs), formulated at the United Nations Conference on Sustainable Development in Rio de Janeiro in 2012, encapsulated a set of goals to meet the urgent environmental, social and economic issues faced by us today (Hajian & Kashani, 2021).

Socio-environmental Sustainability Agency refers to the capacity and actions of individuals, groups, and communities to influence and drive both social and environmental sustainability (WCED, 1987). It is the intersection of social responsibility and environmental stewardship, emphasising the interconnectedness of social systems and environmental health. This concept is rooted in the understanding that sustainable development cannot be achieved without addressing the social dimensions of environmental issues and vice versa. While sustainability can sometimes be discussed in terms of economic or ecological aspects alone, the socio-environmental perspective ensures that human action and the environmental ecosystem are considered together holistically (Gesthuizen et al., 2025).

Next, we share an example of how a socio-environmental sustainability agency might be developed in an STE*A*⁺M education context.

6.5.1 Developing Socio-Environmental Sustainability Agency—A Case Study

In our Monash Lens piece on "How to build an integrated STEM lesson using Minecraft" (Kidman et al., 2018) and our paper on exploring three case studies where students learn about sustainability and the environment through virtual worlds (Gesthuizen et al., 2025), we illustrate the potential of Minecraft in integrated STE*A*⁺M education. Virtual experiences can be used by teachers to raise socio-environmental awareness in students by instilling a sense of purpose and urgency toward a sustainable future by incorporating ecological sustainability and social justice dilemmas into the activities and discussions. Students design and construct sustainable virtual homes or cityscapes inspired by their local contexts, with consideration of people living in these spaces.

We discussed three cases of how virtual experiences transform students' learning in our paper (Gesthuizen et al., 2025): a young builder designing a sustainable house; a group of isolated students on a coral atoll collaborating to build homes considering their critical water, food, and energy needs; and students at a public event redesigning buildings around a city square where the event is staged at. Minecraft can facilitate critical thinking by representing different environments and biomes and simulated weather conditions like sunlight, rain, and snowfall, which impact resource availability (water, food, energy).

Hands-on experiences, such as building in virtual worlds, might be combined with real-world field trips and activities to create rich learning experiences. For example, in order to create a virtual model of a sustainable home, interviews and/or research can be conducted within the community. Students then brainstorm possible solutions before narrowing their ideas and prototyping their solutions in Minecraft. After building their sustainable house, students can communicate and pitch their designs to their peers. This allows them to reflect on others' designs in order to refine their own designs and evaluate their design process. They can follow up with an investigation of how to make feasible and re-create (realise or implement) some of the features of their design into their own homes.

Using the abovementioned learning experiences as examples, this thinking system incorporates a potential combination of the following neo-twenty-first century skills.

Cognitive Skills

- *Generative thinking* to reason and navigate the vast amount of information and knowledge on socio-environmental issues in order to raise problems for investigation and inquiry;
- *Scenario thinking* to visualise and analyse strategically and imagine opportunities for a better future;
- *Transformative thinking* to explore innovative solutions for sustainable change; and
- *Interdisciplinary thinking* to integrate disciplines (and topics) such as physical and human geography (landscape and habitat, human

impact on the environment), science (biology, ecology, chemistry, physics, environmental science), mathematics (mathematical modelling, measurement, statistics, financial mathematics), technology (systems thinking, designing data systems, design thinking), history (socio-cultural-historical contexts), social science (human activities and the environment), and economics (economic impacts, feasibility and logistics), to explore sustainable solutions to issues identified.

Interpersonal Skills

- *Inter-cultural communication and influence skills* to listen to, communicate with, and advocate for local or regional communities;
- *Transformative leadership skills* to motivate and inspire others toward empowerment in the face of adversity.
- *Interdisciplinary communication and influence skills* to collaborate effectively with an interdisciplinary team and to negotiate during the process of creativity and innovation;

Character Qualities

- a sense of *Social consciousness and social justice* to understand that, in this interconnected world, we are seven billion people travelling in a large "ship" (Earth) with different cabins (countries) rather than "193" individual boats with individual captains (Mahbubani, 2015). Thus, having a *global mindset and socio-environmental agency* means that we can no longer ignore climate change and sustainability issues, both locally and globally.

Technological Literacies

- *Socio-technological literacy* to understand and navigate the complexities of technologies involved in the generation and solution of sustainability issues, including *ethical use of artificial intelligence* in the community.

In the context of solving social issues in the local environment, learners co-develop their neo-twenty-first century skills about socio-environmental sustainability with their peers and the local community. The experiences empower and inspire them to take action and advocate for the future. Role models and the opportunities to try and explore (and fail) are important in this respect.

6.6 Resilience Engineering for Climate Adaptation and Human-Centred Design

Another STE*A*⁺M Integrative Thinking System that combines all of the cognitive skills, interpersonal skills, character qualities, technological literacies, and hands-on thinking is ***resilience engineering for climate adaptation and human-centred design***. Engineering in the twenty-first century requires a blend of technical expertise and integrative thinking to address complex, interconnected challenges. Resilience engineering is an emerging approach to designing systems that can anticipate, survive, and adapt to disruptions such as those caused by natural disasters, abrupt discontinuity of supply, terrorism and political instability events, and financial meltdowns (Madni & Jackson, 2009). It emphasises the importance of human factors in maintaining safety and efficiency, particularly in complex systems like autonomous ships (Veitch et al., 2021) and air traffic control (Lopez et al., 2024). Climate change adaptation is a key application area, requiring the integration of resilience engineering principles into infrastructure planning and design (Rifai et al., 2024). Human-centred design methods can help develop tools for equitable community resilience planning (Fry et al., 2023) and co-designing resilient socio-technical software systems (Lopez et al., 2024).

The concept of resilience has evolved from engineering and ecological models to an evolutionary perspective that embraces transformation and adaptation (Davoudi et al., 2013; Fry et al., 2023). Hollnagel (2014) described different interpretations of resilience in different fields. In material science, resilience is a material's inherent quality, representing a static system. In ecology, resilience is about an ecosystem's capacity to absorb changes before it deviates from a particular steady state and becomes unstable, but it does not anticipate nor have any intentions. In psychology, it involves the reflection of psychological systems on experiences to shape responses and anticipate future events. In business and safety management, it is the ability to anticipate and respond to rapidly developing situations, which is crucial for system survival and sustained operations. In socio-technical systems—which consist of (a) the individuals and teams working on tasks, (b) the organisational efforts to manage and coordinate operations, and (c) the industrial system that designs and produces technologies that are used—the potential for resilience lies within effective behavioural adaptation and compensation by people to manage disturbances in practice. This resilience is rooted in how people adapt and compensate, rather than in managing the risk or consequences of failure (Lopez et al., 2024). In engineering, resilience refers to "the intrinsic ability of a system to adjust its functioning prior to, during, or following changes and disturbances, so that it can sustain required operations under both expected and unexpected conditions" (Hollnagel, 2014, p. 222), and is related to four primary abilities of the system (Hollnagel, 2014):

- *Knowing what to do*: how to respond to regular and irregular disruptions;
- *Knowing what to look for*: how to monitor both events in the environment and the performance of the system itself;
- *Knowing what has happened:* how to learn from the experience, both successes and failures; and
- *Knowing what to expect:* how to anticipate developments, threats, and opportunities.

Additionally, a resilient system should be proactive and adaptive - knowing how to look for ways to enhance its ability to monitor risks and make tradeoffs (Madni & Jackson, 2009).

Developing resilience is now a key element in climate adaptation, environmental management, regional economic growth, and strategic planning (Davoudi et al., 2013). Resilient systems must be able to respond, monitor, learn, and anticipate, with humans playing a crucial role in adjusting to changing conditions (Hollnagel, 2014). It is a multifaceted capability of a complex system that encompasses avoidance, absorption, adaptation to, and recovery from disruptions (Madni & Jackson, 2009). This approach views planning for resilience as involving not just recovery from shocks and failures, but cultivating preparedness (Davoudi et al., 2013) and viewing failures as results of necessary adaptations rather than malfunctions (Hollnagel, 2014).

Human-centred design refers to an iterative, participatory approach that aims to uncover users' values, concerns, ideas, and constraints to develop products and services that meet their needs (Fry et al., 2023). In aiming to create products, services, and processes aligned with user needs, human-centred design approaches often use participatory methods for co-generating solutions. Working with communities living in proximity to contaminated lands, Fry et al. (2023) used human-centred design to focus on meeting the needs of those aiming to tackle resilience and equity in climate risk management, engaging and empowering them through public decision-making. They co-developed the Equitable Resilience Builder, a tool designed for interdisciplinary teams of local resilience planners, practitioners, and community members. It guides users through various sections and activities with the consideration of equity issues.

In applying resilience engineering for climate adaptation, we need to integrate human-centred design. A co-design approach, which includes methods of participatory design, can be used to structure and organise resilience engineering analyses (Lopez et al, 2024). This helps to address the limitations of resilience engineering designs that may lose the nuanced details of practice within a profession and discipline. When integrated with resilience engineering, human-centred design approaches can help engineers design systems that are not only resilient but also equitable, accessible, and usable, particularly for underserved and vulnerable communities (Fry et al., 2023). Examples might be designing adaptive and smart infrastructure, such as modular and flexible infrastructure that evolves with climate change, using nature-based solutions like green roofs, permeable pavements, and living shorelines, and enhancing regenerative systems like water-sensitive urban design. With the addition of AI, the integration of human-centred design with resilience engineering for climate adaptation can lead to *resilient interaction design* (IxD), which recognises that humans are required to interact with automation and model-based software in ways that influence and direct the behaviour of systems (Veitch et al., 2021).

6.6.1 Developing Resilience Engineering for Climate Adaptation and Human-centred Design—A Case Study

In early 2025, we conducted a three-day workshop for a group of Indonesian civil engineer lecturers from a university, to review and renew their curriculum for tomorrow's engineers. Through this experience, we realised that engineering education at the professional level is considerably more complex and demanding than the design thinking model in STEM/STEA^{+}M education in schools. During the workshop, we asked the civil engineers, who are also lecturers, to envision the nature of the work that their future civil engineering graduates will be doing. We did this using a higher-order thinking skills (HOTS) strategy of See/Hear, Think, Feel, and Wonder to prompt their creativity via envisaging a future scenario and engaging the senses vividly (Fig. 6.2).

Based on their future scenarios envisioned, the civil engineers discussed and listed the following skills that civil engineers need to be successful:

- Artificial Intelligence and technological skills
- Critical thinking
- Communication with people in other disciplines and with different perspectives
- Leadership and teamwork
- Forecasting skills
- Problem-solving skills to solve complex problems
- Climate change disaster knowledge and skills to solve the challenges
- Skills to investigate and analyse environmental conditions (e.g., soil, water)
- Research skills to find the best way to solve problems (e.g., review existing literature, conduct experiments and simulation, analyse data)

HOTS Trigger Activity (Predict)

It's 2030, and you are a civil engineer working on a major infrastructure project. The world has changed significantly due to advancements in technology, climate challenges, and societal needs.

What does your workplace look like?

What tools are you using? What challenges are you solving?

Who will you be talking to and working with?

See/Hear	Think	Feel	Wonder
What do you see/ hear in your workplace?	What are you thinking about?	What do you feel emotionally and physically?	What questions or uncertainties do you have?

Fig. 6.2 Higher-Order Thinking Skills (HOTS) prompt for scenario and embodied thinking (*Source* Authors)

- Financial and management skills to manage projects and optimise time, money and resources
- Entrepreneurial skills
- Lifelong learning (e.g., updating oneself by reading journal papers, attending courses, and joining a learning community)
- Driven by values and developing one's personal brand (e.g., curious, global-minded, striving for excellence, innovative, and intellectual).

These skills interact with one another in a system. For example, AI and sophisticated technological tools will aid in investigating environmental conditions with real-time data collection and advanced modelling tools and software. Designing solutions that address climate challenges (resilient engineering) involves consulting different stakeholders (human-centred design) and making decisions to best meet the needs of current and future, yet optimise time, money and resources. These correspond to the STE*A*$^+$M integrative thinking system of *Resilience Engineering for Climate Adaptation and Human-Centred Design*, which incorporates a combination of various neo-twenty-first century skills in Table 6.1.

The three thinking systems described (social entrepreneurial mindset, socio-environmental sustainability agency, and resilient engineering for climate adaptation and human-centred design) involved synergising and transforming twenty-first century skills in thinking, feeling, ethics, social, and technology in different contexts to achieve the goal of making a positive difference in the society. In STE*A*$^+$M education, we aim to provide experiences and opportunities to inculcate neo-twenty-first century skills. To prepare students for the complex challenges of the future, we must teach them STE*A*$^+$M integrative thinking skills—synergising disciplinary knowledge with twenty-first century competencies to solve real-world problems. Just as a dancer must first master the steps before adapting to new rhythms and unpredictable conditions, students need strong foundations in STEM disciplines before they can effectively integrate diverse skills. By progressively increasing the complexity and fluidity of learning experiences, we equip them to "dance to different tunes"—to think critically, adapt swiftly, and apply interdisciplinary knowledge in ever-changing contexts. This approach fosters agility and resilience, enabling graduates to eventually "dance on quicksand" by applying integrative thinking systems to novel and evolving challenges in society.

6.7 Conclusion

In this chapter, we explained the need to re-evaluate our definition of twenty-first century skills in the increasingly complex multipolar world with urgent climate and environmental issues. Just as we need to consider not just disciplines in silos but integrate disciplines into interdisciplinary STE*A*⁺M education, we need to shift our focus from individual twenty-first century skills toward synergising the foundational skills into *meta-skills* and *integrative thinking systems*. Integrated STE*A*⁺M education offers an effective way to nurture and transform students' mindsets and develop *A*gency. We need to research effective pedagogies to develop these in our students (see case study 2). But first, we need to develop these meta-skills and integrative thinking systems in ourselves (see case study 1) to inspire ourselves as teachers and connect with various stakeholders (see case studies 1 & 3) so that we can, in turn, inspire our students. Reflecting on the quote at the start of the chapter by Reynolds et al. (2017), instead of merely observing and fretting over global events, we all need to become "players in the world" and take action for a better future.

> Those who come before (us) plant trees so that those who come after can enjoy the shade.
> 前人种树，后人乘凉 (A Chinese Proverb)

References

Allen, J. P., & van der Velden, R. K. W. (2012). Skills for the 21st century: Implications for education. *Researchcentrum voor Onderwijs en Arbeidsmarkt, Faculteit der Economische Wetenschappen*. ROA Research Memoranda No. 11 https://doi.org/10.26481/umaror.2012011

Anderson, J., Rainie, L., & Vogels, E. A. (2021, February 18). Experts say the 'new normal' in 2025 will be far more tech-driven, presenting more big challenges. *Pew Research Center*. https://www.pewresearch.org/internet/2021/02/18/experts-say-the-new-normal-in-2025-will-be-far-more-tech-driven-presenting-more-big-challenges/

Babu, V., & Hisrich, R. (2023). Negotiation for entrepreneurship: Achieving a successful outcome. *Anthem Press*. https://doi.org/10.2307/jj.168332

Boix Mansilla, V., & Jackson, A. (2011). *Educating for global competence: Preparing our youth to engage the world*. Council of Chief State School Officers and Asia Society. https://asiasociety.org/files/book-globalcompetence.pdf

Boix Mansilla, V., Miller, W. C., & Gardner, H. (2000). On disciplinary lenses and interdisciplinary work. In S. Wineburg & P. Grossman (Eds.), *Interdisciplinary curriculum: Challenges to implementation*, (pp. 17–38). Teachers College Press.

Bozkurt, A., & Sharma, R. C. (2021). On the verge of a new renaissance: Care and empathy oriented, human centered pandemic pedagogy. *Asian Journal of Distance Education, 16*(1), i–vii. https://doi.org/10.5281/zenodo.5070496

Buchanan, R. (2019). Systems thinking and design thinking: The search for principles in the world we are making. *She Ji: THe Journal of Design Economics and Innovation, 5*(2), 85–104. https://doi.org/10.1016/j.sheji.2019.04.001

Bukusi, A. D. M. (2020). Transformative thinking for business leaders in the twenty-first century. *African Journal of Business Management, 14*(12), 551–560. https://academicjournals.org/journal/AJBM/how-to-cite-article/BA74D2465740

Cooperrider, D., Whitney, D., & Stavros, J. (2003). *The appreciative inquiry handbook*. Oakland Berrett-Koehler.

Das, K., Mungra, Y., Ambika, A., & Dhir, A. (2024). Creative meta-skills: Construct, dimensions and implications for marketing professionals. *Australian Marketing Journal, 32*(2), 98–112. https://doi.org/10.1177/14413582231154268

Davoudi, S., Brooks, E., & Mehmood, A. (2013). Evolutionary resilience and strategies for climate adaptation. *Planning Practice & Research, 28*(3), 307–322. https://doi.org/10.1080/02697459.2013.787695

Dodge, A. (2017, July 11). *Every job is a STEA⁺M job: Why your child will use tech in their future career*. https://ozobot.com/every job is a-steam-job-why-your-child-will-use-tech-in-their-future-career/

Dunne, D. D., & Dougherty, D. (2016). Abductive reasoning: How innovators navigate in the labyrinth of complex product innovation. *Organization Studies, 37*(2), 131–159. https://doi.org/10.1177/0170840615604501

Finch, D., Peacock, M., Levallet, N. & Foster, W. (2016). A dynamic capabilities view of employability. *Education + Training, 58*(1), 61–81. https://doi.org/10.1108/ET-02-2015-0013

Fry, M., Maxwell, K., Eisenhauer, E., Julius, S., Kiessling, B., Matsler, M., Ollove, M., & Romanoski, S. (2023). Centering equity in the development of a community resilience planning resource. *Climate Risk Management, 40*, 1–11. https://doi.org/10.1016/j.crm.2023.100520

Gao, Y. (2024). Generative thinking—The way of thinking in modern philosophy. *Scientific and Social Research, 6*(3), 144–149. https://ojs.bbwpublisher.com/index.php/ssr/article/view/6463/5607

Gesthuizen, R., Tan, H., & Kidman, G. (2025). Inspired to build understanding: Learning about sustainability and environmental education from within a virtual world. *International Research in Geographical and Environmental Education, 34*(1), 25–42. https://doi.org/10.1080/10382046.2024.2348265

Griffin, P., McGaw, B., & Care, E. (Eds.) (2012). *Assessment and teaching of twenty-first century skills*. Springer. https://doi.org/10.1007/978-94-007-2324-5

Hajian, M., & Kashani, S. J. (2021). Evolution of the concept of sustainability. From Brundtland report to sustainable development goals. In C. M. Hussain & J. F. Velasco-Muñoz (eds.), *Sustainable resource management: Modern approaches and contexts,* (pp. 1–24). Elsevier. https://doi.org/10.1016/B978-0-12-824342-8.00018-3

Hollnagel, E. (2014). Resilience engineering and the built environment. *Building Research & Information, 42*(2), 221–228. https://doi.org/10.1080/09613218.2014.862607

Iwata, M., Pitkänen, K., Laru, J., & Mäkitalo, K. (2020). Exploring potentials and challenges to develop twenty-first century skills and computational thinking in K-12 maker education. *Frontiers in Education, 5*, Article 87. https://doi.org/10.3389/feduc.2020.00087

Kidman, G., Gesthuizen, R., & Tan, H. (2018). How to build an integrated STEM lesson using Minecraft. *Monash TeachSpace*. https://www.monash.edu/education/teachspace/articles/how-to-build-an-integrated-stem-lesson-using-minecraft

Kidman, G., Gesthuizen, R., Tan, H., & Dielenberg, H. (2023). An entrepreneurial STEM teaching framework: Integrating business and STEM education. In S. Kaya-Capocci & E. Peters-Burton (Eds.), *Enhancing entrepreneurial mindsets through STEM education,* (pp. 93–113). Springer. https://doi.org/10.1007/978-3-031-17816-0_5

Kidman, G., Tan, H., & Macdonald, S. (2024). Reimagining STEM education: Developing agency through STE*A*$^+$M and design thinking. In T. Barkatsas, P. McLaughlin, & W. Goff (Eds.), *Reimagining education for the second quarter of the 21st century and beyond: Dilemmas, challenges, advancements and innovations* (pp. 1–23). Brill. https://doi.org/10.1163/9789004688490_001

Lopez, T., Sharp, H., & Wermelinger, M. (2024). Co-designing resilient socio-technical software systems. In *2024 International Workshop on Designing Software (Designing '24),April 15–14, 2024, Lisbon, Portuga*l. ACM, New York, NY, USA, https://doi.org/10.1145/3643660.3643948

Madni, A. M., & Jackson, S. (2009). Towards a conceptual framework for resilience engineering. *IEEE Systems Journal, 3*(2), 181–191. https://doi.org/10.1109/JSYST.2009.2017397

Mahbubani, K. (2015, June 8). Sailing on the same boat | UN. https://mahbubani.net/sailing-on-the-same-boat/

Meyer, P. (2012). Embodied learning at work: Making the mind-set shift from workplace to playspace. *New Directions for Adult and Continuing Education, 2012*(134), 25–32. https://doi.org/10.1002/ace.20013

Mitsea, E., Drigas, A., & Skianis, C. (2023). VR gaming for meta-skills training in special education: The role of metacognition, motivations, and emotional intelligence. *Education Sciences, 13*, 639. https://doi.org/10.3390/educsci13070639

Monat, J. P., & Gannon, T. F. (2015). What is systems thinking? A review of selected literature plus recommendations. *American Journal of Systems Science, 4*(1), 11–26. https://doi.org/10.5923/j.ajss.20150401.02

Moore, M. G. (2002). First thoughts on metaskills [Editorial]. *The American Journal of Distance Education, 16*(1), 1–13. https://doi.org/10.1207/S15389286AJDE1601_1

Ng, D. T. K., Leung, J. K. L., Chu, K. W. S., & Qiao, M. S. (2021). AI literacy: Definition, teaching, evaluation and ethical Issues. *Proceedings of the Association for Information Science and Technology, 58*, 504–509. https://doi.org/10.1002/pra2.487

Organisation for Economic Cooperation and Development. (OECD). (2019). *OECD future of education and skills 2030: Conceptual learning framework: Student agency for 2030.* https://www.oecd.org/education/2030-project/teaching-and-learning/learning/student-agency/

Partnership for Twenty-First Century Skills. (2009). *P21 framework definitions.* https://eric.ed.gov/?id=ED519462

Penna, K. N., & Campelo, S. P. (2021). Social consciousness and critical thinking: Reconciling social entrepreneurship and transformative education. In J. V. de Carvalho, Á. Rocha, P. Liberato, & A. Peña (eds). *Advances in tourism, technology and systems. ICOTTS 2020. Smart Innovation, Systems and Technologies,* (vol. 208, pp. 159–170). Springer. https://doi.org/10.1007/978-981-33-4256-9_15

PricewaterhouseCoopers. (2004, September). *Megatrends: Five global shifts reshaping the world we live in.* PWC. https://www.pwc.com/gx/en/about/corporate-governance/network-structure.html

Ragusa, S. J. (n.d.). *Thinking about "generative thinking".* https://susanjragusa.com/thinking-generative-thinking/

Rifai A., Handayani, S., Prasetijo J., Isradi, M., & Haqqani, M. H. (2024). Climate-resilient infrastructure planning: Integrating climate change adaptation into engineering design. *Journal of Infrastructure, Policy and Development. 8*(15), Article 8006. https://doi.org/10.24294/jipd8006

Ratten, V. (2020). Coronavirus (COVID-19) and entrepreneurship: Changing life and work landscape. *Journal of Small Business & Entrepreneurship, 32*(5), 503–516. https://doi.org/10.1080/08276331.2020.1790167

Repko, A. F., & Szostak, R. (2021). *Interdisciplinary research process and theory* (4th ed.). SAGE.

Reynolds, R., Ferguson-Patrick, K., & Macqueen, S. (2017). "Players in the world": Action for intercultural competence in classroom pedagogy. In S. Choo, D. Sawch, A. Villanueva, & R. Vinz (Eds.), *Educating for the 21st century* (pp. 47–71). Springer. https://doi.org/10.1007/978-981-10-1673-8_3

Ribeiro, T., Silva, J., Paz, M., Cardoso, A., Teles, N., Nogueira, C., & Ribeiro, T. (2023). Strengthening bridges between STEM education and entrepreneurship: Pathways to societal empowerment towards sustainability. In Kaya-Capocci & E. Peters-Burton (eds.), *Enhancing entrepreneurial mindsets through STEM education,* (pp. 25–47). Springer. https://doi.org/10.1007/978-3-031-17816-0_2

Riggio, R. E., & Tan, S. J. (Eds.) (2013). *Leader interpersonal and influence skills: The soft skills of leadership.* Taylor & Francis.

Sahle-Work, Z., Nóvoa, A., Aoyagi, M., Appadurai, A., Awuah, P., Ben Hassen, A. B., Buarque, C., Guerra, E., Jafar, B., Kim, D.-Y., Lin, J. Y., Morozov, E., Mundy, K., Reimers, F. M., Rivera Zea, T., Thiam, S. M., Vike-Freiberga, V., & Yahya, M. (2021). *Reimagining our future together: A new social contract for education.* UNESCO. https://unesdoc.unesco.org/ark:/48223/pf0000379707

Sarkar, M., Overton, T., Thompson, C., & Rayner, G. (2016). Graduate employability: Views of recent science graduates and employers. *International Journal of Innovation in Science and Mathematics Education, 24*(3), 31–48.

Sarpong, D. (2011). Towards a methodological approach: Theorising scenario thinking as a social practice. *Foresight, 13*(2), 4–17. https://doi.org/10.1108/14636681111126210

Senova, M. (2020). Meta-skills are the key to human potential. *Journal of Behavioural Economics and Social Systems (BESS), 2*(1), 33–38. https://doi.org/10.5278/ojs.bess.v2i1.6463

Skills Development Scotland. (2018, February). *Skills 4.0: A skills model to drive Scotland's future.* https://www.skillsdevelopmentscotland.co.uk/what-we-do/scotlands-careers-services/education-team/meta-skills-toolkit/guidance-and-support

Spencer, E., & Lucas, B. (2021, November). *Meta-skills: Best practices in work-based learning: A literature review*. University of Winchester & Skills Development Scotland. https://winchester.elsevierpure.com/files/12520973/Spencer_and_Lucas_2021_Meta_skills_literature_review_SDS.pdf

Stephen, K., Miir, L., & Hall, H. (2020). Towards a definition of metaskills. In *Proceedings of ISIC, the Information Behaviour Conference, Pretoria, South Africa, 30 September - 1 October, 2020. Information Research, 25*(4), paper isic2010. https://doi.org/10.47989/irisic2010

Tammeaid, M., Virtanen, P., & Meyer, J. (2022). Leadership meta-skills in public institutions. *Springer Nature Business & Economics, 2*, Article 84. https://doi.org/10.1007/s43546-022-00262-x

Tayne, K. (2022). Buds of collectivity: Student collaborative and system-oriented action towards greater socioenvironmental sustainability. *Environmental Education Research, 28*(2), 216–240. https://doi.org/10.1080/13504622.2021.2012129

Trilling, B. & Fadel, C. (2009). *Twenty-first century skills: Learning for life in our times*. Partnership for 21st Century Skills. Jossey-Bass, A Wiley Imprint.

Unstoppable. (2019, August 31). *These are the social entrepreneurs changing the world one village at a time*. YouTube. https://www.youtube.com/watch?v=hmWyO1Nvh0Q

van Mele, P. (2008). The importance of ecological and socio-technological literacy in R&D priority setting: The case of a fruit innovation system in Guinea, West Africa. *International Journal of Agricultural Sustainability, 6*(3), 183–194. https://doi.org/10.3763/ijas.2008.0369

Vaughn, M. (2020). What is student agency and why is it needed now more than ever? *Theory into Practice, 59*(2), 109–118. https://doi.org/10.1080/00405841.2019.1702393

Veitch, E. A., Kaland, T., & Alsos, O. A. (2021). Design for resilient human-system interaction in autonomy: The case of a shore control centre for unmanned ships. *Proceedings of the Design Society, 1*, 1023–1032. https://doi.org/10.1017/pds.2021.102

Wahid, R., Halim, S., & Halim, T. (2024). Positive communication skills in the workplace. *Migration Letters, 21*(S8), 59–70. https://migrationletters.com/index.php/ml/article/view/9229

Waldman, J. D. (2007). Thinking systems need systems thinking. *Systems Research and Behavioral Science, 24*, 271–284. https://doi.org/10.1002/sres.828

Westfall, R. D. (1999). Meta-skills in information systems education. *Journal of Computer Information Systems, 40*(2), 69–74. https://doi.org/10.1080/08874417.1999.11647441

Winowiecki, L., Smukler, S., Shirley, K., Remans, R., Peltier, G., Lothes, E., King, E., Comita, L., Baptista, S., & Alkema, L. (2011). Tools for enhancing interdisciplinary communication. *Sustainability: Science, Practice and Policy, 7*(1), 74–80. https://doi.org/10.1080/15487733.2011.11908067

World Commission on Environment and Development (WCED). (1987). *Report of the world commission on environment and development: Our common future*. United Nations. https://sustainabledevelopment.un.org/content/documents/5987our-common-future.pdf

World Economic Forum (WEF). (2023). *The global risks report 2023 18th edition: Insight report*. https://www.weforum.org/reports/global-risks-report-2023/

World Economic Forum (WEF). (2025). *The future of jobs report 2025: Insight report*. https://www.weforum.org/publications/the-future-of-jobs-report-2025/

Yadollahi, A., & Yazdani, S. (2020). Defining cognitive, higher order thinking, and psychomotor meta-skills: Three-curricular content concept analysis. *Journal of Medical Education for Future Demands, 19*(3), Article e110334. https://doi.org/10.5812/jme.110334

CHAPTER 7

STEA^+M Education and Hidden Curriculum Contexts

7.1 Introduction

Despite calls from scholars and educators for an integrated STEM education since the 2000s, or even earlier for integrated curricula, this educational approach remains poorly implemented and has not gained traction in mainstream schools. Current STEM education models are limited by their focus on subject content instead of fostering interconnected, applied learning experiences and outcomes (see Chapter 3). The notion that reforming the curriculum can transform traditional schooling is inadequate (Giroux & Penna, 2025). So, too, are the rhetorics relating to improving teacher preparation, the "back to basics" movement, and attending to the cognitive needs and capabilities of students to overcome the failures of the reform movement. This is because, according to Giroux and Penna, the recommendations are based on structural–functional educational models of curriculum theory, and fail to acknowledge the complex, intimate relationships between the institution of the school and the broader socio-economic cultural dimensions of the society—that is, the hidden curriculum.

Education is never neutral; beyond explicit learning objectives, curricula convey implicit messages that shape students' experiences, identities, and ways of knowing. This hidden curriculum, defined as "the unstated norms, values and beliefs that are transmitted through the underlying structure of meaning in both formal contexts as well as the social

G. Kidman et al., *Knowledge Generation in STEM and STEAM Education*, https://doi.org/10.1007/978-3-032-03247-8_7

relations of school and classroom life" (Giroux & Penna, 2025, p. 23), plays a crucial role in shaping who is recognised as a legitimate learner, what knowledge is valued, and how social structures are reproduced.

In STEM (Science, Technology, Engineering, and Mathematics) education, the hidden curriculum often reinforces rigid disciplinary boundaries and hierarchical knowledge structures, and while it benefits certain students who are inculcated with certain values and ways of thinking, it also marginalises certain groups. The hidden curriculum inherent in STEM education can significantly impact underrepresented students' experiences, identity, and aspirations within individual fields of science (Hansson, 2018), technology (Darvin, 2019), engineering (Sellers & Villanueva Alarcón, 2023; Shiekh et al., 2023), and mathematics (Ernest, 2024). Integrated STEM education is emerging as a promising approach to enhance student engagement and learning outcomes in science, technology, engineering, and mathematics (Thibaut et al., 2018). However, implementing integrated STEM curricula presents challenges, including the need for effective and equitable instructional practices and teacher preparation (Thibaut et al., 2018), and decolonisation of the curriculum (Nqabeni & Buka, 2023). Traditionally positioned as an extension of STEM, STEAM education, with its integration of Arts, is commonly associated with the narrative of fostering creativity and aesthetics into (what is seen as lacking in) STEM education. In analysing the Victorian Curriculum (in Australia) learning areas for curriculum potential for design thinking, we found that the Humanities curriculum (civics and citizenship, economics and business, geography, and history) has a high potential for promoting *A*gency in STEM education (Kidman et al., 2024). We propose a plexus model of STE*A*$^{+}$M (Science, Technology, Engineering, *A*gency and Mathematics) education for fostering interdisciplinary learning, creativity, and problem-solving in twenty-first century education (see Chapter 3). We argue that this interdisciplinary approach better prepares students to tackle complex, real-world challenges by blurring disciplinary boundaries and promoting a more holistic understanding of knowledge, and the mindset of being resilient in taking action toward a positive impact in the community and society around us.

This chapter explores the contextual nature of the STEM and STEAM curricula, not only as a formal educational framework but also as a site where implicit norms shape students' identities, agency, and access to opportunities. By examining both the interconnected ways of knowing—cultural self-knowing, relational knowing, critical knowing, visionary and ethical knowing, and knowing in action —and the hidden curriculum, this chapter aims to illuminate how STEM and STEAM education can either reinforce traditional hierarchies or serve as a transformative space for equitable learning.

To structure this discussion, the chapter will first consider the evolution from STEM to STEAM education and review some of the challenges and tensions relating to their implementation in the current literature. It will then explore the hidden curriculum from the perspectives of curriculum studies (four curriculum ideologies from Schiro, 2013) and critical pedagogies (three contextual lenses, Giroux & Penna, 2025), highlighting its role in shaping students' experiences, from the ways in which certain disciplines are prioritised over others to implicit biases in assessment. Ultimately, this chapter argues that while STE*A*+M holds the potential to transform learning, its success depends on intentional curriculum design, critical reflection, and educator awareness. By recognising and addressing the hidden curriculum, STE*A*+M education and STE*A*+M integrative thinking can become more than an interdisciplinary framework—it can be a vehicle for equity, innovation, and meaningful social change.

7.2 STEM to STEAM to STE*A*+M

The evolution from STEM to STEAM education represents a significant shift in contemporary education, integrating the Arts with science, technology, engineering, and mathematics (Mei et al., 2023). The narrative of this approach is that it aims to foster creativity, holistic development, and innovative problem-solving skills (Mehta et al., 2019). While STEM education has historically prioritised disciplinary expertise and workforce preparedness (Freeman et al., 2014), STEAM education emphasises the importance of creativity and aesthetics in learning, making complex concepts more tangible and engaging for students (Ghanbari, 2015; Mehta et al., 2019). Although STE*A*+M has become increasingly popular in certain countries, its implementation encounters a variety of challenges (Catterall, 2017; Mehta et al., 2019; Mei et al., 2023; Thibaut et al., 2018):

1. **Curriculum andInstructional Challenges**
   - o Restructuring the curriculum to integrate Arts into STEM requires significant effort, as it demands a nuanced understanding of both domains to ensure seamless integration.
   - o Time constraints make it difficult for teachers to adopt inquiry-based or design-based learning methods, as they must also cover the traditional curriculum.
   - o There is a lack of consensus on how integrated STEAM learning and teaching should be implemented, leading to inconsistent practices.
2. **Teacher Preparedness and Training**
   - o Many teachers feel underprepared to teach integrated STEAM due to insufficient knowledge of both STEM content and the Arts, as well as a lack of understanding of how to effectively teach interdisciplinary concepts.
   - o Limited access to professional development opportunities, teaching resources, and training further exacerbates this challenge.
3. **Resource Allocation and Infrastructure**
   - o Implementing integrated STEM and STEAM education requires significant investment in resources such as construction tools, electronic materials, art supplies, and materials used in design, educator training, and infrastructural modifications.
   - o Collaboration between teachers of different disciplines demands additional human, financial, and material resources, which may not always be available.
   - o Creating a school culture and environment that supports an integrated STEM approach is costly and time-consuming.
4. **Conceptual andPhilosophical Challenges**
   - o The vague conceptualisation of STEAM makes it difficult for educators to implement it effectively, often leading to uncertainty or reluctance.

- o There are concerns about maintaining the focus and rigour of STEM education while incorporating arts, with fears that art may be added superficially or that the core purpose of STEM could be diluted.
- o Competing aims between artistic and scientific creativity, as well as broader societal and environmental goals, create tension in aligning STEAM's objectives.

Critics argue that STEAM may be seen as piggybacking on well-funded STEM initiatives (Ghanbari, 2015). However, proponents advocate for policy changes to recognise arts as core subjects, address equity issues, and promote interdisciplinary learning (Catterall, 2017). The concept of STEAM (and also STEM) remains complex and contested, with various interpretations and applications across different contexts (see Chapter 3 of this book).

The above tensions focused on the structural–functional aspects of STEM and STEAM education. Structural-functionalism perspectives originate from the field of sociology, which views.

> society as a whole that is composed of interrelated parts and analyses social phenomena in relation to its functions that maintain order and harmony and gradually evolves from simple to more complex forms of equilibrium. (Marques, 2021, p. 23)

From such a view, the main role of education is to socialise individuals to their future roles or shared common norms to ensure social cohesion in society (Marques, 2021). The idea that STEM education is to prepare students for the future workplace, and the focus on the interrelated parts of the education system such as teacher education, school structure, and the curriculum, and the need to have shared consensus in order to reach a desired equilibrium of an ideal education, are based on the structural–functional lens of education. However, this is inadequate, as we can see that inequities within STEM education still perpetuate. In Chapter 3, we argue for a *Plexus Model of STEA⁺M Education* that focuses on developing *A*gency in students and teachers to become *A*gents of change. While we do not claim that the model is a panacea to resolve the challenges relating to STEM and STEAM education, we would like to see it (and the ideas in this book) as a way forward in transforming interdisciplinary STEM education, and STEA^{+}M education as a pedagogical framework for engaging with societal issues. In this chapter, we hope to raise awareness of the hidden curriculum contexts surrounding STEM and STEAM (shortened as STEM/STEAM) education and suggest some ways for learners and educators to examine the existing norms, values, and structures. This is essential since the design and implementation of STEM/STEAM curricula are deeply contextual, shaped by cultural, social, and institutional factors that influence what is taught, how it is taught, and who benefits from it. For example, while STEM might teach students how to build a robot, STE*A*⁺M encourages them to consider the robot's design, ethical implications, and potential impact on society. In one of our professional development programs, we modelled a makerspace learning experience with the teachers by asking them to design a bionic hand that can pick up a ping-pong ball. The context is to develop ideas for prosthetic limbs for children, and we shared background information on community-school partnerships in recycling plastic milk bottle caps into filaments to make 3D-printed prosthetic limbs for others (e.g., Thomas, 2019). The challenge of picking up a spherical ping-pong ball is to make sure that the bionic hand is able to perform a delicate manoeuvre, with an appreciation of how our hands move and flex. There were many ideas and discussions between educators from different backgrounds (mainly science and mathematics). STEM knowledges were shared and applied, along with creativity and hands-on skills (see Fig. 7.1a for some of the design models created). One design employed 'claws' using pipe cleaners that are hooked at the end of the 'fingertips' to help with grip. Other designs used strings inside straw sections to simulate the tendons pulling on different joints of the finger. One ethnic Chinese Malaysian teacher's design combined the idea of fingers with a pair of chopsticks, influenced by her cultural heritage (see Fig. 7.1b). The teachers tested and refined their designs, with Aha! moments of insight and inspiration. They were elated when their bionic hands managed to pick up a ping-pong ball, then shared and refined their designs (see Chapter 1 for neurophysiological principles of integrative learning).

7.3 Hidden Curriculum Contexts in STEM and STEAM

There are different definitions of curriculum and ways to conceptualise curriculum. The three levels of the curriculum generally encapsulate the function of the curriculum in the teaching and learning context: the planned or intended curriculum, the enacted or implemented curriculum,

(a)

(b)

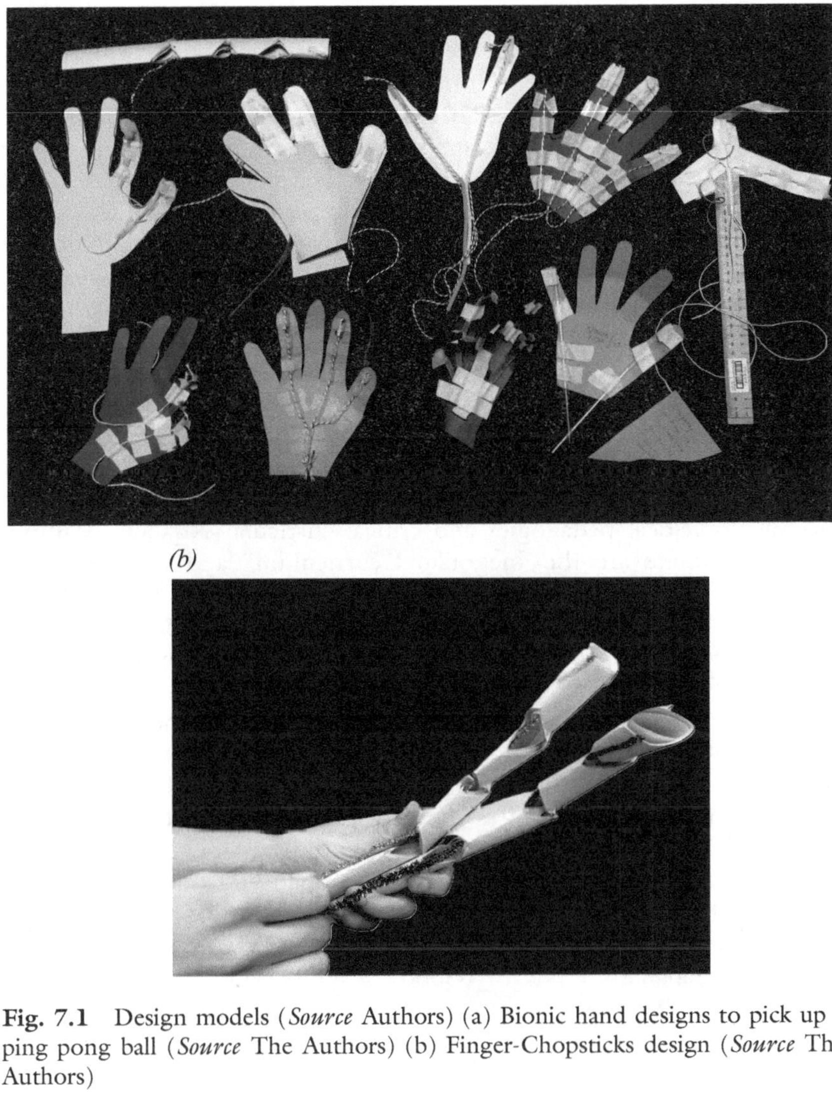

Fig. 7.1 Design models (*Source* Authors) (a) Bionic hand designs to pick up a ping pong ball (*Source* The Authors) (b) Finger-Chopsticks design (*Source* The Authors)

and the learned or assessed curriculum (see e.g., Ziebell & Clarke, 2018). The fourth level, the hidden curriculum, also known as the "implicit curriculum" or "unwritten curriculum", refers to "the unspoken or implicit values, behaviors, procedures, and norms that exist in the educational setting" (Alsubaie, 2015, p. 125). These underlying expectations, competencies, insights, and social mechanisms can either support or obstruct student success by influencing their belief systems (Alsubaie, 2015). Sometimes referred to as "hidden in plain sight", values, norms, and beliefs are usually taken for granted, tacitly transmitted, and in process at all times (Hansson, 2018). What is obvious is prone to obliviousness.

While the hidden curriculum is explored more extensively in the science, mathematics and technology education within their own disciplines (e.g., issues on gender, textbooks, socio-economic status, philosophy, values and attitudes; see Darvin, 2019; Ernest, 2024; Hansson, 2018), it is emergent in integrated STEM and STEAM education. We recognise that the effects of hidden curriculum are not always negative (Alsubaie, 2015) and that the examination of hidden curriculum paves the way to critical pedagogies and critical curricula (Nqabeni & Buka, 2023). Examples are the "negotiated curriculum" and the "emergent curriculum". The negotiated curriculum is created when the curriculum is negotiated between the educators and learners, parents, and the community (Gobby, 2017), focusing on the needs of the learner and building a meaningful and relevant curriculum for the learners. It takes advantage of learners' out-of-school experience, knowledge, and expertise to support school-based learning (Gobby, 2017). Another example is the emergent curriculum, which is used mostly in early childhood settings. The curriculum is developed as the educator builds upon the emergent experiences of the learner (e.g., admiring the rainbow formed after a rain) for learning. This learning process is open and ongoing, without a fully fixed set of learning goals in advance (Gobby, 2017). In the autobiographical memoir titled *Totto-Chan: The Little Girl at the Window* written by Japanese UNICEF Goodwill Ambassador Tetsuko Kuroyanagi (1981/1982), Kuroyanagi recounts her unconventional education experience at an experimental and progressive elementary school after being expelled from a public school. The school location is six old train cars, there is no strict timetable for learning, and children are free to choose the order of subjects learnt for the day. The learning experiences occur in real-world contexts, emphasising outdoor learning and nature, and going beyond textbooks. Fun, creative, adventurous, and engaging experiences were described, such as bringing something from the hills and something from the sea for lunch that the children had to make. Each child's strengths and interests were valued and included, especially those of students with disabilities or non-native Japanese speakers. Socio-emotional learning and empathy were encouraged and developed through storytelling and pastoral nurturing. It is an example of a negotiated and emergent curriculum that is inspiring and humanistic.

With the various curriculum conceptualisations in mind, it is important to note that these conceptualisations are often aligned with particular sets of belief systems about the purpose of schooling and the methods to achieve the objective. These curriculum ideologies, or curriculum philosophies, inform the nature of the hidden curriculum as they are often implicit and assumed. Schiro (2013) identifies four distinct curriculum ideologies that shape how educators approach teaching and learning. These are:

1. *Scholar Academic Ideology*, which views the purpose of schooling as to "help children learn the accumulated knowledge of our culture: that of the academic disciplines" (Schiro, 2013, p. 4). The curriculum is designed to help students master the core disciplines, such as science, technology, engineering, and mathematics, by exposing them to structured, rigorous content. There is a hierarchy of academic scholarship, with the scholars at the top inquiring into the knowledge, the teachers as disseminators to transmit the knowledge to the students, who are at the bottom rung learning established knowledge and being inducted into the discipline.
2. *Social Efficiency Ideology*, which views the purpose of schooling as to "efficiently meet the needs of society by training youth to function as future mature contributing members of society" (Schiro, 2013, p. 5). The curriculum is structured to produce competent workers who can efficiently contribute to society's economic and industrial needs. This ideology is highly reflected in STEM education since science, technology, engineering, and mathematics knowledge and skills are valued as important and critical to the economic, scientific, and technological advancements of today's societies.
3. *Learner-Centered Ideology*, which views the purpose of schooling as to meet the needs and concerns of individuals and to support learners in becoming "agents who must actualize their own capabilities" (Schiro, 2013, p. 5). Learner's interests, experiences, and

developmental needs are prioritised, and learning is seen as an organic process that should be guided by curiosity, self-exploration, and intrinsic motivation rather than rigid curricula. Using makerspace pedagogies (e.g., Gesthuizen et al., 2024) and providing differentiation and metacognitive scaffolding strategies for second language learners in mathematics (Cho et al., 2021) are examples of learner-centred STEM curricula.

4. *SocialReconstruction Ideology*, which views the purpose of education as emancipatory, to "facilitate the construction of a new and more just society that offers maximum satisfaction to all of its members" (Schiro, 2013, p. 6). The goal of this ideology is to empower students to challenge injustices and innovate for a better future. An activity we collaborated with teachers during COVID-19 is an integrated STEM activity where learners design a tool to pick up discarded face masks (Mangao et al., 2022). Here, the problem is the discarded face masks that were gathered on the ground by the roadside. These are a hazard for the local public workers to clean up, and they need a tool to avoid contamination and spread of the coronavirus. Students sketched different designs and shared their designs online. They also attempted to build the tools using everyday materials found at home (Mangao et al., 2022).

The four curriculum ideologies highlighted above characterise the motivations driving STEM/STEAM education and its research. They offer insights into the hidden curriculum by contextualising STEM and STEAM education and curricula. It is an approach to make explicit our underlying implicit values and beliefs when we unpack arguments and perspectives in STEM and STEAM education. With this in mind, we examine three contextual lenses to explore the hidden curriculum.

7.4 Three Contextual Lenses of Seeing STEM and STEAM's Hidden Curriculum

Giroux and Penna (2025) presented three different traditions in educational theory that provide the contextual lens with which we can view the nature of the hidden curriculum. We adapt these three lenses to categorise issues and debates that researchers raise in relation to the contexts in the hidden curriculum of STEM/STEAM education, as seen in Table 7.1. As mentioned earlier, the structural–functional lens of the hidden curriculum sees STEM education as preparing learners for their future roles to ensure social cohesion in society. The social phenomenological lens sees STEM education as being shaped by and, in turn, shaping student experiences. The radical critical lens sees STEM education as a power structure and questions the social justice and equity issues within the curriculum contexts.

There are significant intersectionalities between Schiro's (2013) curriculum ideologies and Giroux and Penna's (2025) lenses, as both theories examine how curriculum (explicit and hidden) transmits values, knowledge, and power. These intersectionalities reveal how STEM/STEAM education implicitly supports or challenges existing structures, through the following prompts:

1. How does STEM/STEAM's hidden curriculum operate within different ideological approaches?
2. How do different ideological approaches reinforce existing contexts or raise tensions about dominant assumptions in STEM/STEAM education?

Note that there are parts where only "STEM" and not STEM/STEAM are highlighted, which is also indicative of the dominant discourse surrounding STEM rather than STEAM. Key intersectionalities are presented in Table 7.1.

It is by no means that all discussions of the contexts and tensions raised in Table 7.1 will be situated within the particular stance and lens; the same issue can be viewed with another lens, raising different questions and discussion points. Multiple lenses and ideologies can be employed in a pragmatic and pluralistic approach to investigate an issue in various dimensions. The contexts and tensions illustrate the complexities and are a springboard of ideas for further discussion and investigations into the hidden curriculum contexts of STEM/STEAM education, with the aim of transforming STEM/STEAM education toward the holistic development of the learner as agents of change and benefiting their community and the world at large.

Table 7.1 Key intersectionalities between three contextual lenses to investigate STEM/STEAM hidden curriculum and Shiro's Four curriculum ideologies, with examples

Lens	*Ideology*	*Examples of Hidden Curriculum Contexts and Tensions in STEM/STEAM Education*
Structural–functional lens (STEM as maintaining social order)	Scholar Academic (STEM as elite disciplinary knowledge)	• *Contexts:* STEM is framed as an objective, hierarchical field (or individual fields) and meritocratic. STEM education reinforces the idea that knowledge is neutral, objective, static, rigorous, logical, and analytic (e.g., Bishop et al., 2006). The inclusion and roles of the different disciplines in STEM/STEAM are presented in ways that are assumed and taken for granted • *Tensions*: Tensions about how disciplines are integrated and function within the STEM/STEAM frameworks, emphasising the roles of disciplines in maintaining the structure and goals of STEM/STEAM education. Scholars seek to analyse the contributions and interactions of each discipline using scientific methods (from this lens), with equity and inclusion in mind

Lens	*Ideology*	*Examples of Hidden Curriculum Contexts and Tensions in STEM/STEAM Education*
	Social Efficiency (STEM as workforce preparation	• *Contexts*: STEM curricula implicitly teach students to value co-opetition (combination of cooperation and competition, see Fountis, 2023), problem-solving, innovation, and technical expertise, which align with the needs of a capitalist economy. There is a body of literature investigating STEM skills needed for the workforce, from the perspective of the industry, tertiary institutions, and students (see Chapter 5) • *Tensions*: STEM curricula need to prepare learners for the challenges of today's workplace, situated within poly-crisis (see Chapter 5, World Economic Forum, 2023), such as climate change, geopolitical tensions, and advancements in AI technology. Agency, social responsibility, values, ethical reasoning, and cross-disciplinary competencies are needed (see Chapters 3 and 6), but how they might be integrated into STEM/STEAM curricula needs to be enabled by overcoming structural challenges (highlighted in the earlier section on challenges of STEAM education)

(continued)

Table 7.1 (continued)

Lens	*Ideology*	*Examples of Hidden Curriculum Contexts and Tensions in STEM/STEAM Education*
	Learner-Centered (STEM as personal development)	• *Contexts*: STEM curricula are made accessible through inquiry-based learning, to allow students to explore and construct their own understanding. STEAM curricula, with the inclusion of the arts, foster creativity, and encourage girls and other underrepresented learners to engage in STEM • *Tensions:* Structural barriers remain (see earlier section on challenges of STEAM education); students with less prior STEM exposure still struggle; and the emphasis on high-stakes assessments and international rankings like PISA and TIMSS can deepen inequities, operating within accountability regimes (e.g., Tomei et al., 2014). More analyses of student data on who is left behind in STEM education do not necessarily translate into beneficial policy changes and systemic reform; or if there are reforms, their policies, implementation, and outcomes may have unintended effects

Lens	*Ideology*	*Examples of Hidden Curriculum Contexts and Tensions in STEM/STEAM Education*
	Social Reconstruction(STEM as social justice and transformation)	• *Contexts:* STEM is framed as a tool for solving real-world problems while still working within existing systems (e.g., sustainable energy within a capitalist market). Also, see the contexts and tensions in the examples above for assumptions about how hidden curriculum issues might be explored • *Tensions:* Can STEM truly be transformative if the systems producing inequities remain unchanged? Do S, T, E, M and STEM not have creativity inherently within the disciplines? Must engineering education in schools be just about design thinking? Should STEM education be for developing scholars, or for everyday people as STEM literacy? Who decides?

(continued)

Table 7.1 (continued)

Lens	*Ideology*	*Examples of Hidden Curriculum Contexts and Tensions in STEM/STEAM Education*
Social phenomenological lens(STEM as shaped by individual's interpretations of experiences and interactions)	Scholar Academic (STEM as elite disciplinary knowledge)	• *Contexts*: Each STEM discipline has unique epistemological approaches and traditions (see Chapters 2 and 4). There are common epistemological elements like knowledge construction and evidence-based reasoning, as well as distinct approaches to validation and problem-solving (e.g., Berland & Busch, 2012; Lederman & Lederman, 2020) • *Tensions*: Cross-disciplinary and integrated STEM education can be a challenge when negotiating the various epistemological approaches of STEM/STEAM disciplines. Individuals and groups construct meaning and negotiate differences in epistemological approaches since knowledge and learning are subjective and context-dependent from this lens. The emphasis on rigorous, theory-driven approaches can create hierarchies within disciplines (e.g., more girls taking biology or psychology in STEM compared to "hard" sciences like physics or mathematics, Wong et al., 2023) or marginalise approaches that are more applied or qualitative (e.g., environmental science or science communication)

Lens	*Ideology*	*Examples of Hidden Curriculum Contexts and Tensions in STEM/STEAM Education*
	Social Efficiency (STEM as workforce preparation	• *Contexts*: STEM education may prioritise practical, industry-aligned knowledge (e.g., coding in computer science, financial and actuarial mathematics), which may devalue theoretical or exploratory research in fields like pure mathematics or astrophysics. STEM/STEAM fields continue to exclude certain groups of society, resulting in the "leaky pipeline" and "glass ceiling" in education and the workplace (e.g., Figueiredo, 2023; Grogan, 2019) • *Tensions*: STEM education is supposed to align with current industry trends, but these trends are constantly changing (e.g., the rise of AI and data science). This leads to mismatches between what the industry, tertiary institutions, and learners view as STEM skills needed by graduates for tomorrow's workplace (Sarkar et al., 2016). Defining what counts as STEM fields is problematic and confusing, for example, agriculture is considered STEM in some contexts and not in others (Panizzon et al., 2015). The challenge of plugging the "leaky pipeline" in STEM remains problematic, systemic, and in high need

(continued)

Table 7.1 (continued)

Lens	*Ideology*	*Examples of Hidden Curriculum Contexts and Tensions in STEM/STEAM Education*
	Learner-Centered (STEM as personal development)	• *Contexts*: The hidden curriculum shapes students' self-concept and motivation. In STEM, this might involve implicit messages about who is "smart enough" or "good at mathematics/science", which can influence students' sense of belonging and identity. Students from marginalised groups may interpret tacit messages in ways that reinforce feelings of exclusion or imposter syndrome (Kidman et al., 2013). For example, a need to (re)look at providing diverse role models, current streaming practices, and inclusiveness of teaching practices and teaching materials and their influence on implicit messages about who belongs in STEM (e.g., Wilkie & Tan, 2019) • *Tensions*: Students may experience tension between their personal values (e.g., social justice, personal and connected learning) and the values implicitly promoted in STEM education (e.g., innovation for profit, scientific objectivity and neutrality). A learner-centred perspective recognises that students' conceptions of STEM are shaped by their personal experiences and cultural backgrounds, but curricula often fail to reflect this diversity. Students may feel pressured to conform to the dominant idealised norms of their field, limiting their ability to explore alternative approaches (Wong et al., 2023)

Lens	*Ideology*	*Examples of Hidden Curriculum Contexts and Tensions in STEM/STEAM Education*
	Social Reconstruction(STEM as social justice and transformation)	• *Contexts*: Recognising that the conceptions of STEAM are situated and strategic, and empowering STEAM learners and educators to be agents of positive change (Colucci-Gray et al., 2019). This social phenomenological lens emphasises the situated and strategic nature of STEM/STEAM conceptions, focusing on how individuals and groups construct and enact these conceptions (e.g., Gesthuizen et al., 2020). It highlights the role of agency and context in shaping STEM/STEAM education • *Tensions*: The hidden curriculum perpetuates colonial or Eurocentric views of STEM, marginalising non-Western perspectives and innovations (Kidman et al., 2013). Students struggle to reconcile their personal or cultural understandings of STEM with the dominant narratives presented in education. Reverse stereotyping occurs in the sense that although some STEM fields have increased female participation than others, this influences the labelling of those disciplines as soft (versus hard) sciences, and leads to the devaluing of these fields as less demanding, rigorous and less worthy of funding (Light et al., 2022). There are also tensions in questioning the origin and propagation of perceptions. For example, girls' interest in biology and boys' interest in physics exist in elementary schooling (Toma, 2024) and out-of-school learning contexts (Baram-Tsabari & Yarden, 2008)

(continued)

Table 7.1 (continued)

Lens	*Ideology*	*Examples of Hidden Curriculum Contexts and Tensions in STEM/STEAM Education*
Radical critical lens (STEM as reinforcing or challenging power structures)	Scholar Academic (STEM as elite disciplinary knowledge)	• *Contexts*: The dominant STEM canon is exposed as historically exclusive, privileging Western, male, and positivist perspectives. The hidden curriculum in STEM often privileges certain ways of knowing (e.g., quantitative, reductionist) over others (e.g., Indigenous knowledge systems, qualitative approaches) and reinforces epistemic injustices (e.g., Hansson, 2018) • *Tensions*: The hidden curriculum highlights the need for equitable representation of disciplines in STEAM, critiques the potential marginalisation of certain disciplines (e.g., the Arts) within STEM/STEAM and advocates for equity. Scholars with this lens challenge power dynamics and seek to disrupt hierarchies among disciplines. The selection, organisation, and teaching and learning of curricular content, with respect to knowledge, skills, attitudes and values, are problematised (Baldinger et al., 2020). For example, how textbooks in various countries portrayed Pythagoras' Theorem, arguably one of the fundamental and most well-known theorems in mathematics, needs to be examined and contested. It is important for teachers and students to recognise that nearly all ancient civilisations, including but not limited to Egyptian, Babylonian, Chinese, Indian, and Greek cultures, understood and applied the concept underlying Pythagoras' Theorem (Fan et al., 2018), rather than being accepted unquestioningly as 'invented' and attributed to a Western (white, male) person (as with other named laws and theorems students learn in science and mathematics textbooks). Also, see tensions when using the Structural-Function lens with the Social Reconstruction ideology above

Lens	*Ideology*	*Examples of Hidden Curriculum Contexts and Tensions in STEM/STEAM Education*
	Social Efficiency (STEM as workforce preparation	• *Contexts*: The idea of STEM/STEAM as a "neutral" skill set for economic development is problematised—who benefits from STEM/STEAM education, and who is left out? Whose knowledge, skills, and values are prioritised? Aligning with a critical/radical stance, scholars seek to uncover and challenge systemic biases in curriculum design • *Tensions*: The hidden curriculum may downplay the ethical and social implications of STEM work, focusing instead on technical mastery. This can lead to a lack of critical engagement with issues like climate change, data privacy, or algorithmic bias in STEM education. There is a concern that neoliberalism creates a marketisation of STEM education (Mehta et al., 2020), with the value of STEM education being measured by its contribution to economic growth, innovation, and global competitiveness (Andrée & Hansson, 2020). The dominant discourse of STEM education risks becoming the preparation of students for the workforce, at the expense of broader educational aims like holistic development or civic engagement

(continued)

Table 7.1 (continued)

Lens	*Ideology*	*Examples of Hidden Curriculum Contexts and Tensions in STEM/STEAM Education*
	Learner-Centered (STEM as personal development)	• *Contexts*: STEM/STEAM education is perceived as a leverage to transform learning. Applying radical critical lens to the hidden curriculum from the perspective of the learner includes examining the systemic structures that perpetuate exclusion and inequality, and analysing how STEM/STEAM curricula prioritise Western scientific knowledge over Indigenous or community-based knowledge, reinforcing colonial hierarchies. The implications are to decolonise STEM/STEAM curricula and empower learners • *Tensions*: There is a risk toward framing educational choices for STEM studies and fields as a matter of individual factors such as interest, identity and values; few studies examine the impact of STEM initiatives/programs (usually funded by the industry and/or governments) and the implicit hidden curriculum within them (Andrée & Hansson, 2020). Other tensions include investigating how high-stakes assessment practices in STEM/STEAM (e.g., standardised tests, international rankings of student achievements) send implicit messages to learners about what is valued (Erickson, 2022), and influences from the community, industry, society, and learners who benefit from the status quo. Yet another is the investigation into the disruptive forces of social media and AI technologies (and their intersectionalities) on schools, for example, the discourse on the rise of manosphere and misogyny impacting learners and teachers in schools (Wescott & Roberts, 2025), which might influence gendered perceptions and identity of learners, educators, researchers relating to STEM/STEAM (e.g., Riley, 2018)

Lens	*Ideology*	*Examples of Hidden Curriculum Contexts and Tensions in STEM/STEAM Education*
	Social Reconstruction(STEM as social justice and transformation)	• *Contexts:* The hidden curriculum contexts relating to social reconstruction are highlighted in the contexts for different ideologies in earlier examples. These are the headways made by critical scholars investigating the hidden curriculum in STEM fields. There are emergent diverse perspectives on the different aspects of the social structure (within which education sits) being investigated • *Tensions*: Problematising the discourse of the 'STEM crisis' (Andrée & Hansson, 2020) and STEM education (strategic focuses, practices, ideologies) with respect to cultural reproduction, equity, and inclusion. Scholars with this lens critique the broader ideologies and practices of STEM/STEAM education, particularly their role in cultural reproduction and inequities. One example is the commodification of STEM education by industry—STEM educational toys (usually robots), STEM courses (usually coding)—to market their products. Another is the discourse of 'STEM crisis', depicted in several ways such as the decline of STEM knowledge and competencies, low interest and participation, the failure of schools and educators to "foster an interest in STEM and to educate citizens that are scientifically literate, employable and internationally competitive" (Andrée & Hansson, 2020, p. 552). To what extent is STEM/STEAM education seen and implemented as a tool for activism and transformation—addressing climate change, social inequalities, and ethical dilemmas in science and technology—remains to be explored. Can STEM education challenge the systems that fund it? How will STEM/STEAM education and curriculum contexts shift in the multipolar, multilateral, and multi-civilizational world (e.g., Mahbubani, 2022)?

7.5 Addressing the Hidden Curriculum in STEM/STEAM

The hidden curriculum plays a critical role in shaping students' experiences and engagement in STEM/STEAM education, making it essential to first become aware of its presence within the contexts surrounding our educational environments. Implicit messages regarding identity, ability, and belonging can either empower or dishearten students, particularly those from marginalised backgrounds (Ireland et al., 2018). For instance, if students consistently see certain groups underrepresented among STEM/STEAM role models or in leadership positions, they may internalise the belief that these fields are not for them (Bock, 2022; Ireland et al., 2018). Similarly, assessment practices that prioritise standardised testing over creative problem-solving and growth mindsets can disadvantage students without the "right kind of capital (broadly construed)" (Erickson, 2022, p. 398). STEM teachers' curriculum ideologies influence what they emphasise in their teaching, learning, and assessment practices (Tan & Wei, 2024). By recognising and understanding these hidden influences, we can take meaningful action to address issues to foster more inclusive and empowering STEM/STEAM programs for all students. This is why we highlight the A^+ as *A*gents of change in the Plexus Model of STEA^+M Education (see Chapter 3). We hope to also empower the learners and educators in taking steps to recognise the hidden curriculum contexts and tensions in STEM and STEAM education. Next, we discuss considerations when designing inclusive and transformative STEM/STEAM curricula, taking into account the tensions discussed in the hidden curriculum.

7.6 Designing Inclusive and Transformative STEA^+M Curriculum

There are several aspects to designing an inclusive and transformative STEA^+M curriculum, with consideration of the hidden curriculum contexts and tensions. In this section, we highlight the following: (1) consideration of the inclusion, roles and representations of the integrated STEM/STEAM curriculum, (2) non-content aspect of STEM/STEAM curriculum, (3) interconnected ways of knowing, and (4) assessment—hidden curriculum and alternative assessment.

7.6.1 *The Inclusion, Roles, and Representations of Disciplines in Integrated STEM/STEAM*

Researchers seek to find out and raise issues based on what is observable, focusing on the structure and function within STEM/STEAM education (see Table 7.1). Examples are the issues of the inclusion and roles of the different disciplines in STEM/STEAM, and the need for equitable representation of disciplines, raised by STEM and STEAM researchers (e.g., English, 2016; Martín-Páez et al., 2019).

The disciplines included in STEAM/STEM vary among researchers and educators (Panizzon et al., 2015). Some researchers use "STEM" to refer to a single subject, such as science, mathematics, technology, and engineering (English, 2016). This view is widespread in the society. Indeed, there is a myriad of science educational kits and educational robotic toys touting themselves as "STEM enabled". Note, however, that researchers do not use the acronym "STEAM" to refer to a single Arts or Humanities subject, signalling the subservient or supporting nature of Arts in STEAM. Other researchers use "STEM" to refer to a combination of two or more STEM subjects. The acronym "STEAM" is used when an Arts subject is incorporated, where a range of Arts and Humanities subjects might be what the *A* refers to (see Chapter 4 for a discussion). In this book, we highlight the teacher and student *A*gency in STEM as STE*A*⁺M (see Chapter 3), where the emphasis is on conceptual understanding and interdisciplinary connections.

Researchers have increasingly called for equitable representation of disciplines within STEM and STEAM education, as current practices often exhibit significant imbalances favouring certain fields. Systematic literature reviews highlight the persistent dominance of disciplines like science and engineering (English, 2016; McLure et al., 2022). These disciplines frequently play specialised roles, with engineering, for example, acting as a "hinge" that connects other STEM components. However, this uneven focus sidelines other critical areas, such as mathematics and the arts, which are essential for fostering interdisciplinary learning. Mathematics is often underrepresented or treated as secondary (Baldinger et al., 2020), while the arts in STEAM are frequently marginalised rather than integrated as a core element. This disciplinary bias not only limits the potential for holistic educational experiences but also risks perpetuating inequities in how students engage with and perceive different fields. There seemed to be an implicit message and assumption that "*there is nocreativity inSTEM education;* therefore, we need to add the Arts to promote creativity". We need to challenge this hidden assumption. There are different understandings of what constitutes creativity in the different disciplines of S, T, E, and M. For example, in mathematics, Hadar and Tirosh (2019) proposed a framework of nine characteristics of creative thinking tasks: (1) tasks requiring alternative solutions for a given solution, (2) tasks with more than one solution, (3) tasks requiring more than one pathway to be solved, (4) tasks taking mathematical knowledge outside the context of mathematics, (5) tasks requiring identification, and implementation of mathematical principles, patterns, or structures, (6) tasks requiring connections between mathematical ideas, (7) tasks using mathematical procedures to solve problems from other contexts, (8) tasks requiring the posing of mathematical problems, and (9) tasks requiring the exploration of mathematical ideas. Evidently, these characteristics are heavily situated in the mathematics education discipline and do not all translate easily to S, T, and E, which have their own epistemological conceptions of creativity. These probably lead to the perceived lack of creativity in STEM. Incidentally, when reviewing STEM and STEAM papers from 2010–2020, Aguilera and Ortiz-Revilla (2021) found that evidence from empirical studies does not seem to support the argument that STE*A*⁺M education is more effective than STEM in developing student creativity. Both approaches have positive effects. They also cautioned that student creativity was more commonly measured via questionnaires and ad-hoc instruments, indicating a need for more research and clarity in the design and conceptualisation of creativity frameworks.

Compounding these challenges, integrated STEM/STEAM curricula developed by practitioners vary widely in content and approach, often reflecting disciplinary bias and an imbalanced focus (Sgro et al., 2020). There is considerable ambiguity about what constitutes different levels of integration and where the boundaries between disciplines lie. Additionally, curriculum developers' conceptions of dissolving disciplinary borders remain unclear, further hindering efforts to achieve meaningful integration (Sgro et al., 2020). To address these issues, a more comprehensive review of praxis and conceptualisation of integrated STEM/STEAM education by various stakeholders is urgently needed (Aguilera & Ortiz-Revilla, 2021; English, 2016; Sgro et al., 2020). There are various models and frameworks for how the disciplines should be integrated, such as applying a set of "big ideas" within and between disciplines

(Chalmers et al., 2017), building cohesion across representations (Nathan et al., 2013), and problem-based learning approaches (Daugherty & Carter, 2018). For example, Kelley and Knowles (2016) conceptualised STEM learning as a set of interrelated pulleys that leverage one another. However, science and engineering seemed to represent larger pulleys than technology and mathematics, which could be conjectured as conceiving science and engineering as domains of enquiry, and mathematics and technology as instrumental domains (Sgro et al., 2020). The representations of engineering (as design thinking) as "central in most definitions of integrated STEM" (Roehrig et al., 2021, p. 6) and science seemed to be unchallenged, although mathematics and technology integration are given more consideration than other frameworks, seen in their seven key characteristics of integrated STEM (Roehrig et al., 2021):

1. Focus on real-world problems
2. Engage in engineering design
3. Context integration
4. Content integration (using M and T beyond tools to service S and E)
5. Engage in authentic STEM practices
6. Foster twenty-first century skills of collaboration, critical thinking, creativity, communication, problem-solving
7. Learn about STEM careers.

As we raised in Table 7.1, the focus on the ways STEM/STEAM disciplines are represented and how meanings are construed and constructed depends on the strategic intent of the curriculum developers. It is interesting that while the design thinking process is seen as representing E in STEM, it is seen as just one of the ways of thinking in engineering. In our professional development experience with a group of civil engineering lecturers in Indonesia, we realised that the commonly applied design process (as understood in school-based integrated STEM frameworks) is representing only part of the entire engineering process. From the perspectives of the lecturers, who are themselves civil engineers by profession, the entire process consists of problem definition, understanding its specification, brainstorming conceptual designs, exploring alternatives, selection of design, detailed design planning, developing methods to execute the design, planning budget and schedule to implement the design, and reporting the proposal. The latter steps of project implementation planning are certainly not explicit in the usual frameworks of Design Thinking (e.g., Kidman et al., 2024). Lande and Leifer (2010) described four ways of thinking accessed by mechanical engineering design students: *Design Thinking* to solve a problem with an idea as outcome; *Engineering Thinking* to create a solution with an artefact or product as outcome; *Production Thinking* to recreate the product or solution with facsimiles of product or plans for making copies as outcomes; and *Future Thinking* to reset the problem with the outcome being a question. *Systems Engineering* is also seen as separate from *Design Thinking*, with the former being more data-driven and analytical compared to human-centred and creative for the latter, both critical to engineering (Greene et al., 2019). Thus, the STEM/STEAM curricula designers should be aware of the dominant conception of E in STEM education as Design Thinking. They should also consider the possibilities of alternative conceptions of E when integrating STEM into their contextualised curricula.

From the radical critical view, some researchers contended that STEM is a "social construct" (Akerson et al., 2018, p. 5), "a twenty-first century invention" (Ostler, 2015, p. 17), and even "a slogan" (Bybee, 2010, p. 30). There is a struggle to reconcile and make sense of the epistemological traditions and non-content aspects of the STEM curriculum. Ostler (2015) highlights that we need to contend with various *tactical* or *strategic definitions* to serve the needs of the users of the acronyms. Colucci-Gray et al. (2019) refer to STEAM as a "boundary object" that is "a knowledge-practice construct that can be shared by different communities and networks, yet with each one holding a particular understanding of the theory or representation" (p. 4). In light of the strong focus and funding for developing STEM capabilities (to a lesser extent STEAM capabilities), it makes sense for researchers and educators in science, mathematics, technology and engineering education to stake a claim for themselves in the STEM/STEAM pie for pragmatic purposes, so to speak (refer to Akerson et al., 2018). Nonetheless, we hope the purposes are transformative and inclusive.

7.6.2 Non-content Aspects of STEM/STEAM Curriculum

Besides the structural issue of what disciplines to include and the roles of the disciplines, we problematise the epistemological approaches of individual STEM/STEAM disciplines and their integration, through the social phenomenological lens (refer to Table 7.1). In inter and trans-disciplinary STE*A*$^{+}$M education, how can we negotiate and reconcile the various disciplinary epistemological traditions? Related to the point of epistemological traditions is the point of: Problematising the selection, organisation, and teaching and learning of curricular content, with respect to knowledge, skills, attitudes, and values. What knowledge, attitudes, values, and dispositions do we see and value in S, T, E, A (or A^{+}), and M (see Chapters 5, 6, 9, & 10), and how can we incorporate them into integrated STEM/STEAM programs? How do the learners, teachers, and teacher educators perceive their own STEM identity and values? Inadvertently, we make strategic choices about the cognitive, affective, and conative aspects of the STEM/STEAM curriculum, pedagogy, and assessment, taking into account the current multipolar, technologically accelerated, and climate-endangered state of the world.

Perhaps the question is not about how STEM/STEAM educators and researchers reconcile with the different epistemological traditions, but how we can flexibly *amalgamate* the different traditions and their corresponding pedagogical approaches into an integrative STE*A*$^{+}$M curriculum. This amalgamation is purposeful and strategic, based on the values and agenda of the context and times, with an emphasis on sociocultural factors and social justice to challenge hegemonic or assumed contexts—the 'water' in which we (as fishes) swim in. Of course, with such strategic intent comes the hidden curriculum relating to power and politics. Questions such as stereotypes about who excels in STEM (and to a lesser extent STEAM) fields (and intersectionalities between gender and minority groups), whose knowledge is valued (Western science vs. Indigenous ways of knowing), how assessment and grading are designed and implemented, have been raised and discussed (see Table 7.1).

Furthermore, STEM/STEAM education can provide authentic contexts for students to develop their academic and disciplinary identity and self-efficacy, as well as empower marginalised groups (Fei et al., 2024). Understanding one's cultural identity and heritage is crucial for personal and academic growth. In the context of STEM/STEAM education, cultural self-knowing involves exploring how cultural backgrounds influence one's perspective and approach to learning. University lecturer Neni Mariana (2019) examined her own and her participants' (who are primary mathematics teachers) experiences and struggles with embedding spiritual and cultural values into mathematics as part of the Indonesian curriculum reform initiative. She utilised Taylor's (2015) five transformative ways of knowing—cultural self-knowing, relational knowing, critical knowing, visionary and ethical knowing, and knowing in action—to empower herself and her participants to critically reflect on their "frames of reference", which included sets of beliefs and values, prior experiences, and cultural understandings. Some of these ways of knowing are explored next.

7.6.3 Valuing Interconnected Ways of Knowing

To create inclusive and transformative STEA^{+}M curricula, educators must embrace interconnected ways of knowing that go beyond technical skills. These approaches recognise the diverse ways students engage with and contribute to STEM/STEAM fields.

- **Cultural Self-Knowing: HowStudents' Identities Shape TheirSTEM/STEAM Engagement**

  Students bring their cultural backgrounds, lived experiences, and personal identities into the classroom, all of which shape how they engage with STEM/STEAM (Beals, 2021). Culturally responsive teaching practices, such as incorporating students' cultural knowledge into lessons and celebrating diverse role models, can help students see themselves reflected in STEM/STEAM fields (Rocha et al., 2024).

  A STEM/STEAM curriculum can value interconnected ways of knowing by incorporating local community knowledge systems and perspectives, particularly in areas with Indigenous student populations (Mariana, 2019). Integrating these systems can create a more inclusive learning experience for those students, improving their academic achievement and sense of cultural identity (Rocha et al., 2024). Rocha et al. (2024) used a Community Cultural Wealth (CCW) model by Yosso (2005) to highlight how an academic-university-hospital-high school partnership nurtured students' cultural wealth in a STEM health/medicine program. The

CCW model frames community cultural wealth as six types of assets or capitals:

- *AspirationalCapital*: hopes, dreams and goals that motivate students to STEM careers;
- *ResistanceCapital*: self-empowerment, ability to advocate and secure students' rights in STEM educational settings;
- *SocialCapital*: diverse relationships and contacts students have to navigate STEM educational settings;
- *NavigationalCapital*: culture of learning and study skills to enable students to navigate STEM educational settings;
- *FamilialCapital*: family support and asset-based resources students have from their families;
- *LinguisticCapital*: diverse communication and language skills that students bring to diverse settings.

The program builds a culture of sustained support through: role models and professionals sharing their experiences as an under-represented group, engaging with diverse members of the STEM community, and rigorous curriculum with high-impact, evidence-based practices. Students developed their career knowledge, communication skills, confidence, research skills, critical thinking, scientific writing, self-reflection and awareness, help-seeking, time management, and study skills (Rocha et al., 2024). This suggests that attending to the different capitals in STEM/STEAM curricula can empower students to "examine various forms of oppression" (Rocha et al., 2024, p. 1036), identify themselves as agents of change, and resist negative stereotypes and deficit notions about gender, ethnicity, and race.

Overall, recent research highlights the importance of cultural awareness and inclusivity in STEM education. Studies show that embracing students' cultural assets and nurturing their cultural wealth can enhance self-efficacy, engagement, and persistence in STEM fields, particularly for students in marginalised groups (Rocha et al., 2024). STEM education, when combined with intercultural competence, can improve students' cross-cultural consciousness and self-efficacy (Fei et al., 2024). Culturally relevant pedagogies, such as incorporating hip-hop in STEM teaching, can address cultural agnosia and reveal the "science genius" of marginalised youth (Emdin et al., 2021). However, caution is needed when framing the

STEM design thinking problem as supporting other (marginalised) groups, as limited cultural representations of these groups can lead to deficit-based views (Keratithamkul et al., 2020). Keratithmakul et al. described a STEM unit where six-grade US students were tasked to design a cheap solar oven to cook the fish Ecuadorian fishermen sold at the markets. Students were not familiar with the Ecuadorian context, and the curriculum provided limited perspectives. The researchers found that students held elitist views, narrow views of another culture, and were indifferent toward the embedded cultural context in the lessons. Unfortunately, the negative perceptions were "imprinted and solidified through limited portrayal of the embedded culture in the curriculum" (Keratithmakul et al., 2020, p. 1). Hence, it is not just the identities of students belonging to marginalised groups but also the identities and perspectives of students in the dominant group that need to be attended to, in order to develop empathy. The teachers' own perceptions and cultural epistemologies are critical factors in how the curriculum is designed and implemented (e.g., Ma, 2015).

- **Relational Knowing:Learning Through Collaboration, Peer Networks, andCommunity Engagement**
  STEM/STEAM education thrives on collaboration, both within the classroom and beyond (Roberts et al., 2022; Rocha et al., 2024). This corresponds to the social capital and navigation capital in Yosso's (2005) framework above. By fostering relational knowing—learning through interactions with peers, mentors, and community members—educators can create supportive networks that enhance student engagement and success (Bang & Medin, 2010). For instance, partnering with local organisations on community-based projects can help students see the real-world impact of their work while building connections that extend beyond the classroom.
- **CriticalKnowing: Challenging Assumptions and QuestioningPower Structures inSTEM/STEAM Fields**
  Critical knowing encourages students to question the assumptions, biases, and power dynamics that shape STEM/STEAM fields. This corresponds to the Resistance Capital in Yosso's (2005) framework. It might involve analysing the historical exclusion of certain

groups from scientific research or examining how technologies can reinforce social inequalities. By developing a critical lens, students can become more informed and ethical practitioners who challengethe status quo and advocate for equity.

We need to support learners in navigating various epistemologies. Bang and Medin (2010) used the example that urban Native Indian children were able to distinguish between what a science teacher would say is "alive" (e.g., animals, plants), and what an elder would say is "alive" (rocks and water are considered alive). They encourage engaging with the learning community to explore the meanings and value systems, recognising that science and its teaching are not acultural but reflect (often Western) values, and to reimagine learning environments that "support students' navigation among multiple ways of knowing, including their community-based ways of knowing" (Bang & Medin, 2010, p. 1019).

- **Visionary and EthicalKnowing: Thinking Beyond Technical Solutions to Ethical and Societal Impacts**

  STEM/STEAM education should encourage students to consider the broader implications of their work, including its ethical, social, and environmental impacts. Visionary knowing involves imagining alternative futures and designing solutions that prioritise sustainability and equity, akin to Future Thinking in engineering (Lande & Leifer, 2010). For example, a STEM/STEAM project on renewable energy could explore not only the technical aspects of solar panels but also their accessibility to low-income communities, with the caution to avoid perpetuating deficit views (see Keratithmakul et al., 2020). Developing empathy involves genuine engagement with the stakeholders and their rich cultural contexts and not pitying them (Keratithmakul et al., 2020). This is why we recommend solving real-world issues that are grounded in the local contexts, where students can investigate and understand themselves and the communities and world around them. Ethical knowing involves understanding the moral and ethical implications of knowledge and its application in the S, T, E, and M disciplines. It requires individuals to consider how their actions and decisions affect others, and to act in ways that are just, fair, and responsible. Emdin et al.

(2021) proposed a Reality Pedagogy framework that helps educators centre the realities and cultures of young people, addressing cultural agnosia by encouraging teachers to connect with students in meaningful ways. Reality Pedagogy involves dialogues between students and teachers to co-develop instructional strategies, co-planning and co-teaching, incorporating students' contexts and cultures, and using curation to study student-generated artefacts of in-class and outside-class realities to inform teachers and teaching. Emdin et al. shared the Science Genius Battles program for urban youths with deep connections to hip-hop culture. Schools send representatives (selected by peers) to a competition, performing a science-themed rap or poem in front of a group of peers, scientists, hip-hop artists, and community members. Assessment criteria include connections to lived and emotional experiences, quality of songs, and the science content's quality, depth, rigour, and accuracy. Hip-hop provides a platform for narrative therapy for learners to sort through complex feelings, thoughts and stressors—for example, describing how one's respiratory system reacts when specific tensions are experienced—while deepening their STEM knowledge (Emdin et al., 2021).

- **Knowing inAction:Learning Through Real-World Projects, Activism, and Applied Problem-Solving**

  STEM/STEAM curricula should emphasise knowing in action—learning by doing and applying knowledge to real-world challenges (e.g., Fei et al., 2024; Tan & Kidman, 2021). This approach not only deepens understanding but also empowers students to see themselves as problem-solvers and change-makers (Roberts et al., 2022). For instance, students visit a polluted river in the community and conduct investigations about the human needs related to water and the challenges to meet these needs. An interdisciplinary STEM-IMREAL model proposed by Nugraha et al. (2024) involves six foundational components:

  - ***I*****ntegratingdisciplines:** with transparent and equitable connections through interdisciplinary approaches;
  - ***M*****ultiplerepresentations:** through various communication and representation modes and the translations across these modes;
  - ***R*****ealistic and relevant problem:** using projects or challenges that relate to real-life problems as a learning context;

- ***E*ngineering design process:** to link each STEM discipline to obtain comprehensive and innovative solutions;
- ***A*ctive collaboration:** to work collaboratively as a community of learners to develop twenty-first century skills; and
- ***L*earning isstudent-centred:** to place students at the centre of learning through hands-on investigations using grade-appropriate content, challenges, and contexts.

The use of hands-on makerspaces and virtual worlds like Minecraft (e.g., Gesthuizen et al., 2025) taps into this knowing-in-action to activate powerful learning experiences for the learner.

7.7 Hidden Curriculum in Assessment

Assessment is a critical component of education, shaping what is valued, how learning is measured, and who is deemed successful (Erickson, 2022). In STEM/STEAM education, traditional assessment methods often reinforce hidden norms and biases, while failing to capture the interdisciplinary and creative nature of learning. This section explores the relationship between assessment and the hidden curriculum, the challenges of evaluating STEM/STEAM learning, and alternative strategies that promote equity, creativity, and deeper understanding.

- **How Traditional Assessment Reinforces or Disrupts Hidden Norms**

Traditional assessment methods, such as standardised tests and rigid grading systems, often prioritise narrow definitions of success, such as memorisation and procedural accuracy (Erickson, 2022). These methods can reinforce hidden norms, such as the privileging of Western scientific knowledge over other ways of knowing or the assumption that certain groups are inherently better at STEM subjects (Bang & Medin, 2010). For example, different culture's educational systems might value different assessment approaches differently, such as Western privileging of open-ended assessment tasks and collaborative learning over multiple-choice examinations, which are seen as closed-ended and encouraging rote learning (Seno-Alday & Budde-Sung, 2022). However, in Chinese language literature, multiple-choice tests have high credibility and are seen as a format that could "objectively evaluate several learning outcomes at varying levels of difficulty across several questions on a wide range of topics" (p. 278) and allow faster feedback to support students' learning. These advantages are also why multiple-choice questions are used in large-scale assessments (nationally and internationally), despite critiques. Alternative assessments that emphasise diverse skills and perspectives can disrupt hidden norms and uncover assumptions, creating more inclusive and equitable learning environments (Emdin et al., 2021; Rocha et al., 2024).

- **The Challenge of Measuring InterdisciplinaryLearning andTwenty-First Century Skills**

Two of the most significant challenges in STEM/STEAM assessment are assessing interdisciplinary learning and twenty-first century skills such as creativity, which are inherently complex and multifaceted (Aguilera & Ortiz-Revilla, 2021; Gao et al., 2020). Most current assessments on integrated STEM/STEAM programs focus on mono-disciplinary knowledge and affective domains, often misaligning with the program aims to improve interdisciplinary understanding (Gao et al., 2020). Different integrated STEM approaches also resulted in different associated outcomes. For instance, STEM project-based learning in the science curriculum aims to enhance students' academic achievement and higher-order thinking skills, whereas out-of-school STEM project-based activities are primarily geared toward fostering students' interests in STEM careers (Le et al., 2023). To address the challenge of assessing interdisciplinarity and skills, researchers propose new frameworks and approaches, such as cross-disciplinary learning (Borda et al., 2022) and targeted assessment of interdisciplinary work (Mansilla & Duraising, 2007). Multiple types of assessments such as quantitative and qualitative approaches for formative, diagnostic, and summative instructional purposes (Borda et al., 2022), and authentic assessments (Tan & Kidman, 2021), can be used. Additionally, assessments need to capture the depth and breadth of STEAM learning, such as student agency in the learning process and product, thinking critically about ethical implications, or approaching problems with innovative solutions. Implementing effective interdisciplinary STEM curriculum and assessment requires overcoming institutional challenges, including teachers' knowledge, mindset and beliefs; lack of professional development and support; and resource limitations (Grangeat et al., 2021).

7.7.1 *Alternative Assessment Strategies*

To move beyond the limitations of traditional assessment, educators can implement alternative strategies that are authentic, metacognitive, and culturally responsive (Grangeat et al., 2021; Tan & Kidman, 2021). These approaches not only provide a more comprehensive picture of student learning but also align with the values of equity and inclusion in STEM/STEAM/STE*A*⁺M education.

- **Authentic Assessment: Project-BasedLearning, Portfolios, and Real-World Applications**
  Authentic assessment focuses on real-world tasks and applications, allowing students to demonstrate their learning in meaningful ways (Tan & Kidman, 2021). For example, project-based learning (PBL) assessments might involve designing a prototype, conducting a community-based research project, or creating a multimedia presentation. Portfolios, which compile students' work over time, provide a holistic view of their growth and achievements. These approaches not only allow assessment of technical skills but also creativity, collaboration, and problem-solving, making them well-suited to STE*A*⁺M education. However, the implication is that assessment for inquiry-based approaches is challenging to design and requires careful planning and skill on the part of the educator (Barron & Darling-Hammond, 2010; Grangeat et al., 2021). On one hand, if there is too much focus on disciplinary and traditional learning goals, then inquiry-based and traditional learning approaches appear to yield similar results (Barron & Darling-Hammond, 2010), or focused on mono-disciplinary outcomes (Gao et al., 2020). On the other hand, assessments can swing the other way, such as the example of Chemistry course summative assessments assessing mathematics and skills and facts instead of underlying Chemistry concepts of making connections between atomic/molecular behaviour to reason about phenomena (Stowe et al., 2021).

- **MetacognitiveAssessment: EncouragingStudents to Reflect on TheirLearning Processes**

  Metacognitive assessment emphasises self-reflection and awareness of one's own learning processes (Grangeat et al., 2021). By encouraging students to reflect on how they approach problems, overcome challenges, and integrate knowledge from different disciplines, educators can foster deeper understanding and lifelong learning (e.g., Shen et al., 2014). For instance, students might write reflective journals, participate in peer feedback sessions, or create learning narratives that document their STEM/STEAM journey. This approach not only assesses learning outcomes but also helps studentsdevelop critical thinking and self-regulation skills (Tan & Kidman, 2021).
- **Culturally Responsive Assessment: Validating Different Ways of Knowing and Problem-Solving**

  Culturally responsive assessment recognises and values the diverse ways students approach learning and problem-solving (Walker et al., 2023). This might involve incorporating students' cultural knowledge into assessment tasks, such as using traditional storytelling to explain scientific concepts or designing solutions that reflect community values (Bush & Cook, 2019; Greene et al., 2019). Culturally responsive assessments also challenge biases in grading and evaluation, ensuring that all students have an equal opportunity to demonstrate their abilities. Walker et al., (2023, p. 4) proposed five principles of cultural responsiveness for assessment design:

  1. share power across all parties at all stages of the assessment process;
  2. foster engagement, use of productive cognitive strategies, and feelings of belonging in the academic learning environment;
  3. reflect expectations that all students have the potential to perform at high levels through promoting higher-order thinking and self-reflection;
  4. allow for flexibility to account for individual differences in culture, interests, and identities for all learners; and
  5. measure what students know and can do by using asset-based perspectives that disrupt traditional deficit narratives.

As can be seen, these assessment strategies go hand-in-hand with the teaching and learning approaches and the transformative curriculum, valuing interconnected ways of knowing to address the hidden curriculum in traditional STEM assessment modes.

Assessment in STEM/STEAM education is not just a tool for measuring learning; it is a reflection of what is valued and who is included. By moving beyond traditional methods and embracing authentic, metacognitive, and culturally responsive strategies, educators can create assessments that align with the interdisciplinary, agentic, and creative nature of STE*A*⁺M. In doing so, they can disrupt the hidden curriculum, promote equity and inclusion, and empower all students to reach their full potential (Walker et al., 2023).

7.8 Developing a STE*A*⁺M Learner-Centred Curriculum

We need to have a STE*A*⁺M curriculum that values local wisdom: values, knowledges, attitudes, and beliefs, to foster students' agency, identity, and efficacy in STEM/STEAM learning. For deep learning to take place, we should engage in both the disciplinary knowledge and content, but yet be responsive to the local context and situate learning within it to see opportunities to develop interdisciplinary knowledge and skills. In earlier parts of this chapter, we discussed the different curriculum ideologies, which implicate "a 'spectrum of agendas' of the nature of the interlink between content and context" (Venkat, 2020, p. 214).

Table 7.2 shows a comparative framework for different approaches to STEM/STEAM curriculum design, ranging from content-focused models to learner-driven approaches, inspired by Venkat's (2020) "spectrum of agendas" that were seen in mathematics classrooms, Giroux and Penny's (2025) three lenses, and Schiro's (2013) four curriculum ideologies. It outlines key dimensions of curriculum structure, including purpose, focus, implementation, learner engagement, assessment, hidden curriculum, and integration. The table highlights how STEM/STEAM education can shift from a traditional, discipline-centred approach toward a more interdisciplinary and transdisciplinary model that is responsive to real-world and community needs.

The *Content-Focused* approach emphasises the acquisition of disciplinary knowledge, treating subjects as distinct areas of study with

Table 7.2 Shifting priorities in STEM/STEAM education approaches from content-focused to context- and learner-driven

Dimension	*Content-Focused*	*Application-Focused*	*Context- and Inquiry-Focused*	*Context- and Learner-Driven*
Curriculum Purpose	To develop deep disciplinary knowledge, skills, and conventions for further studies	To apply disciplinary knowledge in structured real-world problems	To explore real-world contexts to deepen understanding of both disciplines and their applications	To engage with real-world and community needs, fostering critical citizenship and future-oriented problem-solving
Curriculum Focus	Strongly disciplinary, focusing on structured knowledge acquisition and foundational skills	Focused on applying disciplinary knowledge to solve problems in STEM/STEAM contexts	Emphasises interdisciplinary inquiry and systems thinking to tackle complex, real-world STEM/STEAM issues	Integrates inter and transdisciplinary thinking, inquiry, and design within authentic, learner-relevant contexts
Curriculum Implementation	Teaching follows a structured, sequenced progression of discipline-based knowledge and skills, with problem-solving embedded within subjects	Teaching involves carefully designed contexts that enable disciplinary learning and application, often linking multiple subjects	Teaching balances discussion of real-world contexts with revising curricula and learning new disciplinary knowledge through inquiry-based methods	Teaching prioritises the co-construction of knowledge through critical discussions and integrative problem-solving within societal and community contexts
LearnerEngagement and Agency	Learners engage in structured exploration and problem-solving within disciplines, with varying degrees of autonomy	Learners apply knowledge to real-world problems, often in guided or structured ways	Learners engage in inquiry-based learning, connecting disciplinary and interdisciplinary perspectives	Learners co-construct knowledge, working with educators, experts, and communities to address real-world challenges
Assessment Approach	Focuses on standardised testing, mastery of disciplinary knowledge, and procedural skills	Uses structured tasks to assess the application of disciplinary knowledge in problem-solving	Incorporates formative and performance-based assessments, emphasising process, inquiry, and interdisciplinary thinking	Uses authentic, co-constructed assessments, including self- and peer-assessment, project-based evaluation, and real-world impact measures

Dimension	*Content-Focused*	*Application-Focused*	*Context- and Inquiry-Focused*	*Context- and Learner-Driven*
CurriculumIntegration	Primarily mono-disciplinary	Mono- or multi-disciplinary, depending on the problem context	Multi- or interdisciplinary, with increasing integration of perspectives	Inter or transdisciplinary, emphasising emergent knowledge, systems thinking, and co-designed learning experiences
Hidden Curriculum	**Power Dynamics:** Knowledge is positioned as external, controlled by institutions and experts **CulturalPerspectives:** Dominant disciplinary traditions are privileged; local and Indigenous knowledge may be marginalised **Ethical Implications:** Ethics is rarely explicit; focus is on correctness, efficiency, and precision	**Power Dynamics:** Problem-solving is valued, but often within predetermined structures defined by institutions **CulturalPerspectives:** STEM is seen as neutral and universal, often overlooking sociocultural and ethical dimensions **Ethical Implications:** Ethics is framed as adherence to guidelines rather than critical reflection on societal impact	**Power Dynamics:** Knowledge is negotiated through inquiry; students gain some autonomy in questioning and connecting disciplines **CulturalPerspectives:** Acknowledges multiple ways of knowing, with some space for local and Indigenous perspectives **Ethical Implications:** Begins to integrate ethics as part of systems thinking and sustainability discussions	**Power Dynamics:** Knowledge is co-constructed; students, educators, and communities contribute to meaning-making **CulturalPerspectives:** Explicitly values diverse epistemologies, including Indigenous, local, and community knowledge **Ethical Implications:** Ethics is central—students critically evaluate power structures, environmental impact, and societal well-being

structured, sequenced learning. As the focus shifts toward *Application-Focused* models, students engage with real-world problems, applying their disciplinary knowledge in structured contexts while maintaining a strong disciplinary foundation. *Context- and Inquiry-Focused* approaches further integrate interdisciplinary thinking, encouraging students to explore problems and concepts through inquiry, systems thinking, and real-world relevance. Finally, the *Context- and Learner-Driven* model places learners and their communities at the centre of the learning process, emphasising co-construction of knowledge, transdisciplinary inquiry, and ethical, culturally responsive engagement with societal challenges. The use of the word "focused" implies a primary concern or priority of the curriculum, compared to "driven", which implies that the local context and learners dictate the direction of learning. For example, in an application-focused curriculum, applications are prioritised, but the structure may still be guided by external factors such as institutional goals and standardised curricula.

An important feature of the table is the inclusion of the Hidden Curriculum, which highlights the implicit messages conveyed through different curriculum structures. This dimension addresses power dynamics, cultural perspectives, and ethical implications, demonstrating how each approach shapes students' perceptions of knowledge, expertise, and problem-solving in STEM/STEAM fields. While content-focused models reinforce traditional disciplinary hierarchies and standardised assessment structures, more learner-driven approaches empower students to engage with diverse ways of knowing, co-create knowledge with educators and communities, and critically examine ethical dimensions of STEM/STEAM applications.

By presenting these curriculum models side by side, Table 7.2 provides a tool for researchers, educators, curriculum designers, and policymakers to critically evaluate the design and implementation of STEM/STEAM education. It encourages reflection on the balance between disciplinary depth and interdisciplinary integration, as well as the extent to which learning is shaped by external structures versus learner agency and community-driven inquiry.

7.9 Conclusion

Designing inclusive and transformative STE*A*$^+$M curricula requires a commitment to equity, interconnected ways of knowing, and thoughtful implementation of hands-on learning. Addressing the hidden curriculum in STEM and STEAM education requires a multifaceted approach that begins with educators critically reflecting on their own biases and teaching practices. Educators can take meaningful steps toward creating more inclusive and equitable classrooms by examining how their cultural backgrounds and assumptions influence their interactions with students. Professional development programs, such as workshops on implicit bias and culturally responsive teaching, play a crucial role in supporting this reflective process.

Inquiry-based learning emerges as a powerful tool for challenging dominant narratives and encouraging students to explore multiple perspectives. By designing STE*A*$^+$M programs that address real-world issues of equity, power, and representation, educators can foster critical thinking and empower students to question the status quo. Additionally, integrating counter-narratives—such as highlighting diverse role models and incorporating Indigenous and community-based knowledge—helps disrupt the hidden curriculum and validates diverse ways of knowing.

Empowering students to recognise and reshape the hidden curriculum is equally important. By encouraging student-led inquiry, problem-solving, and critical discussions about the implicit messages in their education, educators can help students become active participants in their learning. Practical steps, such as conducting curriculum audits, collaborating with colleagues, and engaging with communities, provide concrete ways for educators to implement these strategies. They also need to navigate the social and cultural norms and practices in which the classrooms and schools operate.

Finally, assessment practices must be critically examined to ensure they align with the goals of equity and inclusion in STEM/STEAM education. By designing assessments that accommodate diverse learning needs, prioritise valued skills, and empower students, educators can create a learning environment that not only measures achievement but also fosters growth and agency.

In conclusion, addressing the hidden curriculum is an ongoing process that requires intentional reflection, collaboration, and action. By taking these steps, educators can create STE*A*⁺M classrooms that are more inclusive, equitable, and empowering for all students, ultimately helping to dismantle systemic inequities and foster a more just educational system.

References

Aguilera, D., & Ortiz-Revilla, J. (2021). STEM vs. STE*A*⁺M education and student creativity: A systematic literature review. *Education Sciences, 11,* 331. https://doi.org/10.3390/educsci11070331

Akerson, V., Burgess, A., Gerber, A., Guo, M., Khan, T. A., & Newman, S. (2018). Disentangling the meaning of STEM: Implications for science education and science teacher education. *Journal of Science Teacher Education, 29*(1), 1–8. https://doi.org/10.1080/1046560X.2018.1435063

Alsubaie, M. A. (2015). Hidden curriculum as one of current issue of curriculum. *Journal of Education and Practice, 6*(3), 125–128.

Andrée, M., & Hansson, L. (2020). Industrial actors and their rationales for engaging in STEM education. *Journal of Curriculum Studies, 52*(4), 551–576. https://doi.org/10.1080/00220272.2019.1682055

Baldinger, E. E., Staats, S., Covington Clarkson, L. M., Gullickson, E. C., Norman, F., & Akoto, B. (2020). A review of conceptions of secondary mathematics in integrated STEM education: Returning voice to the silent M. In J. Anderson, Y. Li (Eds.). *Integrated approaches to STEM education* (pp. 67–90). https://doi.org/10.1007/978-3-030-52229-2_5

Bang, M., & Medin, D. (2010). Cultural processes in science education: Supporting the navigation of multiple epistemologies. *Science Education, 94*(6), 1008–1026. https://doi.org/10.1002/sce.20392

Baram-Tsabari, A., & Yarden, A. (2008). Girls' biology, boys' physics: Evidence from free-choice science learning settings. *Research in Science & Technological Education, 26*(1), 75–92. https://doi.org/10.1080/02635140701847538

Barron, B., & Darling-Hammond, L. (2010). Prospects and challenges for inquiry-based approaches to learning. In H. Dumont, D. Istance, & F. Benavides (Eds.). *The nature of learning: Using research to inspire practice* (pp. 199–225). Organization for Economic Cooperation & Development. https://www.oecd.org/en/publications/2010/08/the-nature-of-learning_g1ghfff3.html

Beals, R. A. (2021). Transformative education in Science, Technology, engineering, and mathematics: Empowering the next generation of scientists. *Journal of Hispanic Higher Education, 20*(4), 385–401. https://doi.org/10.1177/1538192719877414

Berland, L. K., & Busch, K. C. (2012, June), Negotiating STEM epistemic commitments for engineering design challenges. *Paper presented at 2012*

ASEE Annual Conference & Exposition, San Antonio, Texas. https://doi.org/10.18260/1-2--21726

Bishop, A., Clarke, B., Corrigan, D., & Gunstone, D. (2006). Values in mathematics and science education: Researchers' and teachers' views on the similarities and differences. *For the Learning of Mathematics, 26*(1), 7–11. http://www.jstor.com/stable/40248515

Bock, C. (2022). Diversity, equality, and inclusion in our professions: A thin and leaky pipeline. *The Electrochemical Society Interface, 31*(1), 41–43. https://doi.org/10.1149/2.F08221IF

Borda, E., Haskell, T., & Boudreaux, A. (2022). Cross-disciplinary learning: A framework for assessing application of concepts across science disciplines. *Journal of College Science Teaching, 52*(1), 3–5. https://doi.org/10.1080/0047231X.2022.12315664

Bush, S. B., & Cook, K. L. (2019). Structuring STE*A*+M inquiries: Lessons learned from practice. In M. S. Khine & S. Areepattamannil (Eds.), *STEA+M education: Theory and Practice* (pp. 19–35). Springer. https://doi.org/10.1007/978-3-030-04003-1_2

Bybee, R. W. (2010). Advancing STEM education: A 2020 vision. *Technology and Engineering Teacher, 70*, 30–35.

Catterall, L. G. (2017). A brief history of STEM and STE*A*+M from an inadvertent insider. *The STEA+M Journal, 3*(1), Article 5. https://doi.org/10.5642/steam.20170301.05

Chalmers, C., Carter, M., Cooper, T., & Nason, R. (2017). Implementing "big ideas" to advance the teaching and learning of science, technology, engineering, and mathematics (STEM). *International Journal of Science and Mathematics Education, 15*(Suppl 1), S25–S43. https://doi.org/10.1007/s10763-017-9799-1

Cho, S., Mandracchai, M., & Yang, J. (2021). Nurturing mathematical talents of young mathematically gifted English language learners. In S. R. Smith (Ed.), *Handbook of giftedness and talent development in the Asia Pacific* (pp. 833–855). Springer. https://doi.org/10.1007/978-981-13-3041-4_38

Colucci-Gray, L., Burnard, P., Gray, D., & Cooke, C. (2019). A critical review of STEAM (Science, Technology, Engineering, Arts, and Mathematics). *Oxford Research Encyclopedia of Education.* https://doi.org/10.1093/acrefore/9780190264093.013.398

Darvin, R. (2019). Youth, technology, and the hidden curriculum of the twenty-first century. *Youth and Globalisation, 1*(2), 210–229. https://doi.org/10.1163/25895745-00102002

Daugherty, M. K., & Carter, V. (2018). The nature of interdisciplinary STEM education. In de Vries, M. (Ed.), *Handbook of technology education* (pp. 159–171). Springer. https://doi.org/10.1007/978-3-319-44687-5_12

Emdin, C., Adjapong, E., & Levy, I. P. (2021). On science genius and cultural agnosia: Reality pedagogy and/as Hip-Hop rooted cultural teaching in STEM education. *The Education Forum, 85*(4), 391–405. https://doi.org/10.1080/00131725.2021.1957636

English, L. D. (2016). STEM education K–12: Perspectives on integration. *International Journal of STEM Education, 3,* Article 3. https://doi.org/10.1186/s40594-016-0036-1

Erickson, S. S. (2022). The game of grades and the hidden curriculum. *The Physics Teacher, 60*(5), 398–399. https://doi.org/10.1119/10.0010403

Ernest, P. (2024). The ethics of authority and control in mathematics education: From naked power to hidden ideology. In Ernest, P. (Ed.), *Ethics and mathematics education: The good, the bad and the ugly* (pp. 199–249). Springer. https://doi.org/10.1007/978-3-031-58683-5_12

Fan, L., Mailizar, M., Alafaleq, M., & Wang, Y. (2018). A comparative study on the presentation of geometric proof in secondary mathematics textbooks in China, Indonesia, and Saudi Arabia. In Fan L., L. Trouche, Qi C., S. Rezat, & J. Visnovska (Eds.), *Research on mathematics textbooks and teachers' resources* (pp. 53–65). Springer. https://doi.org/10.1007/978-3-319-73253-4_3

Fei, B., Huang, G., & Cai, X. (2024). Improving the cross-cultural consciousness and self-efficacy of college students through STEM education. *Culture and Education, 36*(2), 479–499. https://doi.org/10.1177/11356405241261299

Figueiredo, D. (2023). Walking through the leaky academic pipeline in STEM: Equity, not equality needed for women and under represented minorities (URMs). In F. Bhatti & E. Taheri (Eds.), *Gender inequality—Issues, challenges and new perspectives.* https://doi.org/10.5772/intechopen.111538

Fountis, A. (2023). Co-opetition reloaded: Rethinking the role of globalization, supply chains, and mechanism design theory. In S.-L. Peng, N. Z. Jhanjhi, S. Pal, & F. Amsaad (Eds.), *Proceedings of 3rd International Conference on Mathematical Modeling and Computational Science. ICMMCS 2023. Advances in Intelligent Systems and Computing, vol 1450* (pp. 525–539). Springer. https://doi.org/10.1007/978-981-99-3611-3_44

Freeman, B., Marginson, S., & Tytler, R. (Eds.) (2014). *The age of STEM: Educational policy and practice across the world in Science, technology, engineering and mathematics.* Routledge. https://doi.org/10.4324/9781315767512

Gao, X., Li, P., Shen, J., & Sun, H. (2020). Reviewing assessment of student learning in interdisciplinary STEM education. *International Journal of STEM Education, 7*, Article 24. https://doi.org/10.1186/s40594-020-00225-4

Gesthuizen, R., Kidman, G., Tan, H., Mangao, D., & Macdonald, S. (2020). STEM inspiration: A phenomenological investigation exploring beyond the solution. In E. Creely, J. Southcott, K. Carabott, & D. Lyons (Eds.),

Phenomenological inquiry in education (pp. 225–242). Routledge. https://doi.org/10.4324/9780429285646-15

Gesthuizen, R., Tan, H., & Kidman, G. (2025). Inspired to build understanding: Learning about sustainability and environmental education from within a virtual world. *International Research in Geographical and Environmental Education, 34*(1), 25–42. https://doi.org/10.1080/10382046.2024.2348265

Ghanbari, S. (2015). Learning across disciplines: A collective case study of two university programs that integrate the arts with STEM. *International Journal of Education & the Arts, 16*(7), 1–21. http://www.ijea.org/v16n7/

Giroux, H. A., & Penna, A. N. (2025). Social education in the classroom: The dynamics of the hidden curriculum. In H. A. Giroux. *Teachers as intellectuals: Toward a critical pedagogy of learning* (Rev. ed., pp. 21–40). Bloomsbury Academic. https://www.bloomsbury.com/au/teachers-as-intellectuals-9781350458598/

Gobby, B. (2017). What is curriculum? In B. Gobby & R. Walker (Eds.), *Powers of curriculum: Sociological perspectives on education* (pp. 5–36). Oxford University Press.

Grangeat, M., Harrison, C., & Dolin, J. (2021). Exploring assessment in STEM inquiry learning classrooms. *International Journal of Science Education, 43*(3), 345–361. https://doi.org/10.1080/09500693.2021.1903617

Greene, M. T., Gonzalez, R., & Papalambros, P. Y. (2019). Measuring systems engineering and design thinking attitudes. *Proceedings of the Design Society: International Conference on Engineering Design, 1*(1), 3939–3948. https://doi.org/10.1017/dsi.2019.401

Grogan, K. E. (2019). How the entire scientific community can confront gender bias in the workplace. *Nature, Ecology and Evolution, 3*, 3–6. https://doi.org/10.1038/s41559-018-0747-4

Hansson, L. (2018). Science education, indoctrination, and the hidden curriculum. In M. R. Matthews (Ed.), *History, philosophy and science teaching: New perspectives* (pp. 283–306). Springer. https://doi.org/10.1007/978-3-319-62616-1_11

Hadar, L. L., & Tirosh, M. (2019). Creative thinking in mathematics curriculum: An analytic framework. *Thinking skills and creativity, 33*. Article 100585. https://doi.org/10.1016/j.tsc.2019.100585

Ireland, D. T., Freeman, K. E., Winston-Proctor, C. E., DeLaine, K. D., Lowe, S. M., & Woodson, K. M. (2018). (Un)hidden figures: A synthesis of research examining the intersectional experiences of black women and girls in STEM education. *Review of Research in Education, 42*(1), 226–254. https://doi.org/10.3102/0091732X18759072

Kelley, T. R., & Knowles, J. G. (2016). A conceptual framework for integrated STEM education. *International Journal of STEM Education, 3*, Article 11. https://doi.org/10.1186/s40594-016-0046-z

Keratithmakul, K., Kim, J. N., & Roehrig, G. H. (2020). Cultural competence or deficit-based view? A qualitative approach to understanding middle school students' experience with culturally framed engineering. *International Journal of STEM Education, 7*, 26. https://doi.org/10.1186/s40594-020-00224-5

Kidman, G., Tan, H., & Macdonald, S. (2024). Reimaging STEM education: Developing agency through STEA+M and design thinking. In T. Barkatsas, P. McLaughlin, & W. Goff (eds.). *Reimagining education for the second quarter of the twenty-first century and beyond*, (pp. 1–23). Brill. https://doi.org/10.1163/9789004688490_001

Kidman, J., Yen, C. F., & Abrams, E. (2013). Indigenous students' experiences of the hidden curriculum in science education: A cross-national study in New Zealand and Taiwan. *International Journal of Science and Mathematics Education, 11*, 43–64. https://doi.org/10.1007/s10763-012-9365-9

Kuroyanagi, T. (1982). *Totto-chan: The little girl at the window* (D. Britton, Trans.). Kodansha Publishers Ltd. (Original work published 1981).

Lande, M., & Leifer, L. (2010). Incubating engineers, hatching design thinkers: Mechanical engineering students learning design through ambidextrous ways of thinking. *Paper presented at 2010 Annual Conference & Exposition*, Louisville, Kentucky. https://doi.org/10.18260/1-2--16990

Le, H. C., Nguyen, V. H., & Nguyen, T. L. (2023). Integrated STEM approaches and associated outcomes of K-12 student learning: A systematic review. *Education Sciences, 13*, 297. https://doi.org/10.3390/educsci13030297

Lederman, N. G., & Lederman, J. (2020). Nature of scientific knowledge and scientific inquiry. In V. L. Akerson, & G. A. Buck (Eds.), *Critical questions in STEM education* (pp. 3–20). Springer. https://doi.org/10.1007/978-3-030-57646-2_1

Light, A. E., Benson-Greenwald, T. M., & Diekman, A. B. (2022). Gender representation cues labels of hard and soft sciences. *Journal of Experimental Social Psychology, 90*, Article 104234. https://doi.org/10.1016/j.jesp.2021.104234

Ma, H. (2015). Chinese secondary school science teachers' perceptions of the nature of science and Chinese native knowledge. In M. Khine (Ed.), *Science education in East Asia* (pp. 439–458). Springer. https://doi.org/10.1007/978-3-319-16390-1_18

Mahbubani, K. (2022). The Asian twenty-first century. *Springer*. https://doi.org/10.1007/978-981-16-6811-1

Mangao, D. D., Kidman, G., Tan, H., Gesthuizen, R., Macdonald, S., & Sinniah, D. N. (2022). *SEAMEO STEM planning and design learning (PaDL) framework: Towards twenty-first century skills and design thinking*. Southeast Asian Ministers of Education Organization (SEAMEO) Regional Centre for Education in Science and Mathematics (RECSAM). http://recsam.edu.my/index.php/our-resources/in-house-publication/recsam-publication

Mansilla, V. B., & Duraising, E. D. (2007). Targeted assessment of students' interdisciplinary work: An empirically grounded framework proposed. *The Journal of Higher Education, 78*(2), 215–237. https://doi.org/10.1080/00221546.2007.11780874

Mariana, N. (2019). A reflective journey within five ways of transformative knowing. In P. C. Taylor & B. C. Luitel (Eds.), *Research as transformative learning for sustainable futures: Glocal voices and visions* (pp. 207–223). Brill. https://doi.org/10.1163/9789004393349_013

Marques, M. (2021). Structural-functionalism in comparative and international education: Antecedents, developments, and applications. In T. D. Jules, R. Shields, & M. A. M. Thomas (Eds.), *The Bloomsbury handbook of theory in comparative and international education* (pp. 23–36). Bloomsbury. https://doi.org/10.5040/9781350078789.ch-001

Martín-Páez, T., Aguilera, D., Perales-Palacios, F. J., & Vílchez-González, J. M. (2019). What are we talking about when we talk about STEM education? A review of literature. *Science Education, 103*, 799–822. https://doi.org/10.1002/sce.21522

McLure, F. I., Tang, K. S., & Williams, P. J. (2022). What do integrated STEM projects look like in middle school and high school classrooms? A systematic literature review of empirical studies of iSTEM projects. *International Journal of STEM Education, 9*, Article 73. https://doi.org/10.1186/s40594-022-00390-8

Mehta, R., Creely, E., & Henriksen, D. (2020). A profitable education: Countering Neoliberalism in twenty-first century skills discourses. In J. Keengwe & G. Onchwari (Eds.), *Handbook of research on literacy and digital technology integration in teacher education* (pp. 359–381). IGI Global. https://doi.org/10.4018/978-1-7998-1461-0.ch020

Mehta, R., Keenan, S., Henriksen, D., & Mishra, P. (2019). Developing a rhetoric of aesthetics: The (often) forgotten link between art and STEM. In M. S. Khine, & S. Areepattamannil (Eds.), *STEA+M education: Theory and practice* (pp. 117–141). Springer. https://doi.org/10.1007/978-3-030-04003-1_7

Mei, L., Cui, K., Guo, X., & Du, C. (2023). Incorporating the Arts: A comprehensive analysis of the evolution from STEM to STEA+M in contemporary education. *Journal of Research in Social Science and Humanities, 2*(11), 24–28. https://doi.org/10.56397/JRSSH.2023.11.05

Nathan, M. J., Srisurichan, R., Walkington, C., Wolfgram, M., Williams, C., & Alibali, M. W. (2013). Building cohesion across representations: A mechanism for STEM integration. *Journal of Engineering Education, 102*(1), 77–116. https://doi.org/10.1002/jee.20000

Nqabeni, N., & Buka, A. M. (2023). Post Covid-19: Exploring the decolonisation factor in a mathematics classroom in South African schools. *E-Journal of Humanities, Arts and Social Sciences (EHASS), 4*(3), 287–299. https://doi.org/10.38159/ehass.2023439

Nugraha, M. G., Kidman, G., & Tan, H. (2024). Interdisciplinary STEM education foundational concepts: Implementation for knowledge creation. *EURASIA Journal of Mathematics, Science and Technology Education, 20*(10), Article em2523. https://doi.org/10.29333/ejmste/15471

Ostler, E. (2015). The conception and evolution of STEM: A brief historical perspective. In E. Ostler (Ed.), *STEM education: An overview of contemporary research, trends and perspectives* (pp. 11–26). Cycloid Publications.

Panizzon, D., Corrigan, D., Forgasz, H., & Hopkins, S. (2015). Impending STEM shortages in Australia: Beware the 'smoke and mirrors'. *Procedia—Social and Behavioral Sciences, 167*, 70–74. https://doi.org/10.1016/j.sbspro.2014.12.644

Riley, D. M. (2018). Refuse, refute, resist: Alt-right attacks on engineering and STEM education diversity scholarship. *Paper presented at 2018 CoNECD—The Collaborative Network for Engineering and Computing Diversity Conference, Crystal City, Virginia*. American Society for Engineering Education. https://doi.org/10.18260/1-2--29572

Roberts, T., Maiorca, C., Jackson, C., & Mohr-Schroeder, M. (2022). Integrated STEM as problem-solving practices. *Investigations in Mathematics Learning, 14*(1), 1–13. https://doi.org/10.1080/19477503.2021.2024721

Rocha, J., Castillo-Lavergne, C. M., & Yancy, C. W. (2024). Affirming and nurturing students' cultural wealth to enhance self-efficacy: Examination of urban high school students' lived experiences in a STEM-medicine pipeline program. *Urban Education, 59*(4), 1012–1047. https://doi.org/10.1177/00420859211073897

Roehrig, G. H., Dare, E. A., Ellis, J. A., & Ring-Whalen, E. (2021). Beyond the basics: A detailed conceptual framework of integrated STEM. *Disciplinary and Interdisciplinary Science Education Research, 3*(1), 1–18. https://doi.org/10.1186/s43031-021-00041-y

Sarkar, M., Overton, T., Thompson, C., & Rayner, G. (2016). Graduate employability: Views of recent science graduates and employers. *International Journal of Innovation in Science and Mathematics Education, 24*(3), 31–48.

Schiro, M. S. (2013). *Curriculum theory: Conflicting visions and enduring concerns* (2nd ed.). SAGE.

Sellers, V., & Villanueva Alarcón, I. (2023). From message to strategy: A pathways approach to characterize the hidden curriculum in engineering education. *Studies in Engineering Education*, *4*(2), 176–200. https://doi.org/10.21061/see.113

Seno-Alday, S., & Budde-Sung, A. (2022). Teaching a while measuring b: Cultural bias in assessing student performance. *Journal of International Education in Business, 15*(2), 273–289. https://doi.org/10.1108/JIEB-01-2021-0005

Sgro, C. M., Bobowski, T., & Oliveira, A. W. (2020). Current praxis and conceptualization of STEM education: A call for greater clarity in integrated curriculum development. In V. L. Akerson, G. A. Buck (Eds.). *Critical questions in STEM education* (pp. 185–210). Springer. https://doi.org/10.1007/978-3-030-57646-2_11

Shen, J., Liu, O. L., & Sung, S. (2014). Designing interdisciplinary assessments in sciences for college students: An example on osmosis. *International Journal of Science Education, 36*(11), 1773–1793. https://doi.org/10.1080/09500693.2013.879224

Shiekh, K., Nieusma, D., Zhu, Q., & Rea, S. C. (2023). How the hidden curriculum reveals the enculturation experience of underrepresented students in engineering. *Proceedings of the 2023 IEEE Frontiers in Education Conference (FIE), College Station, TX, USA, 2023*, 1–8, https://doi.org/10.1109/FIE58773.2023.10343275

Stowe, R. L., Scharlott, L., J., Ralph, V. R., Becker, N. M., & Cooper, M. M. (2021). You are what you assess: The case for emphasizing chemistry on chemistry assessments. *Journal of Chemical Education, 98*, 2490–2495. https://doi.org/10.1021/acs.jchemed.1c00532

Tan, H., & Kidman, G. (2021). Authentic assessment in STEM education: An integrative review of research. In T. Barkatsas & T. McLaughlin (Eds.), *Authentic assessment and evaluation approaches and practices in a digital era* (pp. 24–52). Brill. https://doi.org/10.1163/9789004501577_002

Tan, L., & Wei, B. (2024). How science teachers deal with STEM education: An explorative study from the lens of curriculum ideology. *Science Education, 109*, 82–105. https://doi.org/10.1002/sce.21904

Taylor, P. C. (2015). Transformative science education. In R. Gunstone (Ed.), *Encyclopedia of science education* (pp. 1079–1082). Springer. https://doi.org/10.1007/978-94-007-6165-0_212-2

Thibaut, L., Ceuppens, S., De Loof, H., De Meester, J., Goovaerts, L., Struyf, A., Boeve-de Pauw, J., Dehaene, W., Deprez, J., De Cock, M., Hellinckx, L., Knipprath, H., Langie, G., Struyven, K., Van de Velde, D., Van Petegem, P. and Depaepe, F. (2018). Integrated STEM education: A systematic review of instructional practices in secondary education. *European Journal of STEM Education, 3*(1), Article 2. https://doi.org/10.20897/ejsteme/85525

Thomas, K. (2019, September 11). Plastic bottle lids inspire 3D-printed prosthetic limbs, artworks, and future leaders. *ABC News*. https://www.abc.net.au/news/2019-09-11/project-to-print-prosthetics-for-kids-from-recycled-plastic/11491280

Toma, R. B. (2024). Elementary school students' interests and attitudes towards biology and physics. *Journal of Biological Education, 58*(4), 972–983. https://doi.org/10.1080/00219266.2022.2147208

Tomei, A., Dillon, J., & Dawson, E. (2014). United Kingdom: An example of the impact of high stakes accountability regimes on STEM education. In B. Freeman, S. Marginson, & R. Tytler (Eds.). *The age of STEM: Educational policy and practice across the world in science, technology, engineering and mathematics* (pp. 161–177). Routledge. https://doi.org/10.4324/9781315767512

Venkat, H. (2020). Mathematical literacy. What is it? And is it important? In G. Ineson & H. Povey (Eds.), *Debates in mathematics education* (2nd ed., pp. 210–223). https://doi.org/10.4324/9780429021015

Walker, M. E., Olivera-Aguilar, M., Lehman, B., Laitusis, C., Guzman-Orth, D., & Gholson, M. (2023). Culturally responsive assessment: Provisional principles. *ETS Research Report Series*. https://doi.org/10.1002/ets2.12374

Wescott, S., & Roberts, S. (2025). Conceptualising school-level responses to sexual harassment of women teachers as institutional gaslighting. *British Journal of Sociology of Education, 46*(1), 1–18. https://doi.org/10.1080/01425692.2024.2409267

Wilkie, K. J., & Tan, H. (2019). Exploring mathematics teacher leaders' attributions and actions in influencing senior secondary students' mathematics subject enrolments. *Mathematics Education Research Journal, 34*(4), 441–464. https://doi.org/10.1007/s13394-019-00264-3

Wong, B., Chiu, Y.-L.T., Murray, Ó. M., Horsburgh, J., & Copsey-Blake, M. (2023). 'Biology is easy, physics is hard': Student perceptions of the ideal and the typical student across STEM higher education. *International Studies in Sociology of Education, 32*(1), 118–139. https://doi.org/10.1080/09620214.2022.2122532

World Economic Forum (WEF). (2023). *The global risks report 2023 18th edition: Insight report*. https://www.weforum.org/reports/global-risks-report-2023/

Yosso, T. J. (2005). Whose culture has capital? A critical race theory discussion of community cultural wealth. *Race Ethnicity and Education, 8*(1), 69–84. https://doi.org/10.1080/1361332052000341006

Ziebell, N., & Clarke, D. (2018). Curriculum alignment: Performance types in the intended, enacted, and assessed curriculum in primary mathematics and science classrooms. *Studia Paedagogica, 23*(2), 175–203. https://doi.org/10.5817/SP2018-2-10

PART III

Advancing Interdisciplinary and Integrative Approaches in STEA+M Education

This Part extends the conceptual and practical scaffolding established in Parts I and II by deepening the exploration of interdisciplinary and integrative approaches in STE*A*⁺M education. Building on earlier discussions of cognitive integration, systems thinking, and equitable curricula, this section shifts the focus toward implementation: how interdisciplinary principles are enacted through teaching practices, professional learning, and curriculum design. Across three chapters, Part III examines design frameworks, teacher-led curriculum initiatives, and the broader transformative purpose of integration.

Chapter 8 introduces the *Planning and Design Learning (PaDL) Framework* as a culturally responsive model for design thinking in STE*A*⁺M contexts. Drawing from and critiquing well-established models of the Stanford d.school, IDEO, and Double Diamond, the chapter examines how Southeast Asian educators have recontextualised design thinking to reflect local pedagogical values. The PaDL Framework positions teachers as designers of inquiry-rich, interdisciplinary learning experiences, highlighting the importance of adapting educational models to diverse cultural and institutional settings. By emphasising empathy, agency, and iterative problem-solving, the framework serves as a practical tool for enacting inclusive and context-sensitive STE*A*⁺M pedagogy. Chapter 9 builds on the design-focused orientation of Chapter 8 by presenting an empirical account of teachers in Southeast Asia who collaboratively developed interdisciplinary STEAM units through a regional professional development program. These units aligned with curricula in science and mathematics while drawing on locally relevant themes. The analysis underscores the significance of curriculum alignment, professional collaboration, and integrative design strategies. It demonstrates how planning enables the cognitive, affective, and values-based dimensions of learning to coalesce, resulting in holistic and meaningful learning experiences for students. Chapter 10 synthesises the core threads of the book while reframing integration as a future-oriented, ethical, and political praxis. Rather than offering a conclusion to the book, the chapter revisits the book's core aims, to interrogate, reframe, and reconceptualise STEAM education, through a series of "un-conclusions" that reflect the ongoing and unsettled nature of educational transformation. Integration is not treated as a technical solution or a curriculum innovation alone, but rather as a mode of engaging with global challenges, such as epistemic injustice, the climate crisis, and socio-ecological change. Central to this reconceptualisation is the *FRAME model* (Futures-oriented, Responsibility-driven, Agency-centred, Meaning-making, Ethics-informed), which synthesises the book's theoretical, cognitive, and affective commitments. The chapter also advances three provocations: navigating epistemological tensions without diluting disciplinary depth, designing for equity in integrated learning, and reconciling local knowledge with global responsibility. These provocations are not intended as conclusions, but as starting points for continued research, dialogue, and educational experimentation.

Together, these chapters present a vision for STE*A*⁺M education that is interdisciplinary, integrative, and contextually relevant. By grounding the principles of collaboration, agency, adaptability, and cultural responsiveness in localised real-world practice and research, Part III ensures that the theoretical shifts proposed in Parts I and II are realised through actionable strategies for teachers and learners. Ultimately, this section contributes to a broader vision of STE*A*⁺M education as a transformative framework, one that equips learners, teachers, and teacher educators with the knowledge, skills, and agency to navigate complex global challenges and drive meaningful societal change.

Gillian Kidman, Hazel Tan, Roland Gesthuizen, and Dominador D. Mangao

CHAPTER 8

Integrating Design Thinking in STE*A*$^+$M Education: The PaDL Framework for Planning and Learning

8.1 Introduction

Design thinking has gained considerable traction as an innovative educational methodology that enhances creativity, problem-solving, and interdisciplinary collaboration. Initially developed within industrial design, the approach has been widely adopted across multiple fields, including business, healthcare, and education (Morales-Doyle, 2017). Applying design thinking within STEAM (science, technology, engineering, arts, and mathematics) education represents a shift toward more integrative and student-centred pedagogies that mirror real-world problem-solving contexts. This chapter explores the theoretical foundations, pedagogical implications, and practical applications of design thinking in STE*A*$^+$M (science, technology, engineering, innovative *A*gency, and mathematics) education, focusing on the Planning and Design Learning (PaDL) Framework (Mangao et al., 2022), a tool for structuring integrative teaching and learning.

The increasing complexity of societal challenges requires educational approaches that move beyond rigid disciplinary boundaries (Liu & Li, 2023). Traditional STEM (Science, Technology, Engineering, and Mathematics) education has often been critiqued for its emphasis on content knowledge and procedural learning, frequently overlooking the sociocultural dimensions of scientific and technological advancements (Garibay, 2015). STE*A*$^+$M education, by incorporating the *A*gency (see Chapter 3)

G. Kidman et al., *Knowledge Generation in STEM and STEAM Education*, https://doi.org/10.1007/978-3-032-03247-8_8

and emphasising innovation, creativity, and design, seeks to cultivate holistic problem-solvers who can navigate complex global issues. As a methodology that prioritises empathy, ideation, and iteration, design thinking provides a compelling framework for achieving these educational goals.

The evolution of design thinking originates from Herbert Simon's contributions in the late 1960s (Lazar, 2018) and Donald Schön's theories in the 1980s (Waks, 2001), who both defined design as a reflective, iterative, and problem-framing process. Their contributions laid the foundation for contemporary models such as the Stanford d.school model, which articulates design thinking in five stages: Empathise, Define, Ideate, Prototype, and Test (ME113, 2012). Similarly, the IDEO model focuses on inspiration, ideation, and implementation, promoting iterative learning and collaborative problem-solving (IDEO Design Thinking, n.d.). The Double Diamond Model, developed by the British Design Council, provides a structured approach emphasising divergent and convergent thinking in problem-solving (Design Council, 2025). While these models have been widely adopted, they largely reflect Western perspectives on innovation and problem-solving, raising questions about their applicability in diverse educational settings. We explore this further in Chapter 7.

Within STEA^{+}M education, design thinking serves as a bridge between disciplinary knowledge and real-world application. It encourages educators to design learning experiences that engage students in hands-on, inquiry-based projects. Rather than presenting knowledge as static, design thinking situates learning within dynamic, problem-based contexts, requiring students to develop, test, and refine their ideas iteratively (McCurdy et al., 2020). The iterative nature of design thinking aligns closely with scientific inquiry and engineering processes, making it a particularly valuable approach for fostering deep learning and metacognitive skills.

Despite its potential, implementing design thinking in STEA^{+}M education is challenging. Many educators lack formal training in design methodologies, and traditional curriculum structures often prioritise standardised assessments over creative exploration. Moreover, the dominance of Western design thinking models has led to concerns regarding cultural relevance and contextual applicability. This chapter presents a case study of Southeast Asian teachers who collaboratively developed the PaDL Framework. This alternative framework integrates local educational priorities with global best practices in design thinking.

The PaDL Framework (Mangao et al., 2022) consists of two complementary models: the *Design Planning Model*, which guides educators in structuring lessons, and the *Design Learning Model*, which provides students with a scaffolded approach to engaging in design-based inquiry. By centring student *A*gency, ethical considerations, and iterative learning, the PaDL Framework seeks to bridge disciplinary silos and cultivate adaptable, innovative learners. This framework offers educators a structured approach and empowers students to take ownership of their learning, fostering a deeper engagement with STE*A*+M subjects.

As design thinking continues to be prominent in educational discourse, its integration into STE*A*+M education must be guided by theoretical and empirical considerations. This chapter critically examines existing models, highlights the need for culturally responsive adaptations, and presents a novel framework for implementation. By exploring the intersection of design thinking and STE*A*+M education, this chapter aims to provide educators with actionable insights into planning and facilitating integrative learning experiences that prepare students for the challenges of an increasingly complex world.

8.2 Understanding Design Thinking in STEA+M Education

Design thinking is a problem-solving approach emphasising creativity, collaboration, and user-centred innovation. It originated in the mid-twentieth century in industrial design but has since been applied across various disciplines, including business, technology, and education, making it essentially multidisciplinary. As alluded to in Chapter 4, a disciplinary cognate that is 'interdisciplinary' requires an integration process that synthesises ideas and methods (National Academy of Sciences et al., 2005). Design thinking is touted to do this in education and has become an essential framework in STEM (Science, Technology, Engineering, and Mathematics) and STEAM (which often includes the 'Arts'; however, we advocate innovative *A*gency—see Chapter 3) or STE*A*+M education, offering a structured yet flexible way for educators and students to engage in meaningful, hands-on learning experiences.

Herbert Simon (Lazar, 2018) and Donald Schön (Waks, 2001) contributed to the educational theoretical underpinnings of design as a cognitive process, emphasising reflective practice and problem-framing as integral to innovative thinking. The methodology was further developed by design firms such as IDEO and academic institutions like Stanford University's d.school, which structured design thinking into iterative stages. These stages—typically defined as Empathise, Define, Ideate, Prototype, and Test—form the foundation of contemporary design thinking models used in education today. Adopting design thinking in educational settings reflects a growing recognition of equipping students with the skills to tackle complex, real-world problems through iterative and human-centred approaches (Morales-Doyle, 2017).

A critical distinction between traditional problem-solving and design thinking lies in their underlying methodologies and objectives. Traditional problem-solving methods, particularly within STEM education, often emphasise linear, algorithmic, and deductive reasoning. These approaches typically involve defining a problem, applying established formulas or procedures, and deriving a correct or optimal solution. Priemer et al. (2019) present a spectrum of linear problem-solving approaches, and while there is a difference in their foci, the similarity of a step-by-step sequence is apparent. While effective for well-structured problems with clear parameters, traditional problem-solving methods may not fully prepare students for the ambiguities and complexities of real-world challenges. In contrast, design thinking embraces open-ended exploration, allowing for multiple perspectives and creative solutions. Rather than seeking a 'correct' answer, design thinking encourages students to engage in divergent thinking, exploring various possibilities before converging on the most viable solution.

Design thinking prioritises empathy as a central component of the problem-solving process (McCurdy et al., 2020). Traditional STEM education often focuses on technical proficiency and abstract problem-solving without considering broader human and societal contexts. In contrast, design thinking begins with understanding users' needs, experiences, and challenges. This human-centred approach ensures that solutions are technically sound but also meaningful and impactful (Micheli et al., 2019). By incorporating empathy-driven inquiry, design thinking fosters a deeper connection between students and the problems they seek to address, enhancing their ability to develop innovative and socially responsible solutions (McCurdy et al., 2020). The human-centred nature of design thinking is fundamental to STEM and STEAM education. By prioritising empathy, design thinking encourages students to consider their work's social, ethical, and environmental implications. This holds significant importance for global issues like climate change, public health emergencies, and the ethics of technology. Design thinking equips students with the ability to address these complex issues by developing solutions that are not only technically sound but also socially responsible and user-focused.

The relevance of design thinking in STEM and STEAM classrooms is increasingly recognised as educators seek to cultivate twenty-first century skills such as creativity, collaboration, critical thinking, and problem-solving (Abaniel, 2021; Kidman et al., 2024). Integrating design thinking into STEM/STEAM education aligns with contemporary pedagogical approaches emphasising inquiry-based, experiential, and project-based learning. These approaches move beyond rote memorisation and standardised assessments, instead focusing on students' ability to engage with complex problems, experiment with solutions, and iterate based on feedback (Bakar, 2021).

Design thinking is particularly valuable in STE*A*$^+$M education because it facilitates interdisciplinary connections (see Chapter 3). Traditional disciplinary silos can limit students' capacity to address multifaceted problems that require cross-disciplinary knowledge and skills (Garibay, 2015). Integrating science, technology, engineering, arts, and mathematics into a cohesive design process allows students to think holistically, drawing from multiple fields to develop comprehensive solutions (Kidman et al., 2024). For instance, design thinking projects in STE*A*$^+$M classrooms may involve students using scientific principles to understand an environmental issue, leveraging engineering skills to prototype solutions, incorporating technology to refine their designs, applying mathematical analysis to assess feasibility, and using critical and creative thinking to communicate their ideas effectively. This integrative approach mirrors the collaborative and multidisciplinary nature of real-world problem-solving in professional fields (Lyons, 2020).

Another key characteristic of design thinking in education is its iterative nature. Unlike traditional problem-solving models that emphasise arriving at a definitive answer, design thinking is cyclical and adaptive (Micheli et al., 2019). The iterative process involves multiple rounds of prototyping, testing, and refining solutions based on feedback. This approach encourages resilience and adaptability, helping students develop a mindset that views failure not as an endpoint but as a valuable learning opportunity. The iterative nature of design thinking also aligns with scientific inquiry and engineering design processes, making it particularly relevant for STEM and STEAM education (Wensveen & Matthews, 2014). This iterative process enhances students' ability to engage in metacognitive reflection, which is essential for deep learning. Through continuous iteration, students must evaluate their assumptions, analyse the effectiveness of their solutions, and refine their approaches based on new insights (Marthaliakirana et al., 2022). This reflective practice helps them understand the problem more deeply and enhances their strategic thinking and decision-making capacity. Additionally, it fosters a growth mindset as students appreciate that innovation and creativity are ongoing processes rather than one-time achievements.

Incorporating design thinking into STE*A*⁺M classrooms requires a pedagogical shift that moves beyond traditional instruction toward a more student-centred, inquiry-driven approach that prioritises the notion of *A*gency. Teachers must create learning environments encouraging experimentation, collaboration, risk-taking, and social *A*gency. This requires a departure from current curriculum structures and a willingness to embrace uncertainty as students navigate the complexities of real-world problem-solving. Design thinking transforms STE*A*⁺M education, fostering creativity, interdisciplinary collaboration, and human-centred problem-solving. Its origins in design and its evolution as a structured methodology make it a practical approach for addressing complex challenges in educational settings. Design thinking distinguishes itself from conventional problem-solving by prioritising empathy, iteration, and user-focused innovation. This approach empowers students with the essential skills to navigate a complex and interconnected world. As educators explore new ways to integrate design thinking into their teaching practices, the potential for fostering a new generation of confident, innovative, socially conscious problem-solvers becomes increasingly evident.

8.3 Teachers as Designers: Integrating Design Thinking into STEM and STEAM Education

Teachers play a crucial role in shaping students' educational experiences, and increasingly, they are recognised as designers of curriculum and instruction. Rather than simply delivering predetermined content, most teachers actively interpret, adapt, and create learning experiences that meet the needs of their students. The concept of teachers as designers is particularly significant in STEM and STEAM education, where innovation, creativity, and problem-solving are essential. Design thinking provides a powerful framework for educators. However, its practical implementation requires a deep understanding of its principles and professional learning opportunities that support teachers in applying these concepts to their practice.

Kim (2019) highlights the role of teachers as curriculum designers, emphasising that teaching is no longer the passive transmission of knowledge but an active and dynamic process. Teachers continuously adapt and refine their instructional approaches based on students' needs, interests, and evolving educational goals. This perspective positions teachers as designers who plan lessons and create meaningful learning experiences that foster student engagement and deep understanding. This role is even more pronounced in STEM and STEAM education, as teachers must design interdisciplinary, inquiry-driven, and hands-on learning experiences that integrate multiple subject areas and that "student identity formation as scientific investigators, thinkers, and inventors" (Kim, 2019, p. 139).

A fundamental aspect of teachers' work as designers is their interpretative and adaptive role in curriculum implementation. While national and regional curricula provide broad guidelines, teachers must interpret these frameworks in meaningful and relevant ways and adapt to their students' needs (Kim, 2019). This interpretative process involves making decisions about instructional strategies, assessment methods, and learning materials while considering students' diverse backgrounds, learning styles, and interests. Adaptability is particularly crucial in design thinking, as the iterative nature of the process requires teachers to be flexible and responsive to students' evolving understanding and creativity (Pozilova & Zulxumorxon, 2024). Designing compelling learning experiences requires teachers to embrace experimentation, innovation, and a willingness to take risks. Unlike traditional approaches that rely heavily on standardised lesson plans, design thinking encourages teachers to view their classrooms as dynamic learning environments (Pozilova & Zulxumorxon, 2024) where new ideas can be tested and refined. By fostering a culture of iteration, teachers help students develop resilience, critical thinking skills, and a mindset that values learning from failure. This approach aligns with the fundamental principles of STEM and STEAM education, which emphasise inquiry, problem-solving, and real-world application of knowledge.

Despite the growing recognition of the importance of design thinking in education, misconceptions about its nature and purpose persist. Lahey (2017) identifies several common misunderstandings that hinder its effective implementation. One prevalent misconception is that design thinking is about aesthetics or artistic creativity. While creativity is essential, design thinking is fundamentally about problem-solving and innovation (Syah & Nasri, 2023). It involves a structured process that includes empathy, defining problems, ideation, prototyping, and testing. Teachers who perceive design thinking as an artistic endeavour may overlook its broader applications in STEM and STEAM education, where it serves as a framework for developing innovative solutions to complex problems.

Another misconception is that design thinking is a rigid, step-by-step methodology that must be followed linearly (Priemer et al., 2019). In reality, design thinking is iterative and flexible, allowing for movement between different process stages as new insights emerge. Teachers who approach design thinking with a rigid mindset may struggle to integrate it effectively into their practice, as successful implementation requires adaptability and a willingness to revise and refine ideas. Encouraging teachers to see design thinking as a dynamic and evolving process can help them develop a more nuanced and practical approach to its application in the classroom.

Syah and Nasri (2023) identified the belief that design thinking is only relevant for specific subjects or gifted students. While it is commonly associated with engineering, technology, and the arts, design thinking has broad applications across all disciplines and educational contexts. Whether used to develop scientific experiments, mathematical models, literary analyses, or social studies projects, design thinking provides a valuable framework for fostering critical thinking, collaboration, and innovation (Micheli et al., 2019). By challenging the notion that design thinking is limited to specific areas, educators can explore its potential to enhance learning across the curriculum.

Teachers require targeted professional learning programs that provide theoretical knowledge and practical strategies to implement design thinking in their classrooms effectively (Johnson et al., 2020). Research indicates that many educators struggle to understand what design thinking means in practice and how to apply it to their teaching (Lahey, 2017). Without adequate support, teachers may lack confidence and feel overwhelmed by the complexity of design thinking or uncertain how to integrate it into existing curricula. Professional learning programs can help bridge this gap by offering hands-on training, collaborative learning opportunities, and ongoing support (Johnson et al., 2020).

One essential component of professional learning for design thinking is experiential learning. Teachers benefit from engaging in design thinking themselves, working through real-world challenges, and experiencing the iterative process firsthand (Christian et al., 2021). By participating in workshops, collaborative projects, and design challenges, teachers gain a deeper understanding of how design thinking can be applied in their classrooms, especially where "the courses were designed to foster teachers' knowledge of engineering by calling upon their pre-existing knowledge of science and mathematics to develop their self-efficacy in integrating STEM in their classrooms" (Christian et al., 2021). Experiential learning opportunities allow teachers to develop confidence in their ability to facilitate design-based instruction and model the problem-solving mindset they seek to instil in their students.

Professional learning programs should emphasise the importance of reflection and continuous improvement. As design thinking is an iterative process, professional learning should be ongoing and responsive to teachers' evolving needs (Peters-Burton & Johnson, 2018). Opportunities for self-reflection, feedback, and iterative refinement of instructional approaches can help teachers integrate design thinking more effectively into their practice. Providing access to resources, case studies, and exemplars of successful design thinking implementation can further support teachers in translating theoretical knowledge into practical application.

Teachers play a vital role as designers of curriculum and learning experiences, particularly in STEM and STEAM education (Kim, 2019). Teachers can create dynamic and engaging learning environments by interpreting and adapting curricula, fostering creativity and problem-solving, and embracing the iterative nature of design thinking. However, misconceptions about design thinking continue to pose challenges, highlighting the need for professional learning programs that provide educators with the knowledge, skills, and support necessary for effective implementation (Christian et al., 2021). Through targeted professional development, collaboration, and experiential learning, teachers can develop the confidence and expertise needed to integrate design thinking into their practice (Peters-Burton & Johnson, 2018), ultimately enhancing student learning and fostering innovation in education.

8.4 Models and Frameworks for STEA$^+$M Design Thinking

So far, we have described design thinking as a structured yet flexible approach to problem-solving, fostering creativity and innovation across various fields, including education, business, and engineering. Several design models have been developed as guides through the iterative ideation, prototyping, and refinement processes. Below, we explore three widely recognised design models: the Stanford d.school model, the IDEO model, and the Double Diamond Model. Each model presents a unique perspective on design thinking as problem-solving, yet all share a common emphasis on iterative thinking, creativity, and user-centred design.

8.4.1 The Stanford d.school Model

The Stanford d.school model is one of the most well-known design thinking frameworks used in education, developed by the Hasso Plattner Institute of Design at Stanford University. The model has become a cornerstone for establishing design thinking as a pedagogy arena (Syah & Nasri, 2023). The model consists of five stages (see Fig. 8.1). However, the stages are not necessarily linear but iterative, allowing continuous refinement and improvement.

Empathise: The first stage involves understanding the targeted users' needs, behaviours, and emotions. Designers engage in interviews, observations, and other research methods, immersing themselves in their experiences to gain deep insights into their targeted user's challenges

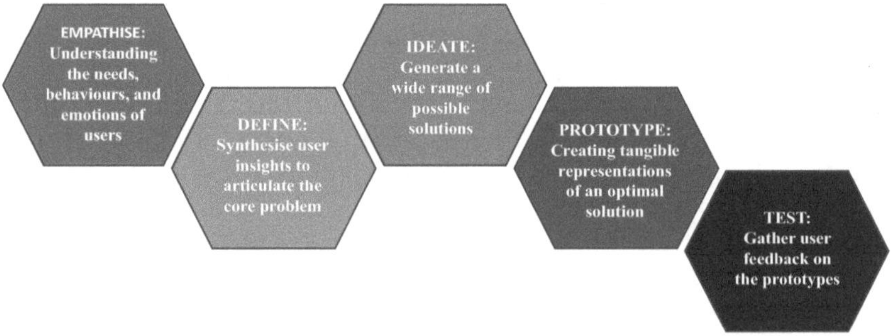

Fig. 8.1 The design thinking model by Stanford d-school (*Source* Authors)

and experiences (ME113, 2012). Empathy is crucial as it ensures that solutions are human-centred rather than solely technology-driven or business-oriented (McCurdy et al., 2020).

Define: After gathering targeted user insights, the next stage is synthesising the information and clearly articulating the core problem that needs addressing. This stage involves distilling targeted user observations into a clear problem statement or "point of view" that guides the design process (ME113, 2012).

Ideate: In this stage, designers generate various potential solutions. Brainstorming, mind mapping, and sketching are standard techniques that encourage divergent thinking, allowing the designer to step beyond the obvious and explore a range of ideas (ME113, 2012). The goal is to explore as many ideas as possible before narrowing them down to the most promising ones.

Prototype: Prototyping involves creating tangible representations of ideas, ranging from simple paper sketches to functional digital or physical models. The goal is to develop a tangible prototype that allows the designer or intended user to engage with the solution, rapidly test concepts in a cost-efficient manner to determine effective elements and areas for enhancement, and foster understanding and empathy (ME113, 2012).

Test: The final stage involves gathering targeted user feedback on the prototypes to refine and enhance the solutions, learn more about the targeted user, and refine the original point of view (ME113, 2012). Testing is an iterative process, often leading back to earlier stages to ensure continuous improvement based on the targeted users' input.

The Stanford d.school model is particularly effective for educators and students. It emphasises creativity, collaboration, and adaptability and provides a structured yet flexible approach to problem-solving, making it applicable across disciplines.

8.4.2 The IDEO Model

IDEO, a global design and consulting firm, developed a design thinking model that aligns closely with real-world innovation practices. This model (see Fig. 8.2) consists of three interconnected phases: Inspiration, Ideation, and Implementation and is sometimes referred to as the 3I's model. It was developed as a tool for non-government organisations and social enterprises (Zamakhsyari & Fatwanto, 2023).

Fig. 8.2 IDEO design thinking process (*Source* Authors)

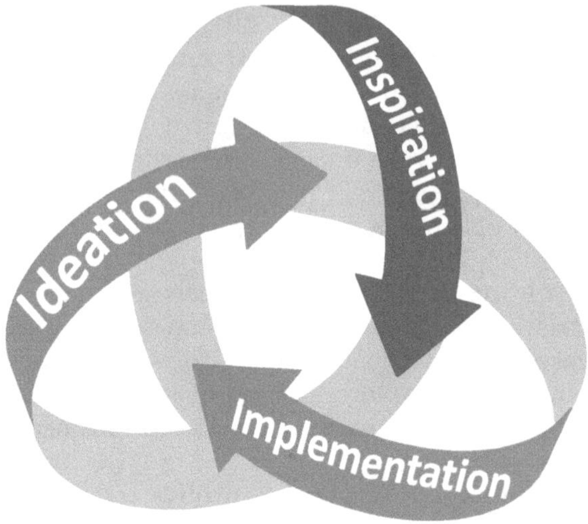

Inspiration: The first phase involves exploring and understanding the problem context through research and user engagement in its daily environment (Verganti et al., 2021). This stage includes conducting ethnographic studies, interviews, and observational research to understand users' needs and challenges. The goal is to determine insights that will drive the ideation process.

Ideation: In this stage, designers generate, develop, and refine ideas based on the insights gathered (Verganti et al., 2021). The emphasis is on divergent thinking and visual representations, where multiple perspectives and creative solutions are encouraged. It is a collaborative approach where multidisciplinary teams contribute to various solutions.

Implementation: The final phase involves bringing the ideas to life through prototyping, testing, and refining. This stage ensures that the solution is feasible, scalable, and impactful. A communication strategy is essential to this stage (Verganti et al., 2021).

The IDEO model is widely used in industry because it emphasises user-centred design and real-world application. It encourages experimentation and continuous refinement, making it a practical approach for tackling complex problems in dynamic environments.

8.4.3 *The Double Diamond Model*

The Double Diamond Design Process, which was developed in 1967 by the UK Design Council to enhance the effectiveness of the UK design industry, offers a structured framework for design thinking. It consists of four stages: Discover, Define, Develop, and Deliver. The model is visually represented as two connected diamonds, illustrating the divergence and convergence of ideas (see Fig. 8.3). As an instructional model, it is used for instructing the design of products, services, and communications (Design Council, 2025).

There are two phases and four stages in the Double Diamond Model.

Phase 1: Problem Phase: This is where a comprehensive understanding of the problem is developed. It requires systematically defining the issue and establishing clear objectives before generating solutions. It involves two stages:

> **Discover:** The first stage focuses on research and exploration. Designers engage in user research, market analysis, and contextual inquiry to uncover insights about the problem space. This stage is about gathering as much information as possible through a divergent thinking process (de Chantal et al., 2020) before narrowing down the focus (convergent thinking).

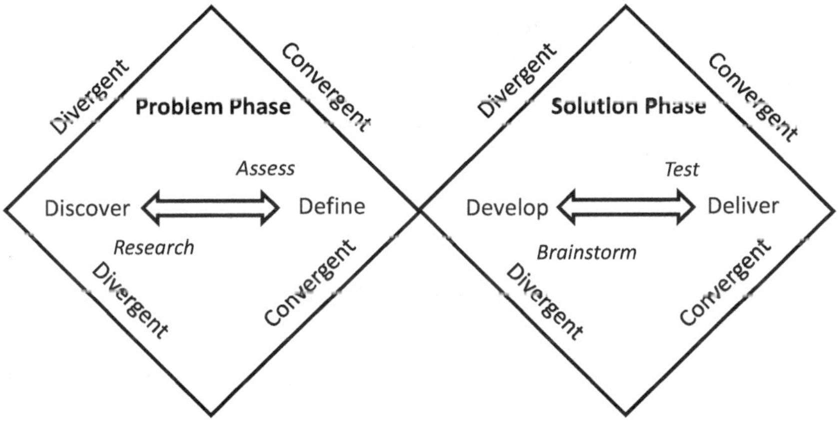

Fig. 8.3 Double diamond model (*Source* Modified from Kidman & Tan, 2025)

Define: In this phase, designers synthesise research findings and define the core problem. This stage involves filtering information, identifying key challenges, and formulating a clear problem statement that guides the design process. The most promising concepts are enhanced via feedback, iteration, prototyping, testing, and aligning with project objectives (Kidman & Tan, 2025).

Phase 2: Solution Phase: This comes after the problem phase and emphasises creating and choosing possible solutions (Kidman & Tan, 2025):

Develop: Designers generate and refine multiple solutions once the problem is clearly defined. This phase consists of brainstorming, creating prototypes, and conducting iterative tests, allowing students to convert abstract concepts into tangible designs (Liu et al., 2020)
Deliver: The final stage focuses on implementing and finalising the solution. Designers test prototypes in real-world settings, refine them based on feedback, and ensure the solution is ready for deployment (Ragni et al., 2018).

The Double Diamond Model benefits large-scale projects and policy design, emphasising a rigorous approach to problem definition and solution development. Its structured yet flexible nature makes it applicable to various fields, including education, healthcare, and technology.

The three design models—Stanford d.school, IDEO, and Double Diamond—offer unique insights and methodologies for approaching complex problems. While they differ in structure and emphasis, all are committed to iterative thinking, creativity, and user-centred design. Understanding these models can help educators, designers, and engineers effectively apply design thinking in their respective fields.

8.4.4 Comparison of the Three Models

The three models reviewed in this section align with STEM/STEAM education by fostering creativity, innovation, and iterative problem-solving. However, their classroom implementation varies in complexity and adaptability. The Stanford d.school and IDEO models offer strong user-centred approaches. The Double Diamond Model, though applicable for large-scale problem-solving, may require adaptations for classroom use. It is, however, advocated for classroom use as an instructional model in design-based learning areas in Australia's Victorian Curriculum (Kidman & Tan, 2025). A key similarity across all models is their emphasis on iteration, encouraging continuous refinement and improvement. However, their fields of origin and primary application areas differ, affecting how they are best integrated into educational settings. The three models are also of 'Western' origin (from the UK and the USA).

8.5 Design Thinking and a "Western" Bias

To further explore this bias, we conducted a Google search for images using the prompt *design thinking model examples* and selected the images tab (see Fig. 8.4). We acknowledge there are possibly hundreds of thousands of similar images available on which to base our discussion. However, we took a screenshot of these 240 images as our data selection. We did not aim to select 240 images; the number appeared on the screen while enabling some readability for easy analysis.

Each image was opened and explored for its relevance as a design thinking model potentially suitable for use in education. Those with relevance were then considered for the country of origin. Table 8.1 summarises the findings. Of the sample, almost half were considered to be links to websites not containing a Design Thinking model suitable for use in an educational setting. For example, one was a Gantt chart for enacting a design thinking process. Of the remaining 140 links, we found 30 sites with more than one design thinking model. These sites provided model comparisons, so each model was counted. Hence, the total number of models exceeded the Google 'images' page count.

Many of the 'images' were repetitive. For example, the popular Stanford d.school model (ME113, 2012) often appeared as an original backlink or a backlink to a site with a Stanford d.school model modification. The Google search on the design thinking model was biased toward images from the United States of America and the United Kingdom to a lesser extent. However, this is unsurprising as many innovative companies are housed in these countries and disseminated globally. However, the United States of America and the United Kingdom are in the Global North. Their education systems differ from those of countries in the Global South. This results in problems translating educational initiatives

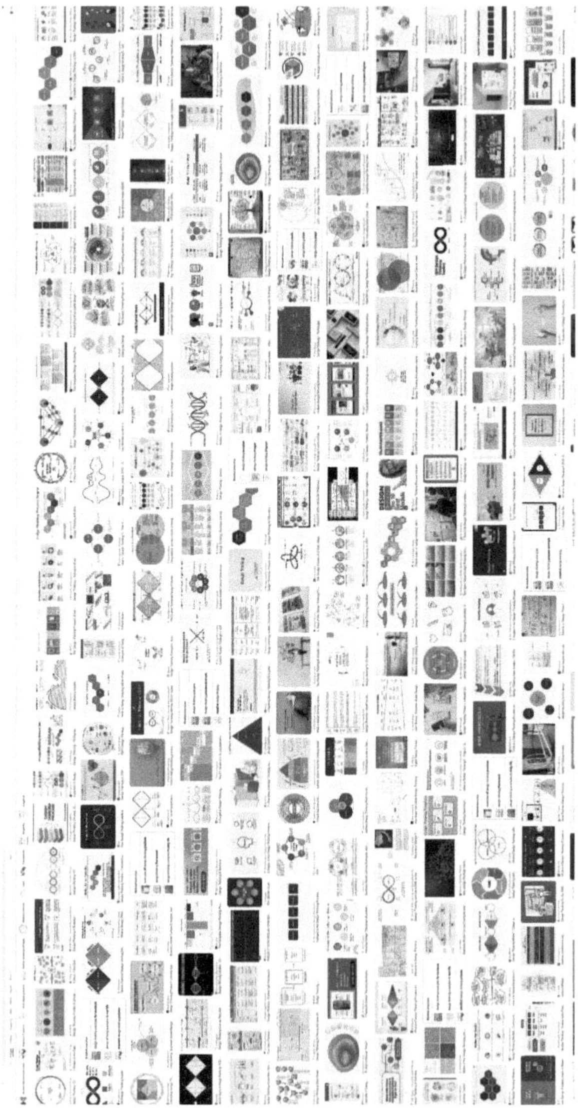

Fig. 8.4 Dominance of 'western' design thinking models (*Source* Authors. Date Collected 22 February 2025)

Table 8.1 Western bias of design thinking models

Criteria of assessment	*Numerical count*
Sites NOT relevant to educational design thinking	100
Sites relevant to educational design thinking	140
Origin of the website or research paper:	125
• United States of America	40
• United Kingdom	2
• Australia	1
• The Netherlands	1
• Spain	1
• Hong Kong	
Total	410

from the global north to the global south. Similarly, we cannot assume we can translate 'Western' initiatives directly into non-Western educational contexts (Kidman & Chang, 2024).

8.6 Developing a Culturally Responsive STEA$^+$M Framework: A Southeast Asian Case Study

This section presents a case study undertaken to overcome some of the Western Global North bias described in the previous section. We conducted a week-long professional development event for 81 science, mathematics, and technology teachers from 10 Southeast Asian countries. The *Teaching to Transform Regional Workshop* was held in Penang, Malaysia, and sponsored by the Southeast Asian Ministers of Education Organization (SEAMEO) Regional Centre for Education in Science and Mathematics (RECSAM).

The Southeast Asian Ministers of Education Organization (SEAMEO) is a regional intergovernmental body created in 1965 by the eleven governments of Southeast Asia to foster regional collaboration in education, science, and culture. Responding to the question of what twenty-first century science and mathematics education entails and identifying the essential skills, knowledge, understanding, values, and attitudes that all Southeast Asian learners should acquire, Mangao et al. (2017) published The SEA-BES Common Core Regional Learning Standards (CCRLS) in Mathematics and Science. These standards are seen as:

> a tool to strengthen regional collaboration on curriculum standards and learning assessment, curriculum and professional development and a mechanism to enhance competitiveness in the era of globalization, nurturing ASEAN values and spirit and in building a harmonious progressive, peaceful and sustainable ASEAN community. (Mangao et al., 2017, iii)

The Common Core Regional Learning Standards define the guidelines and benchmarks for advancing STEM education within the SEAMEO Member Countries. In the Common Core Regional Learning Standards framework for Science, STEM education is featured as a topic entitled "The World of STEM" within the Science Inquiry strand. This section emphasises understanding how STEM operates, including aspects such as STEM careers, applications, the design process, and the importance of evidence. However, the *Science, Engineering and Technology for a Sustainable Society* strand provides more guidance to the teacher and curriculum planner. In this strand, students are engaged in problem-solving and higher-order thinking to develop a deep understanding of sustainability issues through hands-on experiences (Mangao et al., 2017). The Standards are grouped for Key Stage 1 (Grades 1–3), Key Stage 2 (Grades 4–6) and Key Stage 3 (Grades 7–9). Key Stage 3 students are to design products and models, undertake project-based activities, and design ways to use and evaluate technologies and their applications relating to sustainability issues.

Consequently, the Common Core Regional Learning Standards were a guideline for choosing topics and standards while creating STEM activities and materials for a week-long professional development workshop titled Teaching to Transform: 21st Century Skills and Design Thinking in STEM Contexts. This workshop was in line with the SEA-BES project and SEAMEO Action Agenda 2016–2020 Priority #7, which emphasises the implementation of a twenty-first century Curriculum as follows:

> To pursue a radical reform through systematic analysis of knowledge, skills, and values needed to effectively respond to changing global contexts, particularly to the ever-increasing complexity of the Southeast Asian economic, sociocultural and political environment, developing teachers imbued with ideals in building ASEAN Community. (SEAMEO Secretariat, 2016, p. 9)

The Teaching to Transform Workshop systematically analysed knowledge, skills, and values to create a STEM process model suited for Southeast Asia classrooms. This process model was based on analysing readily available online 'Western' STEM education models and the experiences, knowledge, skills, and values of the teachers who attended the Teaching to Transform professional development workshop. In developing the process model, four principles from the SEAMEO Action Agenda (SEAMEO Secretariat, 2016) and the Common Core Regional Learning Standards (Mangao et al., 2017) guided the construction. These were to:

- focus on the students, not the curriculum
- consider the vision, focus, objectives, and student needs
- consider the resources that each SEAMEO Country has available
- consider the tasks a teacher has confidence in developing.

The authors conducted an integrative review of readily available online STEM pedagogy models and then as a workshop activity with the teachers. As the participants were expert teachers and subject specialists in mathematics, science and digital technology drawn from 10 Southeast Asian countries (the 11th SEAMEO country of Timor-Leste was not represented), we were cognisant of their English language proficiency. Therefore, we designed the integrative review workshop to focus on the visual representations found online and not on the research papers or websites supporting the STEM pedagogy visual representations. The workshop activity examined, critiqued, and synthesised each visual representation iteratively, the 'STEM pedagogy models', by considering each inherent concept. The goal was to create a new 'STE*A*+M process framework' based on Southeast Asian perspectives for potential STEM lesson planning. The methodology used in the project called on the participants to have significant education and classroom teaching insights from their respective Southeast Asian country. The participants needed to examine and critique a set of STEM pedagogical models, adopt or discard inherent key concepts and relationships, and then synthesise adopted concepts to create new understandings in the form of the desired new 'STEM process framework'. This STEM process framework was envisaged to be a guide for STEM lesson planning by Southeast Asian teachers (see Chapter 9).

Following an Internet search using the term 'STEM design thinking pedagogy models', 54 visual representations of STEM pedagogy models were downloaded and analysed against the following criteria:

- had a focus on guiding the teacher's lesson planning
  - a step-by-step proccss was cvident
  - promoted student-centred learning
- had a focus on guiding the student toward solution generation
  - a step-by-step process was evident
  - had the potential for hands-on design work

The integrative review was concept-centric; a concept's presence (or absence) became the unit of analysis for the review. Following Webster and Watson (2002), a concept matrix was utilised to assist in categorising the 54 STEM pedagogy models in terms of their key concepts and relationships. The concept matrix enabled us to discover recognisable patterns of essential STEM pedagogy concepts and the subsequent selection of STEM pedagogical models for use in the professional development workshops. Table 8.2 provides a truncated example of the concept matrix developed for the model selection. The concept matrix showed that readily available online visual representations are constructed from differing concepts and depict different relationships, yet all can be considered STEM pedagogy models.

Following an analysis of the 54 visual representations, we selected 14 STEM pedagogy models for the integrated review. These 14 STEM

Table 8.2 Truncated concept matrix

Model	*Concepts*				
	A	*B*	*C*	*D*	*E*
Centre for Research on Learning and Teaching (2019)	✔	✔			✔
Science Buddies (2019)			✔	✔	
Pontis (2015)			✔	✔	
Australian Institute of Management (2017)			✔	✔	
...					

Note A = Disciplinary content; B = Tools; C = Define problem; D = Develop solution; E = Teacher role ...

pedagogy models were also noted to be of two typologies based on the language and terms used: Set A was considered to have the most potential for teacher use to plan STEM lessons and activities; Set B was considered to have the most potential for student STEM design planning and solution generation. Thus, the integrative review was further defined through the lens of the model user—the teacher or the student.

- Set A ($n = 9$) Teacher lesson planning potential
  - Teaching with Technology (Centre for Research on Learning and Teaching, 2019)
  - A TPACK model (Anderson et al., 2017)
  - Conceptual model for teaching arithmetic (Spooner et al., 2017)
  - Team-initiated problem-solving (TIPS II) model (Horner et al., 2015)
  - Conceptual framework of the math problem-solving process (Krawec, 2010)
  - Science and Engineering Practices (Mills, 2015)
  - Young children's inquiry (Worth, 2010)
  - The difference between science and engineering (Science Buddies, 2019)
  - Writing lab reports: Overview (Trent University, 2019)
- Set B ($n = 5$) Student solution generation potential
  - Polya's 4 Step problem-solving (Ruthankoon, 2018)
  - NGSS Engineering Design Process (CDE, 2016)
  - Design thinking revised (Pontis, 2015)
  - MYP Design Tech Cycle (Santiago College, ND)
  - Australian Institute of Management (2017)

Participants were divided into ten table teams. Each team received a variety of STEM pedagogy models to evaluate. Five teams worked with models designed for teacher planning (Set A), while the other five focused on models about student design learning (Set B). Each team was tasked with conducting an in-depth critique of the details within each STEM pedagogy model and examining the merits of each concept. This critical analysis required the participants to deconstruct each STEM pedagogy model into its basic concepts. Over 2 hours, as a team, they identified the strengths and key contributions of each model's concepts.

Torraco (2005) recommended that "any deficiencies, omissions, inaccuracies, and other problematic aspects" (p. 362) of a model should be identified. Once each STEM pedagogy model was thoroughly analysed and critiqued, teams dedicated another two hours to exploring the bigger picture by comparing concepts across the models. Through this iterative critique process, each team developed their own STEM planning model (Set A, which emphasises lesson planning) or STEM design model (Set B, focusing on student-driven solution creation). By conceptually reconstructing models, participants had a clearer understanding of either STEM teacher lesson planning or student STEM solution generation that pertained to their Southeast Asian contexts.

The ten reconstructed models, one from each team, were showcased as posters around the room, each explained by its creators. A concensorgram methodology (Range et al., 2013) illustrated the distribution of participants' attitudes toward specific concepts or relationships featured in each STEM planning or design model. Participants reviewed the reconstructed STEM models using small dot stickers (see Fig. 8.5). In this review phase, they placed stickers on concepts and/or relationships essential for inclusion in the new STEM process model. Analysing the clustering of dot stickers revealed that some concepts and relationships were viewed as more significant and relevant for the STEM process framework. These concepts and/or relationships were discussed in a group setting, leading to a final selection of ideas and relationships.

The concepts selected during the critical analysis were written on Post-it notes and placed on poster paper along with writing materials for creating the relationships. The teachers collaborated with researchers over the next two days to explore concepts and relationships. As a result, two new models were developed, introducing fresh relationships and viewpoints on STE*A*⁺M design thinking that are well-suited for Southeast Asian classrooms. Using the same critique methods applied to evaluate the 'Western' STEM pedagogy models, the concepts and relationships were integrated into the STE*A*⁺M Planning and Design Learning (PaDL) Framework (Mangao et al., 2022). This framework features two models to guide teacher lesson planning and facilitate student learning and solution generation.

The following section presents the STE*A*⁺M PaDL Framework, which consists of two main models: the Design Planning Model and the Design Learning Model. These models make up the STE*A*⁺M PaDL Framework, represented in Figs. 8.6 and 8.7. It demonstrates how the Planning

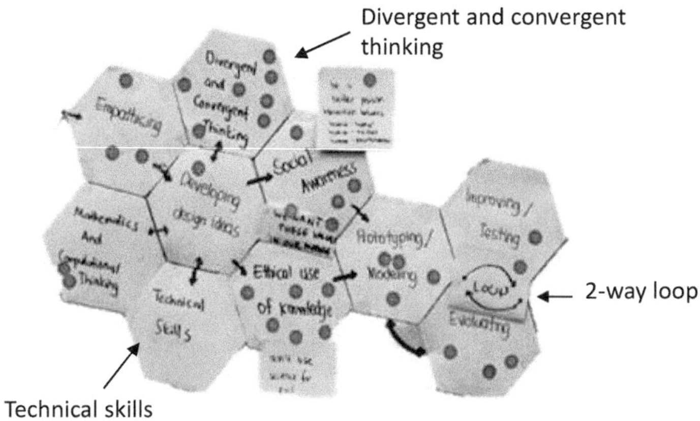

Fig. 8.5 Consensorgram (*Source* Authors)

and Design Learning (PaDL) processes complement each other, serving as a shared starting point for teachers and students (Mangao et al., 2022). The PaDL Framework minimises text and utilises straightforward symbols to depict the actions taken at each step of the STE*A*$^+$M design thinking process and their influence on subsequent decisions and activities. The framework has been tested with novice STE*A*$^+$M design thinkers, revealing that teachers and students could effectively follow the PaDL Framework to organise and teach a sequence of four design lessons. The design and clarity of the PaDL Framework and its two process models enable teachers to estimate better the time required for each lesson task and guide students accordingly.

8.6.1 STEA$^+$M PaDL—Design Planning Model

Figure 8.6 is the *STEA$^+$M Design Planning Model* (Mangao et al., 2022) and indicates that teachers have a starting point for creating enthusiasm in the class. The teachers stressed that this was essential as students from Southeast Asian countries are generally unfamiliar with student-centred learning. As a result, many students would be apprehensive about undertaking a collaborative design challenge with no straightforward 'right or wrong' answer. For many Southeast Asian teachers and students, undertaking a design challenge marks a point of change in how teaching and

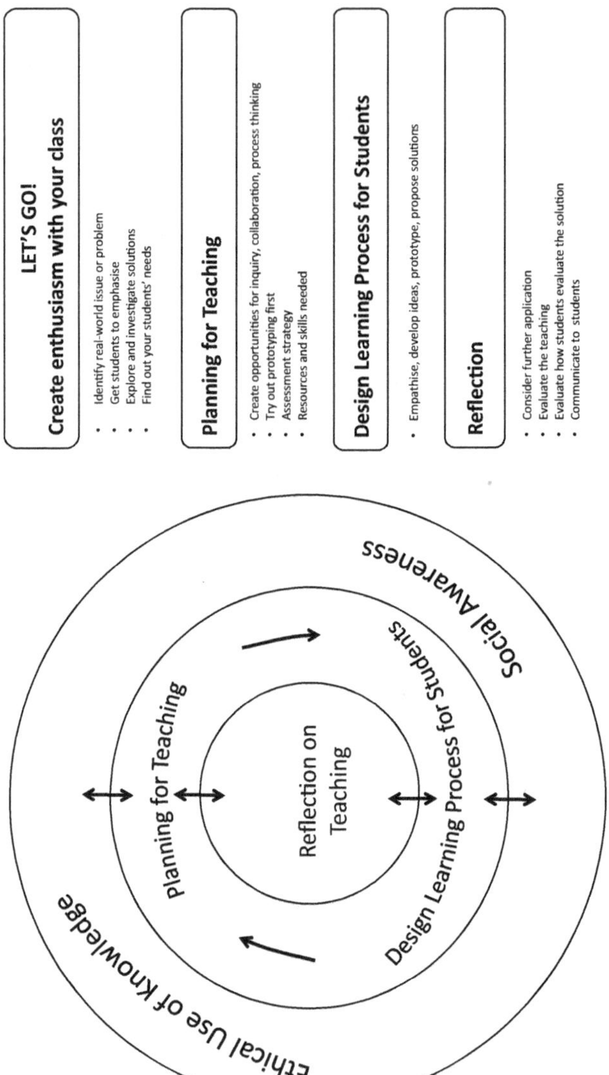

Fig. 8.6 PaDL—design planning model (*Source* Authors)

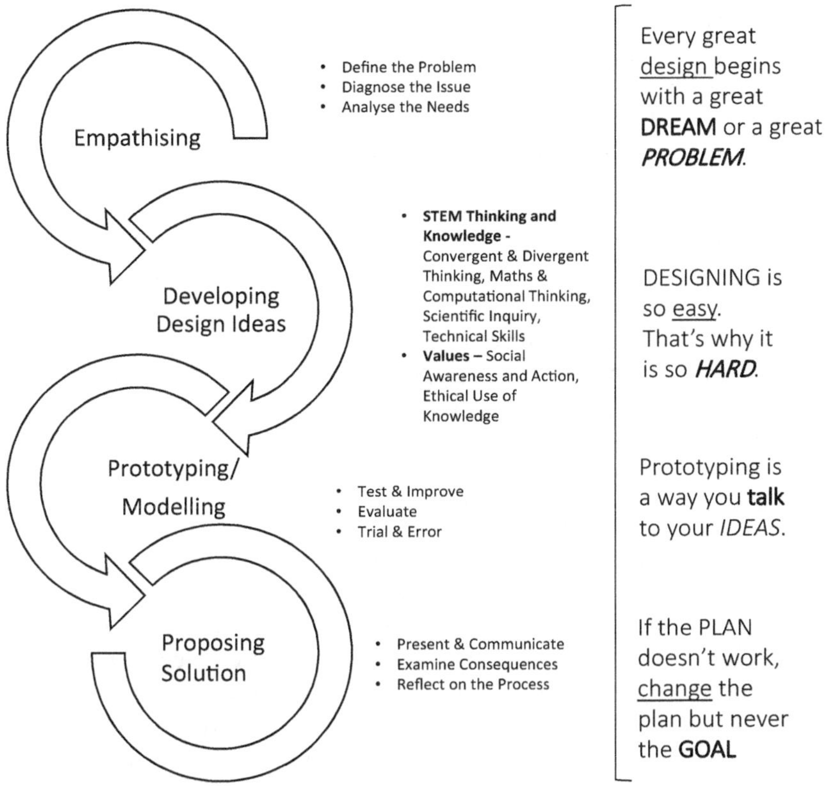

Fig. 8.7 PaDL—design learning model (*Source* Authors)

learning occur. Such challenges were seen as having the potential for the teachers to connect with their students, explore their preferred learning styles, and connect to their local context. The teachers felt it essential that it be explained to the students that they were being challenged in new and unfamiliar ways. It must be emphasised to the students that there are no explicit clues or hidden answers and that rote memory is not required. Instead, the teacher should assist students in developing a sense of purpose and igniting their passion drive. Notably, the teacher needs to prepare students to embrace a sense of frustration, expect mistakes, and try several times to generate a solution.

Teachers believe that students must understand the ethical use of knowledge. As students gather and utilise evidence, moral dilemmas may arise, necessitating a degree of skepticism and openness to alternative explanations and potential data manipulation. Since some students may be engaging in group work and sharing their learning ideas for the first time, mastering social norms becomes essential. Skills such as social awareness and empathy toward the perspectives of others are vital, enabling students to communicate effectively with their peers and resolve conflicts when they occur. An essential goal for Southeast Asian students is to recognise the needs of their peers and communicate accordingly. The teachers felt that only after addressing the ethical and social aspects relevant to the task can they guide students in contemplating the content and nature of the problem, ultimately leading them to seek solutions.

During the planning phase for the teaching step, the teacher selects a real-world problem or issue for students to solve. This choice aligns with the Common Core Regional Learning Standards and the local curriculum. The teacher then organises learning activities and appropriate scaffolding, resources, and assessment methods. These planned activities guide students through the design learning process, which includes stages like empathy and proposing solutions. Importantly, the teacher should allocate time for students to reflect on topics such as the ethical use of knowledge, the implications and consequences of their solutions, teamwork in groups, and the development of other twenty-first century skills. One science educator from the Philippines described the STEA^{+}M Design Planning Model for learning as "a pedagogical approach which scaffolds the lesson structures will develop students' collaborative, problem-solving, twenty-first century skills while learning and applying their disciplinary knowledge and skills". This description emphasised that the teacher had to incorporate collaboration, problem-solving and twenty-first century skills into their lessons, not just disciplinary knowledge. A mathematics educator from Malaysia identified the importance of the students constructing their knowledge, of being able to "discover knowledge and to realise this discovery is related to prior learnings, only then does it become learned knowledge". Inquiry-based processes are critical to STEA^{+}M design learning. Most Southeast Asian teachers are unfamiliar with inquiry-based approaches, and the solution offered was to suggest that in the planning for the teaching step, the lesson sequence was designed to follow the 4-step layout of the STEA^{+}M Design Learning Model.

Throughout the planning process, the teachers felt the need to be cognisant of the Common Core Regional Learning Standards and local curriculum, as the real-world problem or issue was challenging. Suppose classroom time were to be devoted to STE*A*⁺M design learning. In that case, the STE*A*⁺M design learning activities of solution generation must be closely related to the mandated curriculum, and this requires regular checking by the teacher during the teaching and planning. Thus, reflection was central to the STE*A*⁺M Design Planning Model. A mathematics educator from Malaysia commented that when reflecting, it is hoped that the "teachers begin to exercise the possibility of transforming their everyday classroom life, which benefits their being professional and improves the learning support they provide their students". However, it was widely acknowledged that this would be difficult to achieve due to the rare opportunity for STE*A*⁺M learning in some classrooms due to the high-stakes testing in most countries.

8.6.2 STEA⁺M PaDL—Design Learning Model

Figure 8.7 is the *STEA⁺M Design Learning Model* (Mangao et al., 2022), which was designed for students' use. It was envisaged that students would be working in small design groups of 4 to 6 students. Each group would have a copy/ies of the STE*A*⁺M Design Learning Model. Large copies might be printed as posters on the classroom walls or projected onto a screen. The teacher would be pacing the student groups through the 4-step process over several lessons (or over a single lesson for a short, simple design sprint). By deliberately directing the students back to the STE*A*⁺M Design Learning Model during the design lesson, the "students would learn the steps to think like a designer and problem solver", as an Indonesian technology educator explained.

Figure 8.7 indicates that the *PaDL—Design Learning Model* (Mangao et al., 2022) for students has a starting point of empathising with real-world problems or issues. The student design process's empathy stage involves cultivating a human-centred perspective on problem-solving. Teachers wanted the students to observe other people as they valued the skill of examining the local people and listening to their needs and requirements. One Vietnamese mathematics educator explained, "This will help to have empathy and become more deeply understanding". The empathy stage focuses on placing others at the forefront of every activity or project. This human-centred approach was deliberately placed at the beginning of the design process so the solution becomes anchored in the student's real world. A Malaysian Science educator considered empathy essential to leading students and teachers along "the path to making harmonious the right decisions and avoiding societal design failure when solving problems". This notion of a societal design failure was defined to be a solution that harmed society. The students needed to understand that design failure is good when generating solutions—as this is part of learning—but societal design failure is not good as it causes harm. Without empathy, there is no solution. Thus, the initial step of the STE*A*⁺M Design Learning Model involves the student relating to the STE*A*⁺M real-life problem or issue, exploring it from the perspective of others, and becoming aware of what may be considered a societal design failure. Once the student group has collected their insights, they can synthesise them and define the problem. The teachers emphasised that students needed to understand that their prototype or model had people and society at the centre of their solution and that the solution did not contribute to an undesirable outcome for people or the environment.

In the developing design ideas step of the STE*A*⁺M Design Learning Model, students are coached through brainstorming activities to identify different ideas based on real-world problems or issues. These ideas are based on the empathising work completed in the previous step. The students are encouraged to "go wild with different ideas! Write or draw them out. Link different things together to create new ideas" (Malaysian science teacher). Students must work as a group to "find the courage to explore with their hands and with friends. Work not from memory and not from the textbook" (Cambodian mathematics teacher).

The teachers believe it is essential for students to recognise the importance of social awareness and the ethical application of knowledge at this stage. Divergent thinking was highlighted during discussions as the primary brainstorming skill that should be cultivated. Following this, student groups were expected to refine their choices to identify the optimal solution for the prototypes or models, showcasing convergent thinking. It was noted by a Malaysian mathematics teacher that the terms divergent and convergent thinking were to be used on the STE*A*⁺M Design Learning Model and be explicitly taught to the students, but that students are also to be told to use their "conscious and unconscious mind, logical thinking, computational thinking, and imagination".

The next step in prototyping and modelling should accompany a design idea. This process quickly transitions from an initial paper design to a functional solution, physical prototype, or 3D model. The teachers were concerned that the drawn design would be discarded during the excitement of the build and felt a link between the 2D drawing and the 3D model should be evident. This was linked back to the earlier comment by the Indonesian technology educator, who said that students were to learn to think like designers. The essential learning from this step and the subsequent construction step involves deep thinking that reflects the student group's understanding of the problem and their proposed solution. Moving from idea generation to hands-on action and construction requires mental activity that constructs new knowledge. Students must collaborate in groups, draw design plans, communicate, explore alternative ideas, and build new knowledge to do this.

The prototype and modelling phase of the student design process entails creating a distinctive model that presents a potential solution to a real-world problem or challenge. Students should recognise that the final prototype or model:

- serves as a depiction of a suggested solution—meaning it doesn't need to function completely
- aims to portray what the solution could look like, how it may operate, or how it can be utilised
- might not match the size or materials of the proposed final solution
- reflects your creativity, imagination, and playful experimentation

The teachers felt strongly about using an inquiry approach for STE*A*⁺M design lessons. "Students need to embrace failure and learn from mistakes as engineers do. The teacher must inform the students that they learn from their failures" (Singaporean mathematics teacher).

Proposing a solution is the feedback stage of the design learning lessons. The students have the opportunity to evaluate their prototype/model. They need to be encouraged to consider the consequences of their prototype/model and then communicate their group's solution to the real-life problem or issue. Students are required to 'pitch' their solution to an audience to convince the audience that their solution is the best. The teachers felt classroom teachers might not plan time for the 'pitch', so it was necessary to link the pitch to the twenty-first century skill components of the local curricula to which teachers are accountable.

In preparing the 'pitch', the teachers felt students should reflect on the entire design process and improve their 2D drawing and the corresponding prototype/model. The 'pitch' should consider:

- the different components of the prototype/model and how they might be improved or modified to fit a new context
- the consequences of the prototype/model when used. Does the prototype/model solve the problem?
- the most effective method to convey the solution is by using words, diagrams, graphs, and various media to illustrate your approach and persuade others that it is the optimal solution.

When designing the STE*A*⁺M PaDL Framework, the teachers considered conceptual-contextual links. The concepts and contexts were separation-sensitive. The collaborative teamwork called upon disciplinary expertise to generate an educational model for interdisciplinary STEM planning and learning. To do this, conceptual and contextual thinking coincided, indicating an interdependence where one existed independently from the other. Further, each teacher utilised critical thought processes as they argued for each inclusion or exclusion of concepts and relationships in the STE*A*⁺M PaDL Framework.

Conceptual thinking integrates a set of abstract considerations based on a central idea (Yuan & Zhang, 2021). Conceptual thinking requires connecting abstract, disparate ideas. Conceptual thinking must be considered to justify the arrangement of the PaDL Framework as it represents the "theoretical glue that welds the model together" (Whetten, 1989, p. 491). The teachers had disciplinary-based knowledge of the disparate concepts in the 'Western' STEM pedagogical models. Still, it was through their collaborative conceptual thinking that they were able to connect the ideas to generate an understanding of STEM education—an educational approach that these educators had not experienced previously as a learner or as an educator. Contextual thinking focuses on the social, physical, and cultural dimensions of the classroom and educational systems (Czafík et al., 2021). Contextual consideration was essential as it influences the future use of the STE*A*⁺M PaDL Framework, designed to reflect the values and traditions of Southeast Asian societies.

Thus, the STE*A*⁺M PaDL Framework results from a critical symbiosis of a creative blend of design concepts and the context of Southeast

Asian classrooms. Through the STE*A*$^+$M PaDL Framework, a Southeast Asian identity for STE*A*$^+$M design education has become a reality, as it was created according to the characteristics of external features—in this case, Southeast Asian classrooms. Teachers need support in developing their instructional planning knowledge and skills across their careers. This is particularly true when new educational directions such as STE*A*$^+$M emerge. The situation is further complicated when diverse populations are considered—additional language, gifted, and special needs learners. There is a need to refine further and improve the support provided to these teachers. Support is also required in differentiated assessment based on student needs. Thus, the STE*A*$^+$M PaDL Framework must be implemented with diverse populations.

8.7 Conclusion

Integrating design thinking within STE*A*$^+$M education represents a transformative shift in pedagogical practice, emphasising creativity, interdisciplinary collaboration, and problem-based learning. This chapter has examined the theoretical underpinnings of design thinking, traced its evolution across various disciplines, and analysed its application within STE*A*$^+$M education. The review of established models, including the Stanford d.school model (ME113, 2012), IDEO model IDEO (Design Thinking, n.d.), and the Double Diamond Model (Design Council, 2025), has highlighted their contributions to fostering iterative, user-centred problem-solving. However, the dominance of Western frameworks underscores the need for culturally responsive adaptations that align with diverse educational contexts. Chapter 7 further discusses hidden curriculum contexts in STEM and STEAM education.

The case study of Southeast Asian teachers developing the PaDL Framework demonstrates the potential for localised innovation in design thinking pedagogy. Educators can create frameworks that reflect their students' unique needs and priorities by critically engaging with existing models and synthesising new approaches. The PaDL Framework (Mangao et al., 2022), with its dual emphasis on planning for teaching and structuring student learning, provides a valuable tool for bridging theoretical knowledge with practical application. The framework emphasises student *A*gency, ethical considerations, and iterative learning, providing students with essential skills to tackle complex, real-world challenges.

One of the key strengths of design thinking lies in its emphasis on iteration and reflection (de Chantal et al., 2020). Unlike traditional problem-solving approaches that seek definitive answers, design thinking encourages learners to embrace uncertainty, experiment with multiple solutions, and refine their ideas based on feedback. This iterative process fosters resilience, adaptability, and critical thinking—skills essential for success in STEM fields (McCurdy et al., 2020). Additionally, emphasising empathy ensures that students develop socially responsible solutions considering technological and scientific advancements' broader human and environmental impacts.

Despite its many advantages, the widespread implementation of design thinking in STE*A*$^+$M education faces several obstacles. Institutional constraints, such as rigid curriculum structures and high-stakes assessments, often limit opportunities for creative exploration. Additionally, educators require professional development and support to effectively integrate design thinking into their teaching practice. Addressing these challenges requires systemic changes prioritising innovation, interdisciplinary collaboration, and student-centred learning. As educational landscapes continue to evolve, the role of design thinking in shaping future-ready learners cannot be overstated. This chapter advocates for a more inclusive and contextually relevant approach to design thinking in STE*A*$^+$M education, recognising the importance of adaptability and cultural responsiveness. By fostering creativity, collaboration, and problem-solving, design thinking has the potential to transform education and prepare students to engage meaningfully with the complexities of the twenty-first century. Continuing research and dialogue will be essential in refining and expanding the applications of design thinking, ensuring its effectiveness across diverse educational settings.

References

Abaniel, A. (2021). Enhanced conceptual understanding, 21st century skills and learning attitudes through an open inquiry learning model in Physics. *Journal of Technology and Science Education, 11*(1), 30–43. https://doi.org/10.3926/joste.1004

Anderson, S., Griffith, R., & Crawford, L. (2017). Preservice teacher decision making while integrating iPads into instruction. *Contemporary Issues in Technology and Teacher Education, 17*(1), 97–127. https://www.learntechlib.org/primary/p/173323/

Australian Institute of Management. (2017). *Getting to grips with design thinking.* https://www.aim.com.au/blog/getting-to-grips-with-design-thinking

Bakar, S. (2021). From investigating the dynamics of contemporary pedagogical approaches in higher education through innovations, challenges, and paradigm shifts. *Social Science Chronicle, 1*(1), 1–19. https://doi.org/10.56106/ssc.2021.009

California Department of Education (CDE). (2016). *The CA science framework, author.* http://workshops.sjcoe.org/Uploads/72520171234574417 4.pdf

Centre for Research on Learning and Teaching. (2019). Getting started with technology. As cited from Zhu, E., & Kaplan, M. (2014). *McKeachie's teaching tips* (14th ed.). Wadsworth Cengage Learning. http://www.crlt.umich.edu/teaching-technology/getting-started

Christian, K. B., Kelly, A. M., & Bugallo, M. (2021). NGSS-based teacher professional development to implement engineering practices in STEM instruction. *International Journal of STEM Education, 8*, Article 21. https://doi.org/10.1186/s40594-021-00284-1

Czafík, M., Puškár, B., & Vráblová, E. (2021). Conceptual-contextual thinking in architectural education. *Global Journal of Engineering Education, 23*(2), 106–111. http://www.wiete.com.au/journals/GJEE/Publish/vol23no2/05-Czafik-M.pdf

Design Council. (2025). *The double diamond.* https://www.designcouncil.org.uk/our-resources/the-double-diamond/

de Chantal, P. L., Gagnon-St-Pierre, É., & Markovits, H. (2020). Divergent thinking promotes deductive reasoning in preschoolers. *Child Development, 91*, 1081–1097. https://doi.org/10.1111/cdev.13278

Garibay, J. (2015). STEM students' social agency and views on working for social change: Are STEM disciplines developing socially and civically responsible students? *Journal of Research in Science Teaching, 52*(5), 610–632. https://doi.org/10.1002/tea.21203

Horner, R. H., Newton, J. S., Todd, A. W., Algozzine, B., Algozzine, B., Cusumano, D., & Preston, A. (2015). *Working with data—An advanced step in team-initiated problem solving (tips) training.* https://slideplayer.com/slide/12026975/69/

IDEO Design Thinking. (n.d.). *IDEO design thinking*, https://designthinking.ideo.com/

Johnson, J., Macalalag, A., & Dunphy, J. (2020). Incorporating socioscientific issues into a STEM education course: Exploring teacher use of argumentation in SSI and plans for classroom implementation. *Disciplinary and Interdisciplinary Science Education Research, 2,* Article 9. https://doi.org/10.1186/s43031-020-00026-3

Kidman, G., & Chang, C. H. (2024). Divergent futures—A geography of unequal concerns. *International Research in Geographical and Environmental Education, 34*(1), 1–4. https://doi.org/10.1080/10382046.2025.2442172

Kidman, G. & Tan, H. (2025). Exploring cognitive principles of instructional models and problem-solving processes in the STEM disciplines. In T. Barkatsas & W. Goff (Eds.), *Advances in mathematics and STEM education: Incorporating developments in neuroscience, brain science and neurodiversity* (pp. 1–37). Brill. https://doi.org/10.1163/9789004722651_001

Kidman, G., Tan, H., & Macdonald, S. (2024). Reimagining STEM education: Developing agency through STEA^+M and design thinking. In T. Barkatsas, P. McLaughlin, & W. Goff (Eds.), *Reimagining education for the second quarter of the 21st century and beyond: Dilemmas, challenges, advancements and innovations* (pp. 1–23). Brill. https://doi.org/10.1163/9789004688490_001

Kim, M. S. (2019). A systematic review of the design work of STEM teachers. *Research in Science & Technological Education, 39*(2), 131–155. https://doi.org/10.1080/02635143.2019.1682988

Krawec, J. L. (2010). *Problem representation and mathematical problem solving of students of varying math ability*. Open Access Dissertations. 455. https://scholarlyrepository.miami.edu/oa_dissertations/455

Lahey, J. (2017, January 4). How design thinking became a buzzword at school. *The Atlantic*. https://www.theatlantic.com/education/archive/2017/01/how-design-thinking-became-a-buzzword-at-school/512150/

Lazar, L. (2018). The cognitive neuroscience of design creativity. *Journal of Experimental Neuroscience*. https://doi.org/10.1177/1179069518809664. PMID: 30450006; PMCID: PMC6236478.

Liu, S., & Li, C. (2023). Promoting design thinking and creativity by making: A quasi-experiment in the information technology course. *Thinking Skills and Creativity, 49*, Article 101335. https://doi.org/10.1016/j.tsc.2023.101335

Liu, H., Guo, D., Sun, F., Yang, W., Furber, S., & Sun, T. (2020). Embodied tactile perception and learning. *Brain Science Advances, 6*(2), 132–158. https://doi.org/10.26599/BSA.2020.9050012

Lyons, T. (2020). Seeing through the acronym to the nature of STEM. *Curriculum Perspectives, 40*(2), 225–231. https://doi.org/10.1007/s41297-020-00108-2

Mangao, D. D., Ahmad, N. J., & Isoda, M. (2017). *SEAMEO Basic Education Standards (SEA-BES): Common Core Regional Learning Standards (CCRLS) in mathematics and science*. SEAMEO RECSAM, Penang, Malaysia.

Mangao, D. D., Kidman, G., Tan, H., Gesthuizen, R., Macdonald, S., & Sinniah, D. N. (2022). *SEAMEO STEM planning and design learning (PaDL) framework: Towards 21st century skills and design thinking*. Southeast Asian

Ministers of Education Organization (SEAMEO) Regional Centre for Education in Science and Mathematics (RECSAM). http://recsam.edu.my/index.php/our-resources/in-house-publication/recsam-publication

Marthaliakirana, A. D., Suwono, H., Saefi, M., & Gofur, A. (2022). Problem-based learning with metacognitive prompts for enhancing argumentation and critical thinking of secondary school students. *Eurasia Journal of Mathematics, Science and Technology Education, 18*(9), Article em2148. https://doi.org/10.29333/ejmste/12304

McCurdy, R. P., Nickels, M. L., & Bush, S. B. (2020). Problem-based design thinking tasks: Engaging student empathy in STEM. *The Electronic Journal for Research in Science & Mathematics Education, 24*(2), 22–55. https://ejrsme.icrsme.com/article/view/19925

ME113. (2012). *A design thinking process.* https://web.stanford.edu/class/me113/d_thinking.html

Micheli, P., Wilner, S. J. S., Hussain Bhatti, S., Mura, M., & Beverland, M. B. (2019). Doing design thinking: Conceptual review, synthesis, and research agenda. *Journal of Product Innovation Management, 36*(2), 124–148. https://doi.org/10.1111/jpim.12466

Mills, H. (2015). What happened to the scientific method? *CORElaborate.* https://www.corelaboratewa.org/blog/teacher-leaders/single/~board/teacher-leaders-archive/post/what-happened-to-the-scientific-method

Morales-Doyle, D. (2017). Justice-centered science pedagogy: A catalyst for academic achievement and social transformation. *Science Education, 101*(6), 1034–1060. https://doi.org/10.1002/sce.21305

National Academy of Sciences, National Academy of Engineering, and Institute of Medicine. (2005). *Facilitating interdisciplinary research.* The National Academies Press. https://doi.org/10.17226/11153

Peters-Burton, E. E., & Johnson, T. (2018). Cross-case analysis of engineering education experiences in inclusive STEM focused high schools in the United States. *International Journal of Education in Mathematics, Science and Technology, 6*(4), 320–342. https://eric.ed.gov/?id=EJ1193468

Pontis, S. (2015). *Design thinking revised.* https://sheilapontis.com/2015/06/04/design-thinking-revised/

Pozilova, S. X., & Zulxumorxon, K. (2024). A pedagogical review of creativity integration in academic subjects in international contexts. *International Journal of Pedagogics, 4*(11), 247–252. https://doi.org/10.37547/ijp/Volume04Issue11-48

Priemer, B., Eilerts, K., Filler, A., Pinkwart, N., Rösken-Winter, B., Tiemann, R., & Zu Belzen, A. U. (2019). A framework to foster problem-solving in STEM and computing education. *Research in Science & Technological Education, 38*(1), 105–130. https://doi.org/10.1080/02635143.2019.1600490

Ragni, M., Kola, I., & Johnson-Laird, P. N. (2018). On selecting evidence to test hypotheses: A theory of selection tasks. *Psychological Bulletin, 144*(8), s779-796. https://doi.org/10.1037/bul0000146

Range, B., Carnes-Holt, K., & Bruce, M. A. (2013). Engaging middle-grade students to learn in a caring community. *The Clearing House: A Journal of Educational Strategies, Issues and Ideas, 86*(2), 48–52. https://doi.org/10.1080/00098655.2012.738438

Ruthankoon, R. (2018). *Problem solving and decision making*. http://kctathailand.com/courses/course-problem-solving-and-decision-making-executive-course-1-day/

Santiago College. (n.d.). MYP design tech cycle. *Design Cycle*. https://sites.google.com/a/scollege.cl/gr9mypdesigntech/design-cycle

Science Buddies. (2019). *Comparing the engineering design process and the scientific method*. https://www.sciencebuddies.org/science-fair-projects/engineering-design-process/engineering-design-compare-scientific-method

SEAMEO Secretariat. (2016). *SEAMEO 7 priority areas + action agenda 2016–2020*. SEAMEO Secretariat, https://www.seameo.org/SEAMEOWeb2/images/stories/Publications/Centers_Pub/SEAMEO_Education_agenda/04%20SEAMEO%207%20Priority%20Areas%20Implementation%20by%20Centres.pdf

Spooner, F., Saunders, A., Root, J., & Brosh, C. (2017). Promoting access to common core mathematics for students with severe disabilities through mathematical problem solving. *Research and Practice for Persons with Severe Disabilities, 42*(3), 171–186. https://doi.org/10.1177/1540796917697119

Syah, H. S., & Nasri, N. M. (2023). The implementation of design thinking in education: Will it serve the purpose? *International Journal of Academic Research in Progressive Education and Development, 12*(3), 2346–2363. https://doi.org/10.6007/IJARPED/v12-i3/19208

Torraco, R. J. (2005). Writing integrative literature reviews: Guidelines and examples. *Human Resource Development Review, 4*(3), 356–367. https://doi.org/10.1177/15344843052782

Trent University. (2019). *Writing lab reports*. https://www.trentu.ca/academicskills/resources/mathsciences_labreportoverview.php

Verganti, R., Dell'Era, C., & Swan, K. S. (2021). Design thinking: Critical analysis and future evolution. *Journal of Product Innovation Management, 38*(6), 603–622. https://doi.org/10.1111/jpim.12610

Waks, L. (2001). Donald Schon's philosophy of design and design education. *International Journal of Technology and Design Education, 11*(1), 37–51. https://doi.org/10.1023/A:1011251801044

Webster, J., & Watson, R. T. (2002). Analyzing the past to prepare for the future: Writing a literature review. *MIS Quarterly, 26*(2), xiii–xxiii. http://www.jstor.org/stable/4132319

Wensveen, S., & Matthews, B. (2014). Prototypes and prototyping in design research. In P. A. Rodgers & J. Yee (Eds.), *Routledge companion to design research* (1st ed., pp. 262–276). Routledge. https://doi.org/10.4324/9781315758466

Whetten, D. A. (1989). What constitutes a theoretical contribution? *Academy of Management Review, 14*(4), 490–495. https://www.jstor.org/stable/258554

Worth, K. (2010). Science in early childhood classrooms: Content and process. *Early Childhood Research & Practice, 12*(2). http://ecrp.illinois.edu/beyond/seed/worth.html

Yuan, L., & Zhang, J. (2021). Contextual knowledge sharing strategy based on contextual thinking in service design. In E. Markopoulos, R. S. Goonetilleke, A. G. Ho, & Y. Luximon (Eds.), *Advances in creativity, innovation, entrepreneurship and communication of design. AHFE 2021.* Lecture Notes in Networks and Systems (Vol. 276, pp. 40–46). Springer. https://doi.org/10.1007/978-3-030-80094-9_6

Zamakhsyari, F., & Fatwanto, A. (2023). A systematic literature review of design thinking approach for user interface design. *JOIV: International Journal on Informatics Visualization, 7*(4), 2313–2320. https://doi.org/10.62527/joiv.7.4.1615

CHAPTER 9

Interdisciplinary STEM Curriculum Integration: Exploring Teacher-Developed Curriculum Goals

List of Abbreviations

ASEAN	Association of Southeast Asian Nations
CCRLS	Common Core Regional Learning Standards
K–12	Kindergarten to Grade 12
OECD	Organisation for Economic Co-operation and Development
PD	Professional Development
RECSAM	Regional Centre for Education in Science and Mathematics
SEAMEO	Southeast Asian Ministers of Education Organization
S, T, E, M, STEM or STEAM	Science, Technology, Engineering, Arts, and Mathematics
UNESCO	United Nations Educational, Scientific and Cultural Organization

G. Kidman et al., *Knowledge Generation in STEM and STEAM Education*, https://doi.org/10.1007/978-3-032-03247-8_9

9.1 Introduction: Background Information and Context

Since the turn of the twenty-first century, Science, Technology, Engineering, and Mathematics (STEM) education has been promoted worldwide. However, the current landscape of STEM education research and publications reflects a notable imbalance, with research and theories relating to non-western contexts relatively scarce, especially in publications in high-quality journals. This gap was highlighted in several systematic reviews of STEM education publications and research (e.g., Huang et al., 2022; Li et al., 2020; Perignat & Katz-Buonincontro, 2019; Takeuchi et al., 2020) that provide overviews of, and track, the fast-expanding field. In their critical review of 143 empirical research articles on STEM education published 2007–2017, Takeuchi et al. (2020) found an overwhelming number (88%) of articles based on studies in the USA, and none on STEM education in Asia. Other reviews of general STEM education research (e.g., Lee et al., 2019; Li et al., 2020) and specifically of STEM teacher professional development programs (Huang et al., 2022; Nugraha et al., 2023) in the last two decades found similar trends, with a small but increasing percentage of publications on studies conducted in Asian regions such as China, Indonesia, Korea, Malaysia, Singapore, and Taiwan. This is akin to the USA/Western bias we explored in Chapter 8 regarding design thinking models.

Within the Asian context, there is also unequal development of STEM education research. Reviewing 175 STEM education research publications from researchers in the ten Association of Southeast Asian Nations (ASEAN) from 2000 to 2019 sourced via the Scopus database, Cao et al. (2020) found that the publications were unevenly distributed among countries: Malaysia (85 articles), Thailand (35 articles), Indonesia (24 articles), Singapore (23 articles), Philippines (10 articles), Vietnam (4 articles), and Cambodia (1 article). No authors from Brunei Darussalam, Laos, or Myanmar published in the field of STEM education. We note that the youngest of the 11 Southeast Asian nations, Timor-Leste, is not part of the ASEAN at the time of this writing.

Overall, this overarching trend underscores the need for more inclusive and diverse perspectives to enrich the field and scholarship, particularly STEM education research from Asia and, more specifically, Southeast Asia. We sought to address this gap by first presenting a summary of current STEM education initiatives in Southeast Asia, followed by introducing a regional mathematics and science curriculum. In the second half of this chapter, we report our study on the analysis of the learning goals of the teacher-designed STEM units in a professional development program, with respect to the regional curriculum standards. We were interested in finding out what mathematics and science curriculum links the Southeast Asian teachers refered to in their STEM units, and what as the nature of the STEM units developed, in terms of the representation of disciplines, integration of cognitive and affective aspects within science and mathematics, and their overall strategic STEM foci.

The motivation for the research questions is the geopolitical policy focus on science and mathematics curricula, as part of the regional strategic focus on STEM education.

9.2 STEM Education in Southeast Asia

In Southeast Asia, the eleven countries (as listed in Table 9.1) include STEM in their education initiatives and curriculum policies. Some countries (e.g., Laos PDR, Myanmar, Timor Leste), whose immediate priority is to provide equitable access to quality education, focus on STEM as separate science and mathematics subjects. Countries like Indonesia, Malaysia, the Philippines, Thailand, and Vietnam have begun incorporating *integrated* STEM education into their strategic education plans and formal curriculum policies. Singapore has kept to the core S, T, and M subjects but promoted integrated STEM in school-based programmes. All Southeast Asian countries also have non-formal STEM education opportunities for their students (see Table 9.1).

As seen in Table 9.1, interwoven with the formal STEM curriculum are ideas such as "21st century" and "competencies". There is an overall focus on promoting twenty-first century skills, such as critical and creative thinking, problem-solving, higher-order thinking skills, collaboration and communication skills, in the formal curriculum. The initiatives in informal STEM education seemed to focus on motivation and engagement with STEM through projects and events, on promoting digital technologies such as robotics and coding and on teacher professional development. STEM education may be seen as promoting twenty-first century skills and competencies by providing holistic and quality education for students.

STEM education reform is not without challenges. Studies have been conducted in Indonesia (Nugraha et al., 2023; Nugroho et al., 2019),

Table 9.1 Southeast Asian countries' formal and non-formal STEM education initiatives

Country	*STEM education initiatives*[a]
Brunei Darussalam	*Formal*: S, T, M as separate subjects in the SPN21 National Education System for the 21st century (Ministry of Education, Brunei Darussalam. (n.d.). *Non-formal*: Establishing the Science, Technology and Environmental Partnership (STEP) Centre to promote STEM/ STEAM education (MOE Brunei Darussalam, 2021), Annual National STEM and Innovative Design Challenge (MOE Brunei Darussalam, 2023), Empowering STEM educators (https://bruneiprimarystem.education/)
Cambodia	*Formal*: Cambodia Secondary Education Blueprint 2030 advocates the promotion of S, T, E and M education (Ministry of Education, Youth & Sport, 2021). New Generation Schools (NGS) reform launched in 2015, particularly focussing on STEM, ICT, critical thinking skills, and inter- and intra-personal competencies (Donaher & Wu, 2020; Vann, 2023) *Non-formal*: STEM is mainly promoted through festivals, forums, programs, and competitions (Vann, 2023), such as the Annual Cambodian STEM Festival, Robotics Olympiad, STEM sisters, and Eco-Hero, organised by the STEM Education Organisation for Cambodia (STEMEOC, https://stemcambodia.ngo/)
Indonesia	*Formal*: Emancipated Curriculum or Kurikulum Merdeka, focussing on essential competencies (literacy, numeracy, and character values) and student-centred teaching and learning (Randall et al., 2022). Project-based learning to develop soft skills and values of the Indonesian ideology, Pancasila (P5), (to strengthen graduate profile), are mandated for each subject (20% in elementary, 25% in secondary, & 30% in high school subjects), and can be based on an integrated STEM education approach (Nugraha et al., 2023) *Non-formal*: STEM inquiry and problem-based projects in schools through partnerships with and research by universities (Nugraha et al., 2023; Suratno et al., 2020). Teachers are provided with STEM professional development (SEAMEO QITEP, n.d.)

Country	*STEM education initiatives*[a]
Lao People's Democratic Republic (PDR)	*Formal*: S, T, M as separate subjects. The education sector development plans for 2016–2020 and 2021–2025 focus on equitable access to quality education (e.g., compulsory primary education, children living in remote areas and extreme poverty, as well as girls), with the forthcoming plan 2026–2030 being informed by the 10-year ICT in Education Strategy (Fukami, 2024; Global Partnership for Education, 2020; UNESCO, 2015) *Non-formal*: STEM education for young Southeast Asian leaders' initiative (O'Bee, 2018), 'Khaiy Panya STEM' bootcamps initiative with UNICEF (Sihaphom, 2024)
Malaysia	*Formal*: Malaysia Education Blueprint 2013–2025. Focusing on Science, Technology, Reading, Arts and Music (STREAM) for secondary students from 2020 (Landau & Dermawan, 2019; Ministry of Education Malaysia, 2013). New blueprint for 2026–2036 will include a focus on four key areas S, T, E and M (Bernama, 2025). Kurikulum Standard Sekolah Rendah (KSSR) and Kurikulum Standard Sekolah Menengah (KSSM) emphasised integrated STEM approaches for science teachers to support inquiry and problem-solving in the context of daily life, promoting higher-order thinking skills (Ismail et al., 2019; Shahali et al., 2017) *Non-formal*: Project-based inquiry projects and initiatives in schools to promote STEM education (Haron et al., 2019; Ng & Adnan, 2018). Friends of STEM and STEM content providers network platforms to advocate interest in STEM education, such as the Online STEM Olympiad and other events (https://www.facebook.com/STEMalaysia/)
Myanmar (Burma)	*Formal*: S, T, M as separate subjects in the Basic Education curriculum. Focus on twenty-first century skills, soft skills, and higher-order thinking skills (Ministry of Education, Myanmar, 2016, 2019) *Non-formal*: Myanmar STEM Competition (https://www.facebook.com/myanmarSTEM/), and Hour of Code (https://thatepanhub.org/news/hour-of-code)

(continued)

Table 9.1 (continued)

Country	*STEM education initiatives*[a]
Republic of the Philippines	*Formal*: The senior high school level (Grades 11–12) in the K–12 basic education program has three tracks: academic, technical-vocational-livelihood, and sports and arts. STEM is one of the three strands in the Academic track (SHS-STEM Strand) (Tupas & Matsuura, 2019). Subjects in the STEM strand include single disciplines (e.g., General Mathematics, General Biology) and integrated STEM (e.g., Disaster Readiness and Risk Reduction, Research Capstone Project) (Department of Education, Philippines, n.d.) *Non-formal*: STEAM (Agriculture) innovation program involving 13 STEM schools (Aristorenas & Monterola, 2019; Paderna & Monterola, 2022) partnering with industry and university
Singapore	*Formal*: Separate S, T, M, Design and Technology subjects such as electronics, mobile robotics, computing, design and technology (Ministry of Education Singapore, 2022) *Non-formal*: Co-curricular clubs and societies (e.g., STEM club), competitions such as Robotics competition, Engineering Innovation Challenge (Ministry of Education Singapore, 2019) STEM education is the most popular choice by primary schools for their Applied Learning Programme to promote authentic and practice-oriented learning experiences for students (STEM Inc., n.d.; Teo, 2019)

Country	*STEM education initiatives*[a]
Thailand	*Formal*: Basic Education Core Curriculum B.E. 2551 (A.D. 2008) focused on competencies: communication, thinking, problem-solving, applying life skills, and technological application (Soparat et al., 2015). There is no single definition of STEM education, as schools have the flexibility to determine part of their curriculum (Sumonta et al., 2018). National STEM Education Centre, by the Institute for Promotion of Teaching Science and Technology and STEM network schools (SEAMEO STEM-ED, n.d.; Sritrakul, 2018), 13 STEM education centres which are top schools linked to major universities and network schools (Sumonta et al., 2018) *Non-formal*: Public primary schools teach coding (Ministry of Education Thailand, 2019), the STEM ambassador program (British Council Thailand, n.d.), and partnership between SEAMEO STEM-ED in Thailand and University of Danang in Vietnam on "STEM Career Academies in Central Vietnam Project" (Bangkok Post, 26 July 2022)
Timor-Leste (East Timor)	*Formal*: Separate S, T, M subjects in Basic Education curriculum, focusing on equitable access to quality education, holistic education, twenty-first century skills (Ministry of Education Timor Leste, 2011) *Non-formal*: Teacher professional development in mathematics and science (UNESCO, 2020)
Vietnam	*Formal*: New General Education Curriculum 2018 uses a competency-based approach and strongly emphasises STEM integration (Australian Trade and Investment Commission, 2020). STEM and STEAM programs in some high schools, supported by the Ministry of Education and Training in collaboration with the British Council (Nguyen et al., 2020) *Non-formal*: STEAM initiatives (Nguyen et al., 2020). Many STEM activities in private schools and education centres, such as STEM festivals, robotics competitions, science camps, STEM clubs (Duc et al., 2019; Hanoi National University of Education, 2019). Partnership between Taiwan and Vietnam with the e-Learning Integrated STEM Education Center (eLISE) (Chang et al., 2021)

[a]Formal STEM education initiatives can be an explicitly integrated STEM curriculum, or defined as separate S, T, M, and design and technology subjects. Examples of non-formal initiatives are provided, which can be at the school level, national level (e.g., partnership with universities, national events), regional level (e.g., organised by SEAMEO), or international level (e.g., Hour of Code)

Malaysia (Ismail et al., 2019), the Philippines (Tupas & Matsuura, 2019), and Vietnam (Nguyen et al., 2020) on teachers' and students' perspectives on STEM and STEAM education. Generally, it was found that most Southeast Asian teachers and students viewed STEM or STEAM education to be important. However, teachers face significant challenges, such as a lack of exposure and training on STEM integration and a lack of facilities, time, budget, learning resources and supportive environments. Findings also implied "a lack of coordination among high schools and universities, research institutes, organisations and business" (Nguyen et al., 2020, p. 671). There is "a considerable need for awareness raising at both government and teacher levels to embrace STEM education" (Nugroho et al., 2019, p. 417).

In addition to individual countries' STEM education policies and initiatives, STEM education is also driven by regional cooperation efforts between the 11 Southeast Asian nations. Promoting economic and human resource development is a shared goal, and basic education investments in science, mathematics, and technology are seen as means (Devadason, 2017). Hence, regional organisations and education centres develop curricular initiatives and programs and undertake studies and activities for teachers and students as part of the transformation strategy (Devadason, 2017). A regional mathematics and science curriculum is one such initiative.

9.3 Southeast Asian Regional Curriculum: Common Core Regional Learning Standards in Mathematics and Science

The Southeast Asian Ministers of Education Organization (SEAMEO) is an inter-governmental organisation comprising eleven Southeast Asian countries (Brunei Darussalam, Cambodia, Indonesia, Laos People's Democratic Republic, Malaysia, Myanmar, the Philippines, Singapore, Thailand, Timor-Leste, and Vietnam). Since its establishment in 1965, SEAMEO has developed 26 specialist institution centres to provide training and research in education, culture, science, and the environment (https://www.seameo.org/). The Regional Centre for Education in Science and Mathematics (RECSAM) in Malaysia was set up in 1967 and offers leadership in science and mathematics teaching and learning for member countries (Devadason, 2017). An initiative, the SEAMEO

Basic Education Standards, is a regional curriculum project to improve the quality of education through the comparison and identification of gaps in content and developing a common set of shared standards of "what every learner should know, be able to do and value in mathematics and science" (RECSAM, 2019, para. 1).

Author 4 was the regional curriculum's project coordinator and curriculum specialist. The RECSAM team developed the core regional learning standards for mathematics and science (Mangao et al., 2017), known as the Common Core Regional Learning Standards (CCRLS). The regional curriculum aims to address the area of priority in SEAMEO's Education Agenda (2015–2035), "Adopting a 21st Century curriculum", which focuses on

> Pursuing a radical reform through systematic analysis of knowledge, skills, and values needed to effectively respond to changing global contexts, particularly to the ever-increasing complexity of the Southeast Asian economic, socio-cultural, and political environment, developing teachers imbued with ASEAN ideals in building ASEAN community (SEAMEO, 2021, para. 13)

Following the development of the regional curriculum, RECSAM collaborated with Authors 1, 2 and 3 from Monash University to provide a professional development program. As part of the program, researchers and lead teachers from Southeast Asia jointly developed a pedagogical model for STEM processes in teaching and learning based on Southeast Asian perspectives (Mangao et al., 2019, also see Chapter 8). The regional curriculum comprised separate sets of standards for mathematics and science.

- **Mathematics Curriculum**

The curriculum strands for mathematics in the CCRLS consist of four content strands and one process-humanity strand: (1) Numbers, Operations, and Algebra; (2) Geometry, Space, and Measurement; (3) Statistics, Probability, and Data Handling; and (4) Relations and Functions. The fifth strand, (5) Mathematics Process—Humanity, describes the elements in mathematical thinking and processes, values and attitudes integrated

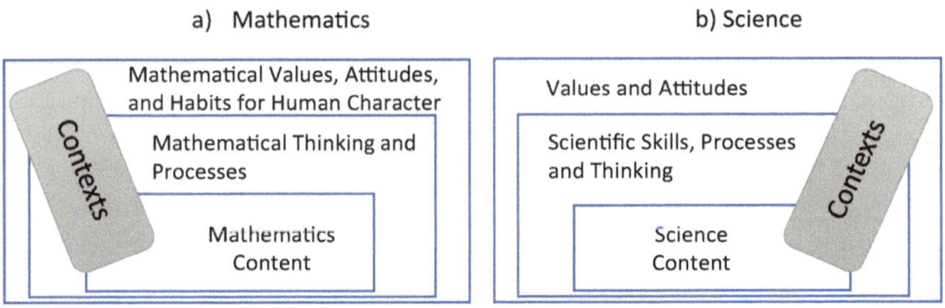

Fig. 9.1 Components of the mathematics and science framework (adapted from Mangao et al., 2017, pp. 5, 76)

across the four content strands (see Fig. 9.1a). Competency in mathematics thinking and processes is seen as a basis for STEM and STEAM education.

- **Science Curriculum**

The science content strands are (1) Life and the Living World, (2) Material World, (3) Energy and Change, (4) Earth and Space, (5) Science, Engineering, and Technology for a Sustainable Society, and (6) Science Inquiry. The last two strands are designated to be integrated into the first four strands. The scientific thinking processes, values and attitudes span the six strands (see Fig. 9.1b), similar to the mathematics curriculum. STEM integration can be seen in strands (5) and (6). A topic, The World of STEM, is housed in (6) Science Inquiry, with standards relating to awareness of the importance, applications and implications of STEM in everyday life and the nature of science and engineering processes and values.

9.4 Related Literature

9.4.1 Strategic Focuses and Representation of Disciplines

As an emerging and rapidly developing field, a myriad of conceptions of STEM education have been developed by STEM education researchers. There are debates and varying understandings about what STEM education is and what constitutes STEM integration. These range from integration within one discipline (or its sub-disciplines) or a collective of STEM disciplines, a combination of subjects, or a comprehensive inclusion of all four disciplines (Huang et al., 2022; Li et al., 2020). Sanders (2009) differentiates between a "business as usual" (p. 21) definition of STEM education that disconnects science, mathematics, and technology education, with an integrative STEM education that "includes approaches that explore teaching and learning between/among any two or more of the STEM subject areas, and/or between a STEM subject and one or more other school subjects" (p. 21). In contrast, Ostler (2015) proposed that "STEM can only be defined and understood by contextualizing it within clearly defined goals, thus creating many small *tactical definitions* rather than one broad conceptual definition" (p. 4). This plurality due to contextualisation was also found by Martín-Páez et al. (2019) in their literature review of STEM education research. They concluded that the theoretical frameworks used by researchers tend to be specific to research goals rather than focus on STEM education per se and that there were multiple interpretations of STEM. Takeuchi et al. (2020) argued that macrosystems and ideological forces (such as the USA's NGSS) shape the dominant discourses on STEM education, epistemology, and ontology. Such a conception of STEM as having diverse contextualised and tactical or strategic focuses reflects the interdisciplinarity and complexity of the field. Being a team of researchers from mathematics, science, and technology education disciplines, our shared understanding of STEM education is aligned with this pluralistic view, with a core definition of STEM being integrated and interdisciplinary in nature.

In addition to the S, T, E, and M disciplines, there are inclusions of other disciplines such as Arts and social sciences (STEAM) (Perignat & Katz-Buonincontro, 2019), and interdisciplinary focuses such as sustainability (Shahali et al., 2017) and disaster education (Sampurno et al., 2015). This multiplicity adds further richness to the field. However, it also adds ambiguity and challenges for researchers to locate and identify STEM education research (Li et al., 2020). Conceptual clarity about STEM is imperative to inform research and practice (Perignat & Katz-Buonincontro, 2019).

Besides the lack of a clear definition of STEM education and of which disciplines to include in STEM, the roles of the disciplines and teaching out-of-field within STEM/STEAM education also raised a concern (Lau & Jong, 2022). Lau and Jong identify that primary teachers, lacking science disciplinary knowledge, are highly concerned with teaching STEM, and secondary school science teachers have a wealth of scientific knowledge, yet they struggle with the knowledge required to teach robotics and coding. English (2016) discusses the *inequitable representation* of mathematics within STEM education. Mathematics, for example, is a core subject yet plays a supporting role in science and engineering (English, 2016; see Chapter 7 on Hidden Curriculum Contexts). In their review of the 2013–2018 literature on STEM education, Martín-Páez et al. (2019) found that out of the 19 studies on STEM interventions, a majority (58%) had engineering as the dominant discipline, followed by science (21%). There was no intervention study with mathematics as the dominant discipline. The researchers concluded that mathematics is too "complex to use as a 'backbone' from which other STEM disciplines begin" (p. 810). Although engineering might be seen as a "hinge" or "contextual integrator" for connecting with other disciplines, through the use of engineering design and engineering-based problems (Martín-Páez et al., 2019, p. 810; Roehrig et al., 2021), engineering education is not a subject in the elementary and middle schools of most countries (English, 2017). Similarly, the role of technology in STEM education has not been well defined (Ellis et al., 2020).

9.4.2 *Integration of Cognitive and Affective Aspects of STEM Education*

In addition to the content knowledge in each discipline, STEM education also integrates various skills and learning processes, such as twenty-first century skills and higher-order thinking skills, as well as the affective aspects, such as attitudes, dispositions, and values. This *integration of cognitive and affective aspects* might be discipline-specific (e.g., formulating and testing hypotheses in science; attitudes toward mathematics) or generic (e.g., twenty-first century skills) (Dare et al., 2021). Along with the multiplicity of STEM conceptions, we found no consistency with the skills, processes, attitudinal variables, and values being investigated in STEM education research. For example, there were different focuses on the twenty-first century skills studied. Dare et al. (2021) outline the 4 Cs—critical thinking, creativity, collaboration, and communication—citing that future STEM professionals need to be adept in these skills due to the rapid advancements in technology and globalisation. Ring-Whalen et al. (2018) highlighted problem-solving, critical thinking, communication, and teamwork as twenty-first century skills in their analysis of STEM units. In contrast, Stehle and Peters-Burton (2019) referred to six twenty-first century skills in their analysis of STEM lessons: knowledge construction, real-world problem-solving, skilled communication, collaboration, use of information and communication technologies (ICT) for learning, and self-regulation. Yet other studies focus on one or two of the twenty-first century skills, such as student collaboration (Herro et al., 2017) and teamwork and communication (Guzey et al., 2016).

Researchers studied general attitudes toward STEM (Kim & Bolger, 2017) and explored culturally relevant values such as uniformity, diversity, harmony, and emotional energy (Davis et al., 2020). The underlying values of performance, accessibility, innovation and creativity in the United States' Next Generation Science Standards (Hoeg & Bencze, 2017) were also identified. Park et al. (2020) compared the disciplinary aims, values and practices of STEM curriculum documents in the United States, Korea, and Taiwan. Park et al. found that features common to science and engineering/technology or specific to science were most frequently mentioned, whereas mathematics features were rarely highlighted. This finding offers another insight relating to English's (2016) notion of the inequitable representation of mathematics discipline in STEM education since curriculum standards documents inform STEM interventions and implementation. While other studies (e.g., Park et al., 2020) focus on the process of STEM integration and classroom practice, we sought insights from the product of a professional development program where teachers designed STEM education units.

9.5 Teachers' Design of STEM Units

Teachers are increasingly seen as "makers" or "designers" (see Chapter 8) rather than "conduits" or "workers" of the curriculum (Clandinin & Connelly, 1992; Schiro, 2013). With curriculum reforms, it is essential that teachers have a voice and agency as part of the reform process (Clandinin & Connelly, 1992), and a strategy to engage teachers in the process is for them to design the curricula (Huizinga et al., 2014). Studies conducted on teachers creating STEM or STEAM curricula have dual aims of providing professional development and investigating teachers' design and enactment of their curricula (e.g., Guzey et al., 2016; Kim & Bolger, 2017; McFadden & Roehrig, 2017; Roehrig et al., 2021). Roehrig et al. (2021) observe that most life science STEM curricula use science as a context rather than content to be applied in developing design challenges incorporating design decisions. Teachers' design of STEM units was found to be dependent on their conceptions of STEM integration and other factors such as their professional identity, confidence, beliefs, and experiences of STEM education among the teacher team members (McFadden & Roehrig, 2017; Ring-Whalen et al., 2018).

The process of teachers designing curriculum collaboratively is complex and dynamic. Teacher design teams have to focus on mapping the organisation and content of the curriculum, as well as selecting and designing the tasks (McFadden & Roehrig, 2017). During the collaboration, team members' values, beliefs and understandings of STEM integration must be clearly articulated to move the design forward (McFadden & Roehrig, 2017). Having facilitators or coaches to support teachers' collaborative design process in drafting and revising ideas was found to be important in developing teachers' design expertise and ensuring that they can view the unit holistically and design a shared integrated curriculum (McFadden & Roehrig, 2017; Quigley et al., 2020; Roehrig et al., 2021). McFadden and Roehrig (2017) highlighted a four-teacher team that "resorted back to a familiar, comfortable pathway" (p. 12). These teachers also struggled to design a STEM-integrated curriculum to meet the teachers' unique needs because they did not have a shared vision for the overall purpose of the unit. The kindergarten teacher team in the study by Quigley et al. (2020) felt constrained by their current prescribed curriculum despite encouragement from their instructional coach and principal. They designed and enacted a STEAM curriculum that was "primarily focused on math and literacy skills that were disconnected to the problem solving" (p. 509).

In addition to investigating teachers' knowledge, perspectives, and attitudes in teacher design studies, some studies used a set of criteria or tools to evaluate the curriculum developed. Kim and Bolger (2017) utilised six criteria to assess integrated STEAM lesson plans developed by 119 Korean elementary pre-service teachers enrolled in a science methods course. In

Table 9.2, we compare Kim and Bolger with two studies that also evaluated STEM curricula. Guzey et al. (2016) assessed 20 integrated STEM curricular materials science teachers (Grades 4–8) developed during a professional development program. They created a tool that has nine criteria. Ring-Whalen et al. (2018) analysed three teacher teams' STEM units using eight criteria emerging from nine science K–12 teachers' conceptions of integrated STEM education. As seen in Table 9.2, these criteria are categorised into the themes of the *Representation of disciplines*, *Integration of cognitive and affective aspects of disciplines*, and the *Strategic STEM focuses* of the researchers, including the focus on specific discipline/s, specific contexts, and/or specific pedagogies.

Guzey et al. (2016) found that the STEM units based on contexts in the physical sciences were rated more motivating and engaging than STEM units based on life sciences and earth sciences, and grade-appropriate mathematics was integrated into the units as data analysis and measurement. When they compared the top five high-scoring STEM units with the remaining 15 medium/low remaining units, there were statistically significant differences in all the criteria except mathematics integration and communication (communicating S, E, and M thinking). Interestingly, the criteria used by researchers vary according to the educational context and focus of the researchers, similar to the tactical definitions of STEM proposed by Ostler (2015). For example, the three studies explored in Table 9.2 (Guzey et al., 2016; Kim & Bolger, 2017; Ring-Whalen et al., 2018) all focused on science teachers and hence had science-focused criteria, followed by the integration and consideration of T, E, and M as criteria. Grade-appropriate STEM content integration and connections to the real world were similarly valued. However, differences are reflected in the criteria being valued between the studies. The role of technology in integrated STEM seemed different: distinguishing T from E is essential in Kim and Bolger's (2017) study, whereas T thinking (e.g., systems thinking, computational thinking) was absent in the criteria in the Guzey et al. (2016) study, and T and E were considered as tools/supports in the Ring-Whalen et al. (2018) study. Additionally, STEM-based pedagogies such as engineering design and organisation of lessons and activities were uniquely included as criteria by Guzey et al. (2016).

The studies by Guzey et al. (2016) and Ring-Whalen et al. (2018) were conducted on in-service science teachers in the United States. Kim and Bolger's (2017) study was on the attitudes of Korean elementary

Table 9.2 Comparison of criteria for analysing STEM Units by Kim and Bolger (2017), Guzey et al. (2016), and Ring-Whalen et al. (2018)

Themes	*Kim and Bolger (2017), six criteria Korean elementary pre-service teacher context*	*Guzey et al. (2016), nine criteria USA in-service teacher context*	*Ring-Whalen et al. (2018), eight criteria USA in-service teacher context*
Representation of disciplines in STEM or STEAM	– Integration of science content – Significant inclusion of all STEAM disciplines – Distinguishing the technology from engineering	– Integration of mathematics content – Interdependence of STEAM	– Connecting the disciplines – Balance of science and engineering – Mathematics and technology as tools/supports in STEM
Integration of cognitive, affective, and conative aspects of disciplines (e.g., thinking processes, twenty-first century skills)		– Teamwork – Communication	– Twenty-first century skills emphasised
Strategic STEM focuses			
Focus on specific discipline/s	– Content appropriate to student level – Science-focused		– Engineering focus – Science focus
Focus on specific contexts	– Creative connections to real world	– Motivating and engaging contexts	– Engineering as contexts – Real-world connections
Focus on specific pedagogies		– Engineering design – Instructional strategies – Performance and formative assessment – Organisation of lessons and activities	

pre-service teachers when they develop STEAM lesson plans. This study explores STEM education research in Asia, specifically in Southeast Asia.

9.6 Professional Development Workshop

A four-day professional development (PD) STEM workshop was conducted in Penang, Malaysia, with 80 teachers, teacher educators, ministry of education delegates and education leaders from Southeast Asia. The authors explored the participants' curiosity, inquiry and STEM practices as they worked collaboratively to solve culturally relevant STEM problems. During phase 1 of the workshop (see Table 9.3), teachers participated as learners in STEM activities to provide first-hand experiences of the collaborative, design-based solving of a real-world problem. This form of PD is essential to develop teachers' confidence in and understanding of STEM pedagogies (e.g., Kim & Bolger, 2017). In phase 2, participants analysed 14 STEM learning and education models and collectively designed a model relevant to the Southeast Asian classroom contexts. Chapter 8 of this book discusses the development process of the *STEM Planning and Design Learning Model* (PaDL) (Mangao et al., 2022).

At the beginning of the session, we organised teachers into groups of eight, representing various disciplines and countries. This group of 10 table teams remained consistent during the workshop, although 19 local teachers and teacher educators only participated in phases 1 and 2. This setup helped them get to know each other better and fostered an environment for sharing and exchanging ideas. During phase 3, the teachers

Table 9.3 Phases in the professional development on STEM curriculum

Phases	*Approach*
1	Teachers experienced six STEM activities in teams during the PD program (n = 80). Post-activity discussions focused on unpacking STEM knowledge, skills, and integration in these activities
2	Teachers work together to develop the model for STEM planning and teaching and the model for the STEM learning process (n = 80, see Chapter 8)
3	Teachers work together in teams to design a STEM unit (n = 61)

collaborated within their table teams to brainstorm and design an integrated STEM unit. The 61 teachers reported here were asked to provide their disciplinary background during the workshop. Most of the teachers identified themselves as having mathematics ($n = 22$) and science ($n = 35$) education backgrounds, and the rest ($n = 4$) mentioned other backgrounds (pre-school, a mix of science and mathematics, engineering). The disciplinary composition of the table teams is shown in Table 9.4. In total, there were three school teachers, 19 teacher educators from universities and teacher training institutions, 29 Ministry of Education delegates from the SEAMEO countries, and 10 education leaders from private education companies and institutions. Collectively, we refer to the participants as "teachers" as they all had teaching backgrounds.

Table 9.3 shows the phases of the PD program. The final task we facilitated in the second half of the workshop was to design a STEM unit of work around a topic selected by teachers in the design team. Teachers in each teacher design team brainstormed the topic for their STEM unit, linked the learning goals to the mathematics and science standards, and designed activities for student investigation of the topic during the six-week learning unit. The teachers incorporated the Pedagogical and Design Learning Model (PaDL) in their units (Mangao et al., 2022). At the end of the workshop, teachers had a draft of their STEM unit, to be completed after the program.

Data collected are the ten teacher teams' draft STEM units. In particular, teachers filled in a template to present the links to the STEM curricula (see Fig. 9.2) and aligned their unit to the CCRLS for mathematics and science. A content analysis (Gray et al., 2007) of the mathematics and science curriculum links was conducted for each of the 10 STEM units. The analysis of the links to Technology and Engineering is beyond the scope of this chapter and will be reported elsewhere. Descriptive statistics were used to analyse and report on the frequency and occurrences of the standards in the CCRLS. Note that the frequency refers to the number of times knowledge or skill of a content area has been mentioned (see Fig. 9.3 in the next section for content areas). Where a team mentioned a topic (e.g., environment and conservation of natural resources), a sub-topic within that topic (e.g., the impact of human activities on the environment), and the curriculum standard within the sub-topic (e.g., conduct research and suggest ways to address the harmful effects of human activities on the environment), we counted these as three mentions. We sought to determine which content strands

Table 9.4 Team number and STEM topics

Team	*Topic*	*Brief description of unit*	*Team composition*[a]
1	Renewable and non-renewable energy	Students are made aware of the input and issues related to the energy crisis due to the depletion of non-renewable sources of energy. They will explore available renewable energy sources, including their advantages and disadvantages. Students will create a low-cost alternative energy model that can replace an existing home appliance at their home (e.g., using palm oil or solar cells). They then present this to their peers and include peer assessment and reflections on their learning	2 Math, 4 Sci
2	Tiger conservation	Students learn about the human-tiger conflict in Southeast Asia and help scientists improve their capture-tag monitor system used in surveying tiger populations. Students will design a capture-tag device to gather tigers' vital statistics, such as size (height, length), blood pressure, and weight. This incorporates empathy and biochemical engineering to create a safe method to capture tigers and implant the tag/chip in tigers. This includes assessing students' presentation, capturing the tag device/method design, and forecasting the tiger population	3 Math, 3 Sci

(continued)

Table 9.4 (continued)

Team	*Topic*	*Brief description of unit*	*Team composition*[a]
3	Sustainable garden	Students learn about healthy and unhealthy food, food safety and security in this interdisciplinary unit. Topics include Biology (plant growth, human body needs and digestive system), Chemistry, Engineering, Technology, Mathematics (geometry, measurement, finance), Physical, Personal and Social Learning, Food Security, Sustainability, and Healthcare. Students design and create a sustainable garden that is a self-sustaining ecosystem (e.g., aquaponics)	2 Math, 5 Sci, 1 mixed Sci and Math
4	Pest management	Students identify and characterise household pests according to their type, habitat, feeding pattern, and natural biological controls. They also integrate technology and engineering concepts in their design and construction of a pest trap or other ways of pest control. This can include light or sound sensors and repellants to deter different pests	1 Math, 6 Sci

(continued)

Table 9.4 (continued)

Team	*Topic*	*Brief description of unit*	*Team composition*[a]
5	Plastic pollution in local rivers	Students will learn about the characteristics and components of plastics and their impact on the environment, humans and animals. After introducing the problem through internet research in class, students will collect and analyse data on local rivers and plastic pollution in fieldwork and then present their findings. They further explore the properties of plastic and its impact on the environment (e.g., fish). Students then brainstorm, design, and prototype STEM solutions to reduce plastic pollution. Students showcase their prototypes to the school and community via an exhibition	2 Math, 3 Sci, 1 pre-school
6	Air pollution	Students learn about air pollution, its causes and effects. The unit includes field trips (e.g., visiting the forest) and experiments (e.g., measuring the amount of CO_2). They explore and work on a project to design a way to reduce air pollution, e.g., living wall, seed bomb, 3R project, bio compost. Students present their projects, peer assess, reflect and refine their projects/products	1 Math, 3 Sci, 1 Engineering, 1 undisclosed

(continued)

Table 9.4 (continued)

Team	*Topic*	*Brief description of unit*	*Team composition*[a]
7	Food hygiene	Students explore the issue of consuming food that is not fresh, from the perspectives of news, social media, legislation/laws, and health. They pose and answer questions about maintaining the hygiene of foods and the consequences of unhygienic foods, learning through role-plays (e.g., selling fish in hawkers or markets). The problem (common in the tropics) is: How do you prevent the flies from gathering around foods? Students need to design and prototype their solutions. Students present their solutions, followed by reflecting on and improving their models	3 Math, 2 Sci
8	Plastic pollution in the ocean	Students will apply STEM knowledge and skills to create a 'seabin' that would remove plastic wastes from our ocean/river, contributing to a better world while actively learning from various disciplines and applying the design process. The unit includes conducting beach clean-up activities and a field trip to a marine animal conservation centre	4 Math, 2 Sci

(continued)

Table 9.4 (continued)

Team	*Topic*	*Brief description of unit*	*Team composition*[a]
9	Clean water	Students will be engaged in discussions about the impact of water pollution and water conservation. They will experiment with filtering various substances from water by designing water filters and pumps. They perform experiments to identify the factors affecting liquid pressure and develop their own water treatment systems. Students will present their design, with justifications. This is followed by feedback and reflections on their learning in STEM	1 Math, 4 Sci
10	Water pollution	Students will develop an understanding of concepts of water pollution and the steps that they can take to help make the environment clean. Using a flipped classroom approach, students share their knowledge of the impact of water pollution. They will also collect and analyse the quality of water samples from various sources. They then brainstorm and design solutions to the issue	3 Math, 3 Sci

[a]Composition of teachers based on their self-identified background; Sci–Science, Math–Mathematics

have the most or least number of *explicit mentions* by teachers in the STEM units. Coding was carried out by Author 1, and any ambiguities were discussed with the other authors. There were very few ambiguities (fewer than 5% of the codes) because the teachers generally used the curriculum descriptors in the CCRLS document provided. Decisions were made about the coding process (e.g., combining the frequency counts for topics and sub-topics as mentioned earlier) and visual representation of data for analysis (e.g., using stacked bar charts). Through this iterative

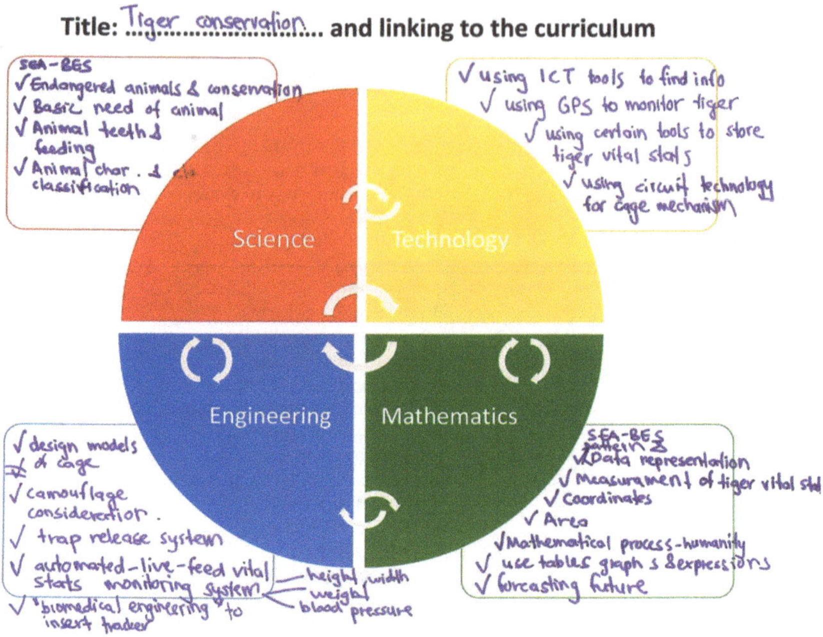

Fig. 9.2 Example of curricula links of a STEM unit (*Source* Authors)

process, the team reviewed and confirmed the analysed codes to ensure reliability and validity.

9.7 Analysis of Science and Mathematics Curricula Links in Teachers' STEM Units

Each of the ten teacher teams designed their STEM unit around a topic relevant to Southeast Asia. The instructions were to use local or regional problems as contexts for learning. The selected topics are shown in Table 9.4 and target upper primary to secondary grade levels (grades 4–10 or ages 10–16).

Table 9.4 highlights that the ten teacher teams created units of work emphasising sustainability concepts. Assessment is primarily formative, utilising peer assessment, reflections, and prototype presentations. The pedagogical emphases favour student-centred problem-solving. The

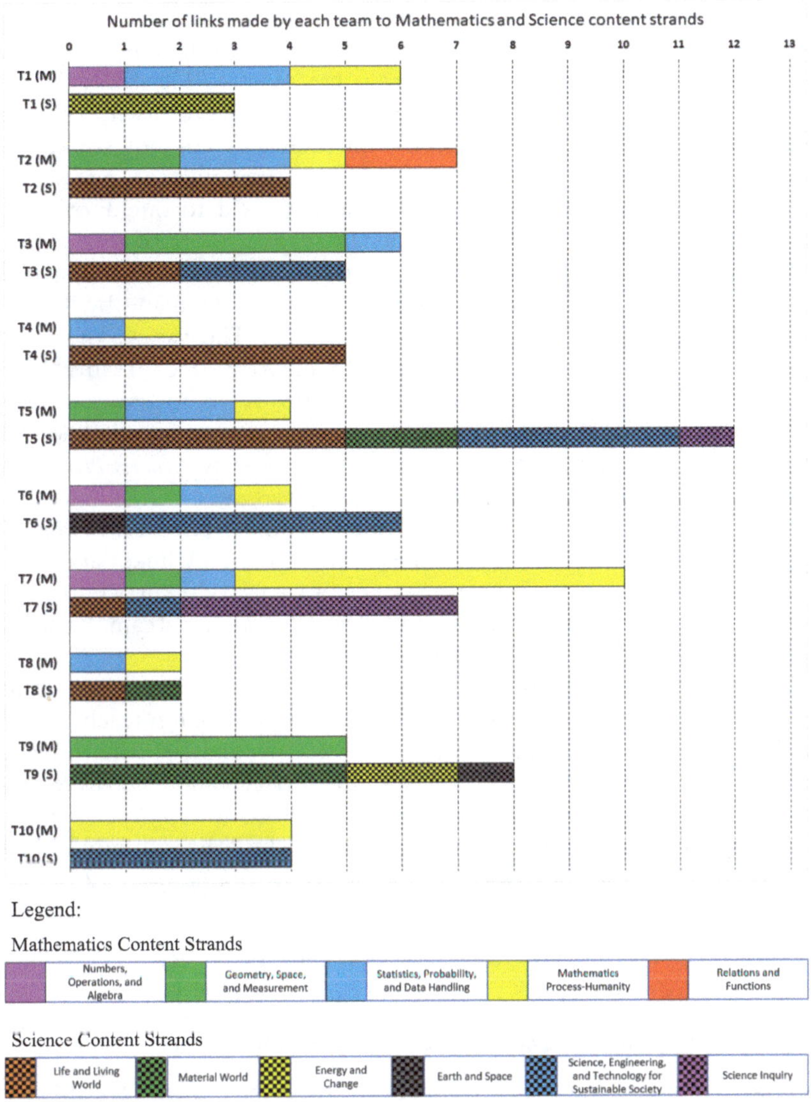

Fig. 9.3 Number and type of mathematics and science curriculum links made by each teacher design team (e.g., T1(M) denotes Team 1: Mathematics) (*Source* Authors)

frequency and occurrences of the mathematics and science curriculum links made by each team are categorised according to the content strands in CCRLS, and a frequency bar chart is created (see Fig. 9.3). The six science content strands are shaded in a checkered pattern with different background colours, whereas the five mathematics content strands are shaded in solid colours. For each team, their mathematics (M) and science (S) curriculum links are plotted as parallel bars next to one another for comparison. As shown in Fig. 9.3, teachers made a range of mathematics and science curriculum links. For Mathematics, a total of 50 links were made by all the teams, with a minimum of two links made by Team 8 and a maximum of 10 links made by Team 7. The variety of strands that teacher design teams referred to ranged from one (Teams 9 and 10) to four (Teams 2, 6 and 7). The frequency refers to the number of times knowledge or skill of a content area has been mentioned. The most frequently cited curriculum strands are *Mathematics Process-Humanity* (18 links), followed by *Geometry, Space, and Measurement* (14 links), and *Statistics, Probability, and Data Handling* (12 links). The least commonly cited strands are *Numbers, Operations, and Algebra* (4 links) and *Relations and Functions* (2 links). For Science, a total of 56 curriculum links were made, with a minimum of two links (again by Team 8) and a maximum of 12 links made by Team 5.

Unlike Mathematics, there is generally a single Science curriculum strand that is the focus of each STEM unit. Four out of ten teacher teams cited one science curriculum strand (Teams 1, 2, 4, and 10), and three out of ten had a strand that was clearly the main focus. The most frequently cited strands are *Life and Living* (18 links) and *Science, Engineering, and Technology for Sustainable Society* (17 links). The least commonly cited strands are *Energy and Change* (5 links) and *Earth and Space* (2 links).

9.8 Discussion on the Nature of the STEM Units Designed

9.8.1 *The Representation of Disciplines*

In this section and the following sections, the nature of the STEM units designed is addressed. As can be seen in the results, more links were made to the science curriculum (56 links) than the mathematics curriculum (50 links). Interestingly, four teacher design teams (Teams 4, 5, 6, and 7) made more science than mathematics links, and the same number of teams (Teams 1, 2, 3, and 7) made more mathematics than science links. Two teams (Teams 8 and 10) have equal numbers of mathematics and science links. Team 5 has a noticeable imbalance, with 12 science and four mathematics links. While the mathematics and science teacher composition in Team 5 (see Table 9.4) is not any different from the other teams, there was a pre-school teacher. The regional curriculum covers elementary to middle school (Grades 1–9), so perhaps the science content for plastic pollution is more familiar to teachers and students alike (relating to animals and the environment) than the mathematics content (data sampling, measurement, and statistics), from the pre-school perspective. It might well be that the teachers in Team 5, including the pre-school teacher, were able to see more connections to the science than mathematics content. The higher number of links might be due to a stronger focus on the science discipline, or it is easier for the Team 5 teachers to connect to the science due to familiarity than the mathematics content.

Although teacher teams made links to the regional curricula for mathematics and science, there are subtle differences between how mathematics and science are treated. Firstly, there are fewer (10%) mathematics than science curriculum links made. Secondly, most teams (all except Teams 3 and 8) referred to one dominant science content area and included a few mathematics content areas. For example, Team 2 had: to measure and design tiger capture and release traps (*Measurement, Geometry, and Space*), analyse and forecast tiger vital statistics (*Statistics, Probability, and Data Handling*), and use coordinates, functions, and various mathematical representations to track and predict tiger population data (*Relations and Functions*). We noted that *Relations and Functions* were only referred to by one table team, Team 2, and "coordinates" and "algebra" were not explicitly mentioned in the curriculum links. This suggests that certain mathematics curriculum strands lend themselves more to integrating content in STEM units, such as measurement and geometry (in engineering design) and statistics (in interpreting, collecting and analysing data), as can be seen in other studies (Guzey et al., 2016; Ring-Whalen et al., 2018) and also discussed in Chapter 7 of this book. The affinity of STEM with certain mathematics curriculum strands found in this study is consistent with the role of mathematics in interdisciplinary STEM education highlighted by past researchers (English, 2016; Martín-Páez et al., 2019). There is a concern that mathematics might be relegated to a support role as the "queen of science" (e.g., Atiyah, 1993), undervalued and under-represented (English, 2016), rather than valued as a discipline in its own right, capable of being foregrounded. We contend that it might be ***numeracy***, rather than mathematics, that is being integrated. The Organisation for Economic Co-operation and Development (OECD, n.d.) defined numeracy as "the ability to use, apply, interpret, and communicate mathematical information and ideas" (p. 1), a skill that parallels reading literacy. According to Goos et al. (2019), numeracy, or quantitative literacy, involves "problem-solving, reasoning, positive dispositions and a critical orientation to mathematics and how it is used in the world" (p. 31). Frankenstein (2009) highlighted the importance of developing critical mathematical numeracy through "*real* real-world problems" (p. 112, emphasis in original), where issues of social justice can be made more explicit and more meaningful. Indeed, the role of mathematics in STEM has been connected to the promotion of twenty-first century skills, mathematical modelling, and responsible citizenship (Ärlebäck & Albarracín, 2019; Maass et al., 2019), and therefore closer to the definition of numeracy. Also, see Chapter 7 for a discussion on this hidden curriculum aspect of STEM.

The nature of M (or N for numeracy?) in STEM needs further consideration and research. Students need to be able to make links and generalisations from the mathematics learnt in the process of designing or making a device for a STEM task, to the abstracted form of mathematics in the curriculum. This process needs to be valued, facilitated and made explicit by the teacher.

9.8.2 *Integration of Disciplinary Thinking Processes, Attitudes, and Values*

In terms of the integrated attitudes and thinking processes strands, it is interesting to note that only two tables had explicit links to the *Science Inquiry* (Teams 5 and 7) strand, despite the application of scientific inquiry in STEM (see Fig. 9.4). In the regional curriculum, *Science Inquiry* involves "science process and thinking skills as well as general inquiry processes such as problem-solving, logical reasoning, decision-making, and reflection" (Mangao et al., 2017, p. 78). The *Science Inquiry* strand is expected to be integrated with the other science content strands and embody attitudes and values (see Fig. 9.1b). Although there were few links (a total of six) to the *Science Inquiry* strand, most of the teams referred to the *Mathematics Process-Humanity* strand (a total of 18 links). Half of them referred to links in the *Science, Engineering, and Technology for a Sustainable Society* strand (a total of 17 links). This finding suggests that, in general, the mathematics and science teachers were aware of and explicitly integrated the twenty-first century skills, thinking processes, attitudes and values into the STEM units.

The emphasis on the *Science, Engineering, and Technology for a Sustainable Society* strand, in contrast to the *Science Inquiry* strand, could be the effects of the STEM professional development program raising awareness, as well as the explicit highlighting of the thinking process, attitudes, and values in the regional curriculum. Some of the

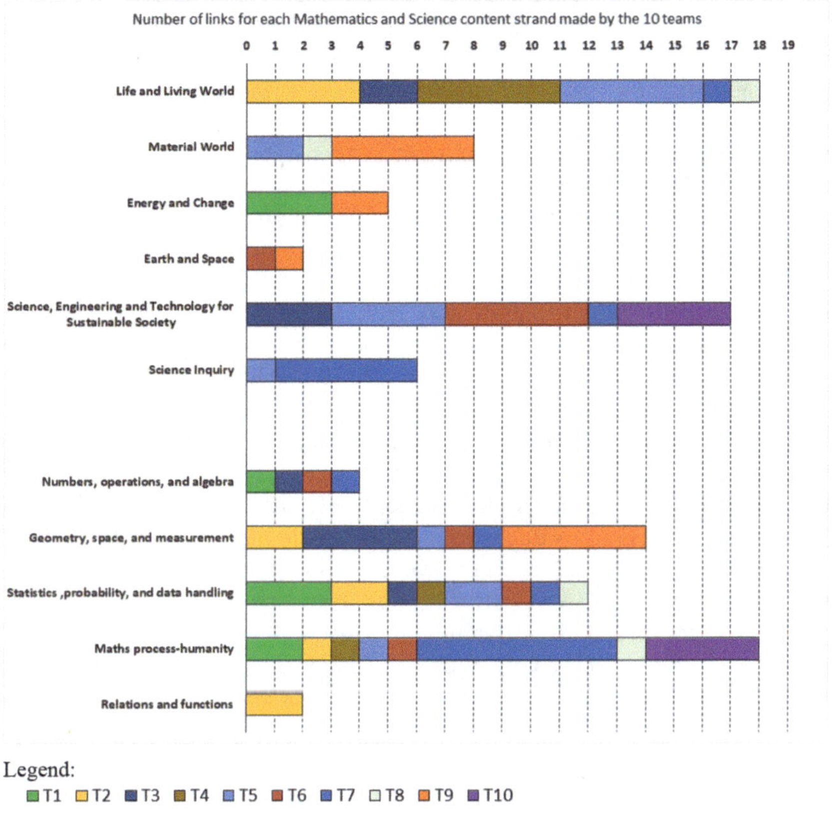

Fig. 9.4 Number of links for each mathematics and science content strand made by teacher design teams (e.g., T1 denotes Team 1) (*Source* Authors)

Mathematics Process-Humanity links referred to by teachers, include: "appreciate others' ideas, meaningful elaboration, usefulness and efficiency", "reasonableness and harmony", "problem solving, promoting creative and global citizenship for sustainable development of the society using mathematics", "estimate for effective use of resources in situations", and "enjoying problem solving through various questioning and conjecturing for extension of operations into decimals and fractions with proportionality and new quantities such as area and volume" indicating the teacher teams were cognisant of linking STEM units with curriculum documents. Similarly, the *Science Inquiry* curriculum in the CCRLS that the teachers mentioned included "inquiry, critical creative thinking", "reasoning, problem solving", "applying, creating, collaboration", "social awareness, responsibility", and "critical reflection". A number of links were made to the topics and sub-topics within the *Science, Engineering and Technology for a Sustainable Society* strand, including "impact of human activities on the environment", "ecosystem and environmental management", "green living", "renewable and non-renewable energy sources", "role of science, technology and engineering to improve the quality of life".

Having explicit curriculum links in the CCRLS to these thinking processes, attitudes, and values may have made it easier for teachers to refer to them in their unit design, compared to creating the curriculum goals from scratch. This supposition is neither confirmed nor refuted in past studies. Researchers tend to evaluate STEM or STEAM units using a set of researcher-designed criteria rather than comparing links from the STEM unit to an official curriculum (see Table 9.2). However, it is clear that Southeast Asian teachers who attended the professional development program value specific twenty-first century skills such as problem-solving, critical and creative thinking, teamwork, and communication. This is consistent with findings from other studies where these twenty-first century skills are valued and integrated into teachers' STEM units (Dare et al., 2021; Guzey et al., 2016; Ring-Whalen et al., 2018; Roehrig et al., 2021).

9.8.3 Strategic Focuses of STEM Integration

In past STEM education research studies, the strategic focuses were found to be on specific disciplines, contexts, and pedagogies (see Table 9.2). In this study, the strategic focus was STEM education pedagogy co-created in the professional development program as a design-based approach, situated in the local real-world context. There was an explicit reference to the regional mathematics and science curricula, and these CCRLS curriculum documents (in both hard and soft copies) were readily available during the PD program. Technology and engineering skills and concepts were also highlighted during hands-on STEM activities to model learning opportunities (e.g., Mangao et al., 2022). The teacher teams were scaffolded to write down curriculum links relating to S, T, E, and M (see Fig. 9.2). As researchers and STEM educators, our strategic focus of the PD was on the meaningful content integration from the S, T, E, and M disciplines rather than deliberately favouring one or two specific disciplines. This chapter presents a partial analysis of teachers' STEM units in relation to the regional science and mathematics curriculum since there was no regional technology and engineering curriculum at the time of the study. Nonetheless, teachers were found to be using the co-created design thinking approach in their STEM unit plans. Digital literacy goals were highlighted by teachers to support general learning and communication skills, as well as skills for using digital tools for data collection in science and data analysis in mathematics.

The STEM units designed by the teacher teams were found to be based on science topics that are highly relevant to the Southeast Asian context, such as pollution (e.g., air pollution, plastic pollution in rivers and ocean) and sustainability (e.g., tiger conservation, sustainable garden, renewable energy). Topics in the *Life and Living World* strand are the most popular, as seen in Fig. 9.4. This finding is interesting when we look at the STEM topics in past studies. Topics in STEM units designed by teachers in other studies are similarly meaningful in their local contexts, such as those by K–5 teachers in Southeastern Wisconsin (Quigley et al., 2020): health, eco-park, water quality, and population growth, as well as STEM units designed by elementary and middle school science teachers in a U.S. Midwest state (Guzey et al., 2016): eco house, pollutants in the pond, greenhouse, and water desalination. Comparatively, the topics in our study were related to life science or environmental science, with no topics in the earth science or physical science that were found in other studies (e.g., soil and earthquakes unit, Guzey et al., 2016; mechanical claw unit, Ring-Walen et al., 2018). There are topics in the *Earth and Space* strand in the Southeast Asian regional curriculum (e.g., volcanic eruptions, solar system, universe and space technology) that teachers in our study did not select. It might be suggested that Southeast Asian teachers prefer certain topics for STEM integration, or that specific topics may lend themselves to STEM integration more readily than other topics in a localised context, especially for novice STEM unit designers.

In the handbook of the environment in Southeast Asia, Hirsch (2017) argued that the environment can be considered a cultural construct and that different countries frame environmental challenges and understandings about environment-related problems contextually. Some key environmental issues in Asia, such as loss of biodiversity, air and water pollution, food security risks, pests and diseases, and natural disasters (Hasnat et al., 2018), might be considered more neutral or less politically sensitive topics in the education setting. These "neutral" or "safe" topics, under the theme of "sustainability", might be considered strategic STEM focuses by teachers, compared to, say, the transboundary haze issue between Indonesia, Singapore, Malaysia and southern Thailand (Hirsch, 2017, p. 6). Further research can examine and unpack the themes or topics of STEM teaching and learning activities in the light of geopolitics and socio-cultural contexts.

9.9 Recommendations

This study is significant because we have 61 representatives from Southeast Asian countries and 10 interdisciplinary teams. The Southeast Asian teachers in this study experienced innovative STEM activities relating to content integration to develop their STEM conceptions collaboratively. In line with McFadden and Roehrig (2017), we aimed to work with the teachers in collaborative teams to develop units of work that provided structure and expertise to prevent the teachers from falling back into their teaching comfort zones. In this chapter, we have presented an analysis of the content integration via curriculum links of teachers' co-designed STEM units, connecting to the regional mathematics and science curriculum (CCRLS). Teachers designed STEM units strategically focusing on local environmental and sustainability issues and integrated disciplinary cognitive, affective, and conative aspects. Having teachers design a STEM unit was an effective professional development approach because it allowed teachers to practise what they learnt during the PD workshops (e.g., Huang et al., 2022). It also enabled the researchers/PD providers to determine the teacher needs and processes as they design for STEM learning. To design STEM units that integrate content, teachers must work synergistically in cross-subject disciplinary teams (e.g., science and mathematics teachers with technology teachers if available in the school). They bring cross-disciplinary knowledge, skills, attitudes and ideas to their work, just like our interdisciplinary research team when we plan and conduct the research and professional development programs. In synergising, a catalyst like professional development workshops or a professional learning team provides teachers with the opportunity, impetus, and momentum to collaborate and design STEM education. Each designed STEM unit was based on a dominant science context, consistent with past findings of science being a dominant discipline in integrated STEM interventions (Martín-Páez et al., 2019). In contrast, various mathematics content strands were applied in a single STEM unit's design and investigative activities. This explains why STEM learning is often based on one science topic, with the mathematics skills required for the STEM tasks or activities spanning various mathematics topics. Thus, STEM units tend to lend themselves more to using, applying, and communicating mathematics knowledge and skills learnt, which might be seen as *numeracy*.

A few implications and recommendations are made based on our research analysis and our experiences conducting professional development and follow-up programs. In countries that do not have integrated STEM or STEAM as a subject (yet), we found it more effective to link STEM education to existing local curricula (e.g., science, mathematics, design and technology) rather than create a brand new curriculum. As discussed in Chapter 7, the curriculum priority can be shifted from disciplinary content-focused toward more focus on content applications to the local context, with increased emphasis on learner engagement and *A*gency (see the Plexus Model of STE*A*$^+$M Education in Chapter 3). A strategy might include the integration or cross-subject connections in the curricula links for science, mathematics, and technology (e.g., the *Science, Engineering, and Technology for a Sustainable Society* strand) as a possible starting point. We recommend that cross-subject integrated S, T, E, M, or STEM curricula explicitly include the cognitive, affective, and conative learning objectives so that teachers can connect to them in their planning (see Chapters 6 and 7 for discussions of STE*A*$^+$M thinking systems and designing inclusive STE*A*$^+$M curricula). There is a need to investigate further the roles of mathematics and science (and technology and engineering) in integrated STEM curricula. Finally, the framework of three dimensions—representation of disciplines, integration of cognitive and affective aspects within each discipline, and strategic STEM focuses—can be made explicit by STEM teachers and researchers to enable the deliberate integration of what is valued by stakeholders in their contexts.

9.10 Limitations

There are limitations to the generalisability of our learnings since we base them on one professional development program and one set of document data (teacher-designed STEM units). Limitations in implementing interdisciplinary STEM curriculum integration remain significant. Despite positive outcomes reported by teachers and students (Mangao et al., 2022), structural barriers such as time constraints, limited professional development, and assessment misalignment continue to challenge consistent implementation. A dominant challenge that many teachers underscored was time constraints. Some detailed their issues with finding sufficient time for collaborative planning and reflection. Others reported that the professional development affordances were inconsistent with their approach to interdisciplinary curriculum integration. Both factors worked to limit the capacity of teachers to adapt and refine their integrative practices effectively. Lastly, given that curriculum assessment models remained heavily grounded in disciplinary silos, this further restricted opportunities for teachers to capture the full scope of student learning outcomes, even when teachers carefully aligned class work and study to exploit the opportunities presented by interdisciplinary contexts. This indicated a missed opportunity, highlighting the need for further research to explore the underlying tensions and confounding factors.

9.11 Conclusion

This chapter explored the implementation of interdisciplinary STEM curriculum integration, examining how teacher-developed curriculum goals influence student learning outcomes. The findings highlight that structural challenges persist despite teachers valuing interdisciplinary integration and reporting positive impacts on student engagement and understanding. Teachers demonstrate adaptive reasoning and problem-solving skills when designing and implementing interdisciplinary learning experiences, drawing on cognitive and communicative dimensions. Unfortunately, we must acknowledge that teaching to include interdisciplinary contexts still faces institutional barriers and varying levels of professional development support. It is hard for schools to move beyond familiar and safe learning silos. As illustrated in the prologue of this compendium, the exercise books in a child's school satchel and their timetabled classes are carefully isolated and compartmentalised. Traditional teaching methods continue to separate subjects into neat compartments. Interdisciplinary learning challenges this notion of teaching and schooling. This chapter shows how breaking these boundaries creates new spaces for creativity and critical thinking.

Our findings underscored the importance of our continued research to develop professional learning opportunities targeting interdisciplinary STEA^{+}M thinking skills and adaptive discourse practices. Teachers require strategies for navigating epistemological boundaries and facilitating cognitive integration across STEM disciplines, for themselves and their students. A *Plexus Model of STEA⁺M Education* (see Chapter 3) is needed to foster collaborative reasoning, adaptive problem-solving, and reflective dialogue among students. Furthermore, assessment practices must evolve to capture the complexity of interdisciplinary learning. Developing adaptable assessment models that demonstrate both disciplinary depth and integrative complexity will strengthen interdisciplinary STEM education. A teacher applying our research and writing with our PADL framework in Southeast Asia reflected on how interdisciplinary learning reshaped her students' engagement. She described how a mathematical modelling task, combined with scientific reasoning and creative problem-solving, allowed students from different backgrounds to contribute unique insights. This shift from compartmentalised thinking to collective problem-solving fostered deeper understanding and increased student agency and confidence.

Future research should explore how professional development models can be adapted to better support teachers in implementing interdisciplinary practices. Identifying the mechanisms through which teachers develop their adaptive expertise and collaborative problem solving skills will provide valuable insights for scaling interdisciplinary STEM education. Ultimately, the uptake and long-term integration of interdisciplinary curriculum goals will require systemic changes at the institutional level, including aligning curriculum structures, providing sufficient time and support for teacher collaboration, and ensuring that assessment models capture the full scope of student learning. It also requires a mindset and shift in culture, to develop the agency for change and transformation.

The challenge we now face is not just to introduce interdisciplinary learning. We must create stimulating and inclusive environments where students can creatively and intellectually thrive. This challenge goes beyond a simple educational shift in schooling; it transforms how teachers help students learn, engage, and take the lead. Ultimately, this pedagogical approach will inspire students to become innovative thinkers and effective change-makers in an ever-evolving world.

> Education is not preparation for life; education is life itself.
> —John Dewey

References

Aristorenas, L., & Monterola, S. L. C. (2019). Creating a culture of innovation and collaboration: STEAM innovation program in the Philippines. In *Proceedings of the Annual Meeting of Japan Society for Science Education, 2019 August 23–25*, Vol. 43, pp. 219–222. https://doi.org/10.14935/jssep.43.0_219

Ärlebäck, J. B., & Albarracín, L. (2019). The use and potential of Fermi problems in the STEM disciplines to support the development of twenty-first century competencies. *ZDM—Mathematics Education, 51*, 979–990. https://doi.org/10.1007/s11858-019-01075-3

Atiyah, M. (1993). Mathematics: Queen and servant of the sciences. *Proceedings of the American Philosophical Society, 137*(4), 527–531. https://www.jstor.org/stable/987071

Australian Trade and Investment Commission. (2020, August 25). Vietnam set to implement STEM education school curriculum.

Bangkok Post. (2022, July 26). *First regional STEM education collaboration in ASEAN*. https://www.bangkokpost.com/thailand/pr/2354624/first-regional-stem-education-collaboration-in-asean

Bernama. (2025, January 9). *New blueprint set to reshape education system, meet future demand—Fadhlina*. https://www.bernama.com/en/news.php?id=2380804

British Council Thailand. (n.d.). *STEM ambassador*. https://www.britishcouncil.or.th/en/programmes/education/our-work-support-higher-education-and-research-sector/NewtonFund/stem-education/ambassador

Cao, T. H., Trinh, T. P. T., Nguyen, T. T., Le, T. T. H., Ngo, V. D., & Tran, T. (2020). A bibliometric review of research on STEM education in ASEAN: Science mapping the literature in Scopus database, 2000 to 2019. *Eurasia Journal of Mathematics, Science and Technology Education, 16*(10), em 1889. https://doi.org/10.29333/ejmste/8500

Chang, C-Y., Lin, P.-L., & Khuyên, N. T. T. (2021). e-Learning Integrated STEM Education center (eLISE) in Asia: A reflection case study of Taiwan and Vietnam research project. *Revista Historia de la Educación Latinoamericana, 23*(36), 15–36. https://doi.org/10.19053/01227238.12833

Clandinin, D. J., & Connelly, F. M. (1992). Teacher as curriculum maker. In P. W. Jackson (Ed.), *Handbook of research on curriculum* (pp. 363–401). Macmillan.

Dare, E. A., Keratithamkul, K., Hiwatig, B. M., & Li, F. (2021). Beyond content: The role of STEM disciplines, real-world problems, 21st century skills, and STEM careers within science teachers' conceptions of integrated STEM education. *Education Sciences, 11*(11), Article 737. https://doi.org/10.3390/educsci11110737

Davis, J. P., Du, J., Tang, J.-H., Qiao, L., Liu, Y.-Q, & Chiang, F.-K. (2020). Uniformity, diversity, harmony, and emotional energy in a Chinese STEM classroom. *International Journal of STEM Education, 7*(44). https://doi.org/10.1186/s40594-020-00232-5

Department of Education, Republic of The Philippines. (n.d.). *Suggested academic track—Science, technology, engineering and mathematics (STEM) strand scheduling of subjects*. https://www.deped.gov.ph/wp-content/uploads/2019/01/Science-Technology-Engineering-and-Mathematics-STEM-Strand.pdf

Devadason, R. P. (2017). Improving basic science and mathematics education in Southeast Asia: The role of SEAMEO RECSAM. In B. Akpan (Ed.), *Science education: A global perspective*, (pp. 125–143). Springer. https://doi.org/10.1007/978-3-319-32351-0_7

Donaher, M., & Wu, N. (2020). Cambodia's new generation schools reform. In F. Reimers (Ed.), *Empowering teachers to build a better world*. SpringerBriefs in Education. Springer. https://doi.org/10.1007/978-981-15-2137-9_6

Duc, N. M., Linh, N. Q., & Yuenyong, C. (2019). Situation of organizing STEM activities in Vietnamese schools. In *IOP Conference Series: Journal of Physics, vol. 1340, presented at the International Annual Meeting on STEM Education (I AM STEM), 2018.* https://doi.org/10.1088/1742-6596/1340/1/012030

Ellis, J., Wieselmann, J., Sivaraj, R., Roehrig, G., Dare, E. & Ring-Whalen, E. (2020). Toward a productive definition of technology in science and STEM education. *Contemporary Issues in Technology and Teacher Education, 20*(3), 472–496. https://www.learntechlib.org/primary/p/214911/

English, L. (2016). STEM education K-12: Perspectives on integration. *International Journal of STEM Education, 3*(3). https://doi.org/10.1186/s40594-016-0036-1

English, L. (2017). Advancing elementary and middle school STEM education. *International Journal of STEM Education, 15*(Suppl 1), S5–S24. https://doi.org/10.1007/s10763-017-9802-x

Frankenstein, M. (2009). Developing a criticalmathematical numeracy through *real* real-life word problems. In L. Vershaffel, B. Greer, W. Van Dooren, & S. Mukhopadhyay (Eds.), *Words and worlds: Modeling verbal descriptions of situations* (pp. 111–130). Brill. https://doi.org/10.1163/9789087909383_008

Fukami, T. (2024, October 29). *Pave the way for a digital future in Lao PDR's education sector*. UNICEF. https://www.unicef.org/laos/stories/pave-way-digital-future-lao-pdrs-education-sector

Global Partnership for Education. (2020). *Endorsement of the education sector strategic plan 2021–2025. Lao PDR*. https://www.globalpartnership.org/content/endorsement-education-sector-strategic-plan-2021-2025-lao-pdr

Goos, M., Geiger, V., Dole, S., Forgasz, H., & Bennison, A. (2019). *Numeracy across the curriculum: Research-based strategies for enhancing teaching and learning*. Routledge. https://doi.org/10.4324/9781003116585

Gray, P. S., Williamson, J. B., Karp, D. A., & Dalphin, J. R. (2007). *The research imagination: An introduction to qualitative and quantitative methods*. Cambridge University Press.

Guzey, S. S., Moore, T. J., Harwell, M. (2016). Building up STEM: An analysis of teacher-developed engineering design-based STEM integration curricular materials. *Journal of Pre-College Engineering Education Research (J-PEER), 6*(1), Article 2. https://doi.org/10.7771/2157-9288.1129

Hanoi National University of Education. (2019). *STEM education in Vietnam* [presentation]. https://sites.dundee.ac.uk/miistem/wp-content/uploads/sites/123/2019/07/vietnam-presentation.pdf

Haron, H. N., Kamaruddin, S. A., Harun, H., Abas, H., & Salim, K. R. (2019). Science, technology, engineering and mathematics initiatives at rural schools and its impact on learning motivation. *Journal of Physics: Conference Series, vol. 1174, presented at the 2nd International Conference on Women in Science, Engineering and Technology, 17–19 July 2018, Kuala Lumpur, Malaysia*. https://doi.org/10.1088/1742-6596/1174/1/01200

Hasnat, G. N. T., Kabir, M. A., & Hossain, M. A. (2018). Major environmental issues and problems of South Asia, particularly Bangladesh. In C. M. Hussain (Ed.), *Handbook of environmental materials management* (pp. 1–40). Springer International. https://doi.org/10.1007/978-3-319-58538-3_7-1

Herro, D., Quigley, C., Andrews, J., & Delacruz, G. (2017). Co-Measure: developing an assessment for student collaboration in STEAM activities. *International Journal of STEM Education, 4*(26). https://doi.org/10.1186/s40594-017-0094-z

Hirsch, P. (Ed.). (2017). *Routledge handbook of the environment in Southeast Asia*. Routledge. https://doi.org/10.4324/9781315474892

Hoeg, D. G., & Bencze, J. L. (2017). Values underpinning STEM education in the USA: An analysis of the Next Generation Science Standards. *Science Education Policy, 101*(2), 278–301. https://doi.org/10.1002/sce.21260

Huang, B., Jong, M. S. Y., Tu, Y. F., Hwang, G. J., Chai, C. S., & Jiang, M. Y. C. (2022). Trends and exemplary practices of STEM teacher professional development programs in K-12 contexts: A systematic review of empirical studies. *Computers & Education*, 104577. https://doi.org/10.1016/j.compedu.2022.104577

Huizinga, T., Handelzalts, A., Nieveen, N., & Voogt, J. M. (2014). Teacher involvement in curriculum design: Need for support to enhance teachers' design expertise. *Journal of Curriculum Studies, 46*(1), 33–57. https://doi.org/10.1080/00220272.2013.834077

Ismail, M. H. B., Salleh, M. F. M., & Md Nasir, N. A. (2019). The issues and challenges in empowering STEM on science teachers in Malaysian secondary schools. *International Journal of Academic Research in Business and Social Sciences, 9*(13), 430–444. https://doi.org/10.6007/IJARBSS/v9-i13/6869

Kim, D., & Bolger, M. (2017). Analysis of Korean elementary pre-service teachers' changing attitudes about integrated STEAM pedagogy through developing lesson plans. *International Journal of Science and Mathematics Education, 15*, 587–605. https://doi.org/10.1007/s10763-015-9709-3

Landau, E., & Dermawan, A. (2019, November 30). New STREAM system a challenge to teachers. *New Straits Times*. https://www.nst.com.my/news/exclusive/2019/11/543385/new-stream-system-challenge-teachers

Lau, W. W. F., & Jong, M. S. Y. (2022). Typology of teachers' stages of concern for STEM education. *Research in Science & Technological Education*. https://doi.org/10.1080/02635143.2022.2064447

Lee, M.-H., Chai, C. S., & Hong, H.-Y. (2019). STEM education in Asia Pacific: Challenges and development. *Asia-Pacific Education Research Journal, 28*(1), 1–4. https://doi.org/10.1007/s40299-018-0424-z

Li, Y., Wang, K., Xiao, Y., & Froyd, J. E. (2020). Research and trends in STEM education: A systematic review of journal publications. *International Journal of STEM Education, 7*(11). https://doi.org/10.1186/s40594-020-00207-6

Maass, K., Geiger, V., Ariza M. R., & Goos, M. (2019). The role of mathematics in interdisciplinary STEM education. *ZDM—Mathematics Education, 51*, 869–884. https://doi.org/10.1007/s11858-019-01100-5

Mangao, D. D., Hamad, N. J., & Isoda, M. (2017). *SEAMEO basic education standards: Common core learning standards (CCRLS) in mathematics and science*. SEAMEO RECSAM. http://www.recsam.edu.my/sub_SEA-BES/

Mangao, D. D., Kidman, G., Gesthuizen, R., & Tan, H. (2019). Development process of SEAMEO STEM planning and design learning (PaDL) framework.

Paper presented at the 8th International Conference on Science and Mathematics Education (COSMED), 11–15 November 2019, SEAMEO RECSAM, Penang, Malaysia. http://cosmed.iconf.me/index.php

Mangao, D. D., Kidman, G., Tan, H., Gesthuizen, R., Macdonald, S., & Sinniah, D. N. (2022). *SEAMEO STEM planning and design learning (PaDL) framework: Towards 21st century skills and design thinking*. SEAMEO RECSAM. http://recsam.edu.my/images/everything/pdf/2022_SEAMEO_STEM_PaDL.pdf

Martín-Páez, T., Aguilera, D., Perales-Palacios, F. J., & Vílchez-González, J. M. (2019). What are we talking about when we talk about STEM education? A review of literature. *Science Education, 103*, 799–822. https://doi.org/10.1002/sce.21522

McFadden, J. R., & Roehrig, G. H. (2017). Exploring teacher design team endeavors while creating an elementary-focused STEM-integrated curriculum. *International Journal of STEM Education, 4*(21). https://doi.org/10.1186/s40594-017-0084-1

Ministry of Education, Brunei Darussalam. (n.d.). *National education system for the 21st century (SPN21)*. Ministry of Education Brunei Darussalam. http://www.moe.gov.bn/Pages/spn21.aspx

Ministry of Education, Brunei Darussalam. (2021). *STEAM programs guidebook*. Science, Technology and Environment Partnership Centre (STEP Centre) https://www.moe.gov.bn/Shared%20Documents/STEP%20Centre%20Information/STEAM%20Guidebook.pdf

Ministry of Education Brunei Darussalam (2023, March 30). *The 5th national STEM and innovative design challenge 2022–2023 prize giving ceremony*. https://www.moe.gov.bn/SitePages/NewsArticle.aspx?AID=1105

Ministry of Education, Malaysia. (2013). *Malaysia education blueprint 2013–2025 (Preschool to post-secondary education)*. Kementerian Pendidikan Malaysia https://www.moe.gov.my/en/dasarmenu/pelan-pembangunan-pendidikan-2013-2025

Ministry of Education, Myanmar. (2016). *National education strategic plan 2016–2021*. The Government of the Republic of the Union of Myanmar Ministry of Education. https://planipolis.iiep.unesco.org/sites/planipolis/files/ressources/myanmar_nesp-english_summary.pdf

Ministry of Education, Myanmar. (2019, April 10). *MoE's reforms: Working on educational development plan on 12 priority sectors*. Ministry of Education (Science and Technology). http://engmoest.moe-st.gov.mm/?p=4686

Ministry of Education Singapore. (2019, July 20). *Welcome remarks by Mr Ong Ye Kung, Minister for Education, at the National Engineers Day 2019 at Devan Nair Institute for Employment and Employability (E2i) Singapore*. https://www.moe.gov.sg/news/speeches/welcome-remarks-by-mr-ong-ye-kung--minister-for-education--at-the-national-engineers-day-2019-at-devan-nair-institute-for-employment-and-employability-e2i-singapore

Ministry of Education Singapore. (2022, January 11). *STEM*. https://www.moe.gov.sg/news/parliamentary-replies/20220111-stem

Ministry of Education Thailand. (2019, August 8). *Primary schools to start teaching coding: Deputy education minister*. http://www.en.moe.go.th/enMoe2017/index.php/news/432-primary-schools-to-start-teaching-coding-deputy-education-minister

Ministry of Education Timor Leste. (2011). *National Education Strategic Plan 2011–2030*. https://www.globalpartnership.org/sites/default/files/2011-08-Ministry-of-Education-Timor-Leste-Education-Plan_processed.pdf

Ministry of Education, Youth and Sport, Kingdom of Cambodia. (2021, May). *Cambodia secondary education blueprint 2030*. https://data.opendevelopmentcambodia.net/en/library_record/cambodia-secondary-education-blueprint-2030

Ng, C., H., & Adnan, M. (2018). Integrating STEM education through project-based inquiry learning (PIL) in topic space among year one pupils. *IOP Conference Series: Materials Science and Engineering, 29*. https://doi.org/10.1088/1757-899X/296/1/012020

Nguyen, H. H., Tran, V. C., & Tran, T. T. (2020). Students and teachers' perspective of the importance of Arts in STEAM education in Vietnam. *Journal of Critical Reviews, 7*(11), 666–671. http://www.jcreview.com/fultext/197-1592981186.pdf

Nugraha, M. G., Kidman, G., & Tan, H. (2023). Pre-service teacher in STEM education: An integrative review and mapping of the Indonesian research literature. *EURASIA Journal of Mathematics, Science and Technology Education, 19*(5), em2262. https://doi.org/10.29333/ejmste/13155

Nugroho, O. F., Permanasari, A., & Firman, H. (2019). The movement of STEM education in Indonesia: Science teachers' perspectives. *Jurnal Pendidikan IPA Indonesia, 8*(3), 417–425. https://doi.org/10.15294/jpii.v8i3.19252

O'Bee, M. (2018, September 15). *STEM education from students in LuangPrabang Laos* [Video]. YouTube. https://www.youtube.com/watch?v=zlHjM0nFEJs

Organisation for Education Co-operation and Development. (n.d.). *The survey of adult skills (PIAAC) Numeracy sample items*. OECD. https://www.oecd.org/skills/piaac/

Ostler, E. (Ed.) (2015). *STEM education: An overview of contemporary research, trends, and perspectives*. Cycloid Publications.

Paderna, E. E. S., & Monterola, S. L. C. (2022). Advancing integrated STEM education in the Philippines through STEM curriculum implementation. In T. W. Teo, A.-L. Tan, & P. Teng (Eds.), *STEM education from Asia: Trends and perspectives* (pp. 127–143). https://doi.org/10.4324/9781003099888-7

Park, W., Wu, Y.-Y., & Erduran, S. (2020). The nature of STEM disciplines in the science education standards documents from the USA, Korea and Taiwan: Focusing on disciplinary aims, values and practices. *Science and Education, 29*, 899–927. https://doi.org/10.1007/s11191-020-00139-1

Perignat, E., & Katz-Buonincontro, J. (2019). STEAM in practice and research: An integrative literature review. *Thinking Skills & Creativity, 31*, 31–43. https://doi.org/10.1016/j.tsc.2018.10.002

Quigley, C. F., Herro, D., King, E., & Plank, H. (2020). STEAM designed and enacted: Understanding the process of design and implementation of STEAM curriculum in an elementary school. *Journal of Science Education and Technology, 29*, 499–518. https://doi.org/10.1007/s10956-020-09832-w

Randall, R., Sukoco, G. A., Heyward, M., Purba, R., Arsendy, S., Zamjani, I., & Hafiszha, A. (2022). *Reforming Indonesia's curriculum: How Kurikulum Merdeka aims to address learning loss and learning outcomes in literacy and numeracy*. INOVASI. https://www.inovasi.or.id/en/publikasi/the-learning-gap-series-two-reforming-indonesias-curriculum-how-kurikulum-merdeka-aims-to-address-learning-loss-and-improve-learning-outcomes-in-literacy-and-numeracy/

Regional Centre for Education in Science and Mathematics (RECSAM). (2019). *Southeast Asian Ministers of Education Organization (SEAMEO) Basic Education Standards* (SEA-BES). http://www.recsam.edu.my/sub_SEA-BES/index.php

Ring-Whalen, E., Dare, E., Roehrig, G., Titu, P., & Crotty, E. (2018). From conception to curricula: The role of science, technology, engineering, and mathematics in integrated STEM units. *International Journal of Education in Mathematics, Science and Technology (IJEMST), 6*(4), 343–362. https://doi.org/10.18404/ijemst.440338

Roehrig, G. H., Dare, E. A., Ring-Whalen, E., & Wieselmann, J. R. (2021). Understanding coherence and integration in integrated STEM curriculum. *International Journal of STEM Education, 8*(2). https://doi.org/10.1186/s40594-020-00259-8

Sampurno, P. J., Sari, Y. A., & Wijaya, A. D. (2015). Integrating STEM (science, technology, engineering, mathematics) and disaster (STEM-D) education for building students' disaster literacy. *International Journal of Learning and Teaching, 1*(1), 73–76. https://doi.org/10.18178/ijlt.1.1.73-76

Sanders, M. (2009). STEM, STEM education, STEMmania. *The Technology Teacher, 68*(4), 20– 26. http://hdl.handle.net/10919/51616

Schiro, M. S. (2013). *Curriculum theory: Conflicting visions and enduring concerns* (2nd ed.). SAGE.

SEAMEO. (2021). *What is SEAMEO?* https://www.seameo.org/Main_about/90

SEAMEO QITEP. (n.d.). Training on STEM implementation in learning for elementary, middle and high school teachers in Alor Regency. In *Southeast Asian Ministers of Education Organization regional center for Quality Improvement of Teacher and Education Personnel.* https://www.qitepinscience.org/blog/author/qispengatur/

SEAMEO STEM-ED. (n.d.). *SEAMEO regional centre for STEM education (SEAMEO STEM-ED) centre profile.* https://seameo-stemed.org/seameo-stem-ed/

Shahali, E. H. M., Ismail, I., & Halim, L. (2017). STEM education in Malaysia: Policy, trajectories and initiatives. *Asian Research Policy, 8*(2), 122–133.

Sihaphom, S. (2024, October 9). *Igniting the spark: Empowering adolescent girls in Lao PDR with the skills needed to thrive in the 21st century knowledge-economy.* UNICEF. https://www.unicef.org/laos/stories/igniting-spark

Soparat, S., Arnold, S. R., & Klaysom, S. (2015). The development of Thai learners' key competencies by project-based learning using ICT. *International Journal of Research in Education and Science, 1*(1), 11–22. https://files.eric.ed.gov/fulltext/ED548501.pdf

Sritrakul, P. (2018). The state of STEM education policy in northern region, Thailand. *Humanities, Arts and Social Sciences Studies, 18*(1), 129–147. http://www.thaiscience.info/Journals/Article/SUIJ/10989786.pdf

Stehle, S. M., & Peters-Burton, E. E. (2019). Developing student 21st century skills in selected exemplary inclusive STEM high schools. *International Journal of STEM Education, 6*, Article 39. https://doi.org/10.1186/s40594-019-0192-1

STEM Inc. (n.d.). *Science, technology, engineering and mathematics, innovation and creativity.* Singapore Science Center. https://www.science.edu.sg/stem-inc/

Sumonta, P., Finley, F. N., & Kaweekijmanee, K. (2018). The evolution and current status of STEM education in Thailand: Policy directions and recommendations. In G. W. Fry (Ed.), *Education in Thailand: An old elephant in search of a new mahout* (pp. 423–460). Springer. https://doi.org/10.1007/978-981-10-7857-6_17

Suratno, S., Wahnono, B., Chang, C.-Y., Retnowati, A., & Yushardi, Y. (2020). Exploring a direct relationship between students' problem-solving abilities and academic achievement: A STEM education at a coffee plantation area. *Turkish*

Science Education, 17(2), 211–224. https://doi.org/10.36681/tused.2020.22

Takeuchi, M. A., Sengupta, P., Shanahan, M.-C., Adams, J. D., & Hachem, M. (2020). Transdisciplinarity in STEM education: A critical review. *Studies in Science Education, 56*(2), 213–253. https://doi.org/10.1080/03057267.2020.1755802

Teo, T. W. (2019). STEM education landscape: The case of Singapore. In *Proceedings of the International Annual Meeting on STEM Education (I AM STEM) 2018 August 13–15 Avanti Khon Kaen Hotel, Thailand.* in *Journal of Physics: Conference Series, 1340*(2019) 012002. https://doi.org/10.1088/1742-6596/1340/1/012002

Tupas, F. P., & Matsuura, T. (2019). Moving forward in STEM education, challenges and innovations in senior high school in the Philippines: The case of Northen Iloilo Polytechnic State College. *Indonesian Journal of Science Education, 8*(3), 407–416. https://doi.org/10.15294/jpii.v8i3.19707

United Nations Educational, Scientific and Cultural Organization (UNESCO). (2015). *Education and sports sector development plan (2016–2020).* Lao People's Democratic Republic Ministry of Education and Sports. https://planipolis.iiep.unesco.org/en/2015/education-and-sports-sector-development-plan-2016-2020-6273

UNESCO. (2020, November 10). *The gaps and challenges faced by teachers in the classroom for science and mathematics primary education in Timor-Leste.* https://en.unesco.org/news/gaps-and-challenges-faced-teachers-classroom-science-and-mathematics-primary-education-timor

Vann, P. (2023). STEM education in Cambodia: Progress and challenges. In K. Heng, K. Sol, S. Kaing, & S. Em (Eds.). *Innovations and challenges in Cambodian education: Youth's perspectives.* https://cefcambodia.com/books/

CHAPTER 10

Interdisciplinary Integration for Future-Making Education in STEAM

10.1 The Purpose of Integration

This chapter returns to the book's aims: to interrogate, reframe, and illuminate the processes of interdisciplinary integration in STEM and STEAM education. From the outset, this book resisted simplistic interpretations of integration as a linear or additive synthesis of disciplines. Instead, it has positioned integration as an entangled, dynamic practice that is pedagogical, ethical, political, and epistemological. At the centre of this conception is the metaphor of the *plexus*: a richly interconnected web of relations in which knowledge systems, contextual realities, and human *A*gency are co-constituted. It is within this plexus that the promise and complexity of STE*A*$^+$M education reside.

In returning to these aims, we acknowledge that interdisciplinary integration is not solely concerned with merging content areas for curriculum efficiency or engagement. Instead, it reflects a more profound commitment to reconfiguring how knowledge is produced, valued, and mobilised in education. This work highlights the importance of disciplinary epistemologies, critical pedagogies, and socio-cultural contexts, revealing that integration is always situated within broader struggles over power, representation, and the future orientation of education. STE*A*$^+$M, in this view, becomes a space for critical inquiry and imaginative possibility, where learners are not simply taught to operate across disciplines but are

G. Kidman et al., *Knowledge Generation in STEM and STEAM Education*, https://doi.org/10.1007/978-3-032-03247-8_10

invited to question the boundaries, assumptions, and implications of those disciplines themselves, and to develop *A*gency.

The chapters in Parts I–III collectively trace this argument across conceptual, practical, and policy-oriented terrains. Yet this concluding reflection is not a synthesis in the traditional sense. It is a recursive gesture that invites reconsideration of the tensions, insights, and provocations explored throughout the book. In doing so, it repositions integration as an ongoing, unfinished, and necessarily pluralistic endeavour, an approach to education that must remain open to diverse knowledges, responsive to global and local urgencies, and committed to cultivating *A*gency within complex, uncertain futures.

Thus, this chapter does not seek to close the discussion, but to reopen it through a new lens: one that views interdisciplinary integration not as a solution to be implemented, but as a relational practice to be lived, negotiated, and continually reimagined. Through this, we reaffirm the critical significance of STE*A*⁺M integration, not as a mere curriculum innovation, but as a transformative orientation toward learning, social justice, and the future of education itself.

10.2 Knowledge, Practice, and Pedagogy

The imperative of interdisciplinary integration in STEM and STE*A*⁺M education emerges not as a technical solution to curriculum design but as a deeply epistemological, ethical, and political project. Throughout the preceding chapters, we have examined how integration is enacted, resisted, and reimagined within contemporary educational contexts. This final chapter seeks not to summarise, but to reflect on and revisit the original questions posed in Part I of the book, while acknowledging that they now appear in transformed form after being entangled with theory, pedagogy, and practice.

Part I of the book, comprising Chapters 1–4, establishes the conceptual and cognitive foundations of interdisciplinary learning. Chapter 1 introduced neuroeducation as a scientific and pedagogical bridge between brain-based learning and classroom practice, offering frameworks (*CEN-STEM* and *CIRCLE*) that align neurocognitive processes with effective STEM teaching. In Chapter 2, we turned to the epistemic structures of disciplinary thinking, mapping the methodological distinctiveness of science, mathematics, and technology and exploring how such differences shape both integration and its limits. Chapter 3 advanced this discussion by presenting the *Plexus Model of STEA⁺M Education*, positioning interdisciplinary integration as an entangled nexus of *A*gency, innovation, and social responsibility. Chapter 4 then explored the cognitive work of integration itself, highlighting the intellectual labour of meaning-making across disciplines through the *Cognitive Integration Framework* and *Crafting Knowledge model*. Together, these chapters underscore the conceptual necessity of integration as more than a pedagogical strategy. We see it is a way of thinking that unsettles disciplinary silos while cultivating new forms of knowledge production.

Part II of the book (Chapters 5–7) delves into the practice of integration, examining the systems of knowledge, skill, and power that circulate through twenty-first century learning. Chapter 5 traced the shifting discourse of future-ready competencies, linking STEM and STEAM to broader socio-educational transformations through a review of global frameworks. Chapter 6 extended this trajectory by elaborating on STEAM as a system of meta-skills, integrative competencies that connect cognitive, affective, and ethical dimensions of learning. Chapter 7 interrogated the hidden curriculum embedded in STEM and STEAM education, making visible the normative assumptions and power relations that shape who is recognised as a legitimate knower. In doing so, it invited us to consider integration not simply as content amalgamation but as an act of disruption, one that can challenge dominant narratives and open possibilities for equity, inclusion, and alternative epistemologies.

In Part III, Chapters 8 and 9 engaged with the design and enactment of integrated curricula. Chapter 8 introduced the *PaDL Framework* for design thinking in STEAM education, showing how interdisciplinary learning can be structured to support iterative, context-sensitive, and student-centred pedagogies. Chapter 9 provided empirical insights into how teachers in Southeast Asia construct interdisciplinary STEM units through collaborative professional development, revealing both the possibilities and constraints of integration in non-Western contexts. These chapters highlight the practical realities of interdisciplinary education, focusing on how teachers adapt, translate, and reimagine curricula in response to regional, cultural, and environmental imperatives.

As we return to the purpose of integration through these cumulative insights, it becomes clear that interdisciplinary practice is not merely a curriculum innovation; rather, it is a fundamental shift in the way we approach education. It is a mode of future-making. It invites us to ask:

- What kinds of futures are we imagining and enacting through STE*A*⁺M education?
- Whose knowledge is included or excluded in the STE*A*⁺M education process?
- What forms of *A*gency are cultivated when learners navigate across disciplinary boundaries?

To answer these questions requires a shift in how integration is conceptualised, not as a synthesis for efficiency or alignment, but as a transformative encounter that reconceptualises what education can be. This chapter, then, begins not with a conclusion but with an invitation to reimagine the work of integration as an ongoing act of ethical and creative responsibility.

10.3 Reconceptualising Integration: From a Cognitive Process to Social Praxis

Building on the preceding narrative, the following foregrounds a pivotal transformation in the conceptualisation of interdisciplinary integration, from a primarily cognitive enterprise to an enactment of social praxis. While the earlier chapters of this book articulated the cognitive, epistemological, and disciplinary dimensions of integration, what emerges from the collective insights of Parts I–III is the inescapable conclusion that transformative STE*A*⁺M education must also be fundamentally ***political***, ***decolonial***, and ***sustainability-oriented***. Interdisciplinary integration, when understood as social praxis, moves beyond the coordination of disciplinary knowledge to become an active engagement with the ethical, ecological, and cultural imperatives of our time.

Political Spaces of STEAM Education: Disrupting Epistemic Hierarchies. Disciplinary thinking and the epistemic traditions of science, mathematics, and technology shape not only what is taught but also whose knowledge is privileged. The CEN-STEM framework (Chapter 1), particularly through Principles 1 and 2, actively engages educators in critical epistemological literacy. Addressing neuromyths (Principle 1) and promoting metacognition and executive function (Principle 2) do more than refine teaching; they empower educators and learners to question dominant assumptions about how knowledge is acquired, who produces it, and whose ways of learning are legitimised. These principles align with our political call to disrupt epistemic hierarchies. They support a shift from passive reception of disciplinary content to active interrogation of power and authority in knowledge systems, especially those rooted in Western, reductionist models. The discussion of epistemological boundaries highlighted how traditional STEM education often perpetuates narrow, Western knowledge systems that prioritise objectivity, abstraction, and disciplinary purity (Chapter 2). The Plexus Model of STEA^{+}M Education (Chapter 3) challenges this hegemony by advocating for an interwoven, *A*gency-driven approach that recognises the interconnectedness of social, environmental, and technological systems. Here, integration is reconfigured as an ethical imperative, one that must attend to how knowledge systems are embedded in histories of power, exclusion, and colonisation. Further critiques of the hidden curriculum are presented (Chapter 7), showing how educational structures reproduce inequities by privileging certain knowledge forms. This demands that STE*A*$^{+}$M actively confronts whose knowledge counts, making pedagogy inherently political.

Decolonisation: Agency and Cultural Relevance. Throughout the book, we position decolonisation as central to the reimagining of STE*A*$^{+}$M education, foregrounding cultural relevance, learner *A*gency, and epistemic justice. The CEN-STEM framework (Chapter 1), particularly through Emotional Safety in Learning (Principle 4), Emotional Engagement in Complex Subjects (Principle 5), and Active Engagement (Principle 9), speaks directly to the pedagogical dimensions of decolonial education. They prioritise emotional connection, culturally situated experiences, and learner voice. These are core aspects of centring *A*gency and resisting the universalising tendencies of Western STEM education. The Principles align with the PaDL Framework (Chapter 8) and Chapter 6's STEAM Thinking Systems, which call for pedagogies that are context-sensitive, culturally adaptive, and co-developed with local educators. CEN-STEM's emotional-motivational principles support the epistemic justice work by recognising learning as relational and affective, not merely cognitive. The PaDL Framework, co-developed with Southeast Asian educators, exemplifies this commitment by adapting dominant Western-centric models to reflect local epistemologies, community priorities, and contextual pedagogical practices. This collaborative work offers a tangible example of how integration can serve as a decolonising act, transforming not only what is taught but how and why it is taught. We extend this approach by reinterpreting twenty-first century skills through the lens of *STEAM Thinking Systems* (Chapter 6), a constellation of meta-skills that includes ethical technological literacy, social consciousness, and global responsibility. Rather than treating these skills as neutral or universally applicable, we argue that they are culturally and contextually embedded capacities, shaped by local realities, social conditions, and ethical dilemmas. This repositioning extends beyond individual competencies to encompass broader educational imperatives, such as global citizenship, ethical and moral reasoning, and sustainability-oriented action. The critique of disciplinary hierarchies (Chapter 7) furthers this decolonial agenda. We interrogate the valorisation of certain forms of knowledge and the marginalisation of others, particularly Indigenous and non-Western knowledge systems. Here, integration becomes not simply a pedagogical strategy but a political act, one that challenges whose knowledge is legitimised in STEM and STEAM domains. Teachers and students are positioned as active agents of cultural and epistemic justice, capable of resisting and reshaping inherited curricula. The book deepens this critique by situating integration within the lived pedagogical realities of classrooms (Part II). We bring the hidden curriculum to light, revealing how implicit norms and structures perpetuate systemic inequities. In doing so, we call for a radical reimagining of integration as a decolonising practice, one that recognises the complexity of learners' socio-political and ecological contexts, and equips them to act within and against these systems. In sum, decolonialising STE*A*$^{+}$M education in this work involves more than content diversification or inclusion; it requires a fundamental transformation of pedagogical assumptions, power relations, and epistemic boundaries. It is a call to centre *A*gency, contextual understanding, and social justice at the heart of integrated education.

Sustainability-Oriented STEAM Education: Ethics and Socio-Ecological Action. We have redefined twenty-first century skills beyond neutrality, tying them to sustainability, global citizenship, and ethical reasoning (Chapter 5). Several CEN-STEM principles (Chapter 1) address systems thinking and complex problem-solving, which are foundational to sustainability education: Neuroplasticity and Targeted Learning (Principle 6), Cognitive Load Management (Principle 7), and Large-Scale Neural Network Interplay (Principle 8). These principles promote adaptive thinking, iterative learning, and flexibility, which underpin the kind of education needed to address socio-ecological crises (as elaborated in Chapters 5–9). For example, the interplay between introspective and task-oriented neural networks (Principle 8) mirrors the need for learners to shift between personal reflection and global responsibility, as urged in this chapter. In this way, the CEN-STEM principles not only serve to optimise the classroom, they prepare learners cognitively and emotionally to navigate complexity, act ethically, and contribute to regenerative, place-based futures. The empirical study (Chapter 9) highlights teacher-developed STEM units that address local environmental crises, demonstrating how integration becomes place-based, ecologically responsive education. The *Plexus Model* of STEA^{+}M Education (Chapter 3) and STEAM Thinking Systems (Chapter 6) further anchor this, framing STEAM as a means to navigate complex socio-ecological challenges and foster regenerative futures. This repositioning of integration reaches its full expression in Part III of the book. The PaDL Framework (Chapter 8) presented an approach to planning and learning that explicitly centres teacher and student agency, cultural relevance, and iterative ethical reflection. Reinforcing this commitment through its empirical study of teacher-developed STEM units (Chapter 9) that directly address environmental issues and sustainability concerns relevant to their communities illustrates the integrated content; they embody integration as an act of place-based, socially responsive education. Teachers were not only knowledge designers but also agents of ecological and cultural renewal. Across the chapters in this book, integration is not conceived solely as a pedagogical method or a curriculum strategy. It is a form of praxis rooted in the pursuit of social justice and sustainability. It calls upon educators to consider how STEA^{+}M education can equip learners to confront ecological crises, redress social inequities, and reclaim silenced epistemologies. Interdisciplinary integration, when grounded in culturally and ecologically responsive pedagogies, becomes a site for decolonial work: unsettling inherited norms, amplifying marginalised voices, and fostering educational futures that are both critically reflective and radically hopeful. Thus, we assert that STEA^{+}M integration must not remain on the periphery of social critique, but instead be reconceptualised as a transformative practice. Integration, in this sense, is not only about what we teach or how we teach it, but also about the world we seek to create through education. It is an active and ongoing reconfiguration of knowledge, identity, and responsibility. It positions learners and educators as co-constructors of just, sustainable, and culturally attuned futures.

Bringing It All Together: Conceptual Coherence. To link the CEN-STEM principles (from Chapter 1) to the political, decolonial, and sustainability-oriented vision of STEA^{+}M education (this chapter), we can trace the conceptual coherence across the book via Fig. 10.1. In Fig. 10.1, we present the ***FRAME Model** for Transformative STEA⁺M Education.* It can be viewed as a conceptual tool for guiding transformative STE*A*⁺M education. FR*A*ME stands for *Futures-oriented*, *Responsibility-driven*, *Agency-centred*, *Meaning-making*, and *Ethics-informed* STEAM education practice. It encapsulates the chapter's central argument: that interdisciplinary integration must move beyond cognitive coordination to become a pedagogical praxis responsive to global challenges and local contexts. The FR*A*ME Model synthesises insights from the CEN-STEM principles, the CIRCLE Framework, and the political, decolonial, and sustainability imperatives of STE*A*⁺M education. It provides educators with a relational and justice-oriented lens through which integration becomes a catalyst for critical hope, ethical learning, and collective educational futures. The CEN-STEM principles offer more than a neuroscience-informed pedagogy; they serve as scaffolds for a deeper political project. They operationalise the theoretical commitments articulated in this chapter: that STEAM education must be relational, ethical, culturally grounded, and justice-oriented. In doing so, they bridge cognitive and emotional development with the social imperative of transformative learning. Thus, the CEN-STEM principles (Chapter 1) are not peripheral to the book's concluding claims; they are the foundational mechanisms through which the political, decolonial, and sustainability goals of STE*A*⁺M education are realised in practice.

The synthesis of the CEN-STEM principles with the political, decolonial, and sustainability-oriented vision articulated in this chapter reveals a structural coherence across the work. While the initial framing of the CEN-STEM principles appears rooted in cognitive science and neuroeducation, a deeper examination reveals that these principles, when situated within broader pedagogical and epistemological commitments, form the cognitive and affective infrastructure necessary for future-oriented, justice-driven STE*A*⁺M education. Rather than being merely technical or instrumental frameworks, CEN-STEM principles can be understood as enablers of the ethical, political, and ecological transformations demanded by a reconceptualised model of integration.

The CEN-STEM principles provide the foundation for a pedagogy that is responsive not only to how the brain learns but also to who the learner is, in what context they are situated, and what futures they are being prepared to navigate. The fostering of principles such as Addressing Neuromyths through Educator Literacy (Principle 1) and Metacognition

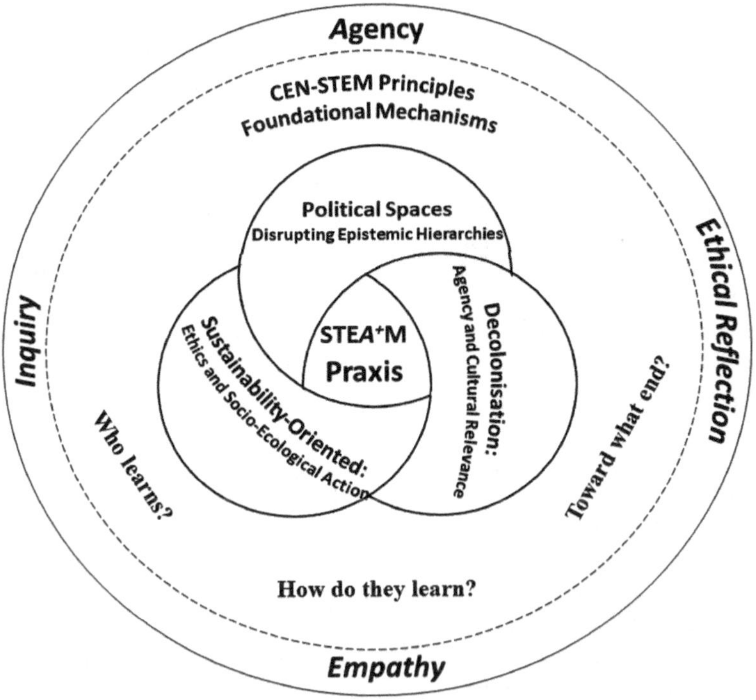

Fig. 10.1 The FR*A*ME model for transformative STE*A*$^{+}$M education

and Executive Function Support (Principle 2) prompts educators to critically examine commonly held assumptions about intelligence, learning, and knowledge production, often rooted in Western models of cognition. In this way, the CEN-STEM principles resist universalising narratives and encourage reflexivity, thereby aligning with the decolonial imperative to disrupt dominant epistemologies and validate diverse ways of knowing. The inclusion of emotional and motivational principles within CEN-STEM, such as Reward and Motivation (Principle 3), Emotional Safety in Learning (Principle 4), and Emotional Engagement in Complex Subjects (Principle 5), further enhances this alignment. These principles support a learning environment that recognises the affective dimensions of learner experience, a key concern in decolonial education where inclusion, belonging, and cultural recognition are central.

The CEN-STEM principles are consolidated in the CIRCLE (Cognitive, Inclusive, Resilient, Creative Learning Environments) Framework for STEM Neuroeducation (Chapter 1), a multilayered, relational model of teaching and learning that emphasises inclusivity, resilience, and creativity, aligning with the transformative aims of STEA^+M as outlined above. The social-emotional and cognitive layers work in tandem to address both the neurocognitive processes of learning and the socio-cultural conditions under which learning occurs. This dual focus enables educators to design learning environments that are simultaneously scientifically informed and socially responsive. The skills-development layer of the CIRCLE Framework also aligns closely with the sustainability-oriented goals of STEA^+M education by emphasising adaptability, ethical judgement, and problem-solving capacities essential for navigating the socio-ecological crises and moral complexities of the contemporary world.

By embedding evidence-based neurocognitive principles within an inclusive, dynamic, and future-facing pedagogical model, the CEN-STEM and CIRCLE frameworks do not merely support effective STEM instruction; they become mechanisms for enacting educational transformation. They operationalise the epistemic and ethical orientations described in this chapter by linking individual cognitive development with collective agency, cultural relevance, and ecological responsibility. In doing so, these frameworks provide a scientifically grounded yet politically potent foundation for educators seeking to reimagine STEAM not as a curricular add-on or integration strategy, but as a form of praxis. This praxis is concerned with who learns, how they learn, and toward what ends, thus foregrounding the political, decolonial, and sustainability imperatives that constitute the book's inescapable conclusion.

10.4 Toward an Ethical, Creative, and Collective STEA^+M Praxis

Here we draw upon the notion of future-making education (Burnard & Colucci-Gray, 2020). Burnard and Colucci-Grey introduce us to the concept of "un-conclusions" (p. 425), the leaning into an agenda of educational innovation. We position STEAM not as a corrective response to disciplinary fragmentation but as an ongoing, generative practice of critique, experimentation, and transformative possibility. We have demonstrated throughout this book that integration is not merely the alignment of learning outcomes or the coordination of subject areas; it is also a speculative, reflexive, and ethically charged orientation toward what education can become when we actively engage with the monumental challenges of our time (for example, climate change, social inequality, technological advancements, and epistemic marginalisation). STEAM education must be reimagined as a framework for cultivating not only knowledge and skills but also the capacities for ethical judgement, collective action, and critical hope.

This reconceptualisation begins by affirming learner *A*gency as a foundational element of future-oriented education (see Fig. 10.1). Across the book's chapters, *A*gency is explored not as a passive trait or individual disposition but as a pedagogical principle that enables learners to see themselves as capable of intervening in the world. In Chapter 3, the Plexus Model of STE*A*⁺M Education places *A*gency at the centre of interdisciplinary practice, emphasising the importance of empowering learners to navigate complexity, engage with social and environmental challenges, and act with purpose. Chapter 8 extends this by situating *A*gency within design thinking practices, showing how structured opportunities for empathy, inquiry, and ethical reflection (see Fig. 10.1) support learners in generating creative, context-responsive solutions. *A*gency in these contexts is not an abstract ideal; it is cultivated through material and relational processes that connect knowledge with responsibility.

Just as learners are repositioned as active agents, so too are communities redefined as educational partners rather than passive beneficiaries of knowledge. Chapter 9 offered compelling illustrations of how teacher-designed interdisciplinary STEM units were developed through a dialogue with local environmental conditions and social concerns. These cases demonstrated how STEAM education can be embedded in community-based learning that is attuned to place, culture, and sustainability. Such approaches enable a more reciprocal relationship between schools and communities, where learning is not only about acquiring disciplinary knowledge but also about co-creating solutions, valuing local knowledge, and fostering collective resilience.

Burnard and Colucci-Gray's (2020) future-making education also demands a renewed commitment to ethical responsibility. As argued throughout Part II and further elaborated in the planning models of Part III, interdisciplinary integration must contend with the ethical and moral dilemmas posed by complex global issues. Chapter 7's critique of the hidden curriculum revealed how traditional models of STEM and STEAM often reproduce normative assumptions and inequities. In contrast, Chapters 5 and 6 made the case for pedagogies that explicitly address socio-political and ecological interdependencies. STEAM integration, therefore, must be anchored in a commitment to social justice, not only in what is taught, but in how and why teaching occurs. This means designing educational experiences that are attentive to power, inclusive of marginalised perspectives, and capable of cultivating learners who are both critically literate and morally engaged.

Reconceptualising the future of STEAM education, therefore, entails a reframing of the concept of integration itself. It should not be seen as a technical solution, but as a pedagogical stance that embraces uncertainty, multiplicity, and contestation. It is an open-ended process of making and remaking educational practice in response to an evolving world. It involves embracing the creative potential of difference, the necessity of collective responsibility, and the urgency of preparing learners for futures that are yet to be known. In this view, the value of integration lies not in its ability to stabilise the curriculum but in its capacity to destabilise fixed assumptions and to enable education to remain responsive, imaginative, and just.

As this chapter and book end, we invite educators, researchers, and policymakers to continue thinking with STEAM not as a fixed and established model, but as a living praxis (as STEA^+M education), a way of working with and through disciplines to imagine new educational futures. These futures are not given; they are made through the ethical, collaborative, and creative work of learners and educators who refuse to see education as separate from the world it seeks to transform. In this sense, STEAM is not the endpoint of integration, but rather a most promising beginning when seen as STEA^+M education.

10.5 Three Parting Provocations

We finish this chapter and book by developing three provocations to extend the generative potential of STEAM education, inviting further dialogue, critical inquiry, and collaborative experimentation, and revisiting the rhetorical questions posed in Sect. 10.2 above. In keeping with the spirit of Burnard and Colucci-Gray's (2020) *future-making education*, our provocations are not presented as problems to be solved, but as ongoing tensions to be engaged with. The tensions are at the heart of interdisciplinary integration and demand sustained attention from educators, researchers, and policymakers alike. The aim here is to foreground STE*A*⁺M as a necessary condition for educational transformation.

Provocation 1 emerges across the chapters and relates to how educators can navigate epistemological tensions in STEAM without undermining disciplinary integrity. As examined in Chapter 2, disciplines are defined not only by their content but by their ways of knowing. Their ontologies, methodologies, and epistemic commitments. Yet interdisciplinary integration often risks flattening these differences in the name of coherence or thematic relevance. The challenge, then, is *to design integrated STEAM learning experiences that preserve the depth, rigour, and specificity of each discipline while simultaneously fostering the cognitive flexibility and synthesis required for transdisciplinary inquiry and Agency.* This necessitates a 'pedagogy of integration', one that attends to the epistemic foundations of knowledge and cultivates student understanding of how disciplines differ, intersect, and co-evolve. The earlier question (Sect. 10.2), "*What forms of Agency are cultivated when learners navigate across disciplinary boundaries?*", is developed through this provocation by identifying the conditions under which learner *A*gency can flourish: when students are supported to recognise, interrogate, and synthesise disciplinary differences. The provocation thus reframes agency not as a generic skill but as a situated practice that arises from grappling with complexity. It calls for pedagogies of integration that enable learners to move reflexively across disciplinary boundaries while maintaining intellectual rigour and epistemological awareness.

Provocation 2 concerns the enactment of equity in integrated learning environments. The chapters in Part II, particularly Chapters 6 and 7, highlighted the need to confront the hidden curriculum, structural exclusions, and normative assumptions that shape students' access to disciplinary knowledge and participation in learning. To 'design for equity' in STEAM education is not only about including diverse content or learners, but also about fundamentally reconsidering how integration can serve as a vehicle for social justice. The challenge, then, is *to create inclusive* STE*A*⁺M *pedagogies that honour different ways of knowing, test dominant discourses, and support learners in recognising their Agency and value within complex systems.* It also demands systemic change and the rethinking of assessment practices, professional development, and institutional priorities to ensure that equity is embedded at every level of educational design and delivery. The earlier question (Sect. 10.2), "Whose knowledge is included or excluded in the STEAM education process?", is further explored through this provocation, which confronts the hidden curriculum and systemic exclusions that persist in integrated learning environments. This provocation reframes the earlier question as a challenge for *A*ction, demanding not just recognition of epistemic exclusion but a radical rethinking of how equity is designed into the architecture of STE*A*$^+$M education. By calling for inclusive pedagogies and systemic change in assessment and institutional priorities, this provocation expands the original question from an inquiry into representation, to a broader agenda for epistemic justice and educational transformation.

Provocation 3 addresses the dual imperative of supporting both local knowledge and global responsibilities within interdisciplinary integration. We highlighted how culturally responsive pedagogies and teacher-designed curricula can effectively integrate local knowledge, environmental stewardship, and community priorities into STEAM learning (Chapters 8 and 9). However, as global crises such as climate change, displacement, and digital surveillance intensify, educators are also called to prepare students to engage ethically with global systems and transnational challenges. The challenge, then, is ***determining how transformative STEA$^+$M education can cultivate learners who are deeply rooted in their local contexts while also being attuned to global interdependencies.*** Addressing this challenge requires pedagogies that are both place-based and globally conscious, offering learning experiences that equip students to act responsibly across various scales, institutions, and knowledge traditions. The earlier question (Sect. 10.2) "What kinds of futures are we imagining and enacting through STEA$^+$M education?" is addressed most directly through this provocation, which explores the challenge of supporting both local knowledge and global responsibilities. This provocation advances the original question by repositioning STEAM education not merely as preparing students for predefined futures, but as STE*A*$^+$M education, enabling learners to participate in shaping those futures. The provocation calls for pedagogies that are place-based yet globally conscious, thus situating STEAM as a space of active world-building. In doing so, it offers a response to the question not in the form of an answer, but in the form of an imperative: to cultivate educational experiences that are responsive to the ecological and geopolitical urgencies of our time.

Taken together, these three provocations reaffirm that interdisciplinary integration is not a finished product, but a living, evolving process. It requires humility, reflexivity, and sustained commitment. As we have argued throughout, STEAM education must not only cross disciplinary boundaries but also bridge divides between theory and practice, tradition and innovation, and individual and collective responsibility. It must engage educators in a practice of continuous learning: questioning assumptions, responding to emerging challenges, and co-creating new possibilities for what education can become.

In this light, as a final chapter, we are not at the endpoint, but at a threshold. This chapter and book offer an invitation to continue the work of integration, not by replicating models, but by inventing, adapting, and reimagining models in dialogue with diverse learners, communities, and futures. The task ahead is not only to understand STEAM integration but to live it as an ethical, relational, and transformative praxis—as STE*A*⁺M education.

Reference

Burnard, P., & Colucci-Gray, L. (2020). Un-conclusions: Disentangling the assemblage of science and arts creativities for future-making education. In P. Burnard & L. Colucci-Gray (Eds.), *Why science and arts creativities matter: (Re-)configuring STEAM for future-making education*. Brill. https://brill.com/display/title/54614?language=en&contents=toc-130902

Epilogue

The phone call arrived on an unassuming summer afternoon early in 2004. I was called from class to go to the school reception. Standing in her doorway, my school principal nodded slightly as muted office staff handed me the telephone receiver. It was an international phone call!

"We need you for an important mission. Can you help?"

Was I biting off more than I could chew? The build instructions were being faxed to me. Only later did I fully grasp that two scientists had called me from the other side of the world: Prof Woody Sullivan from the University of Washington, Seattle, and Bill Nye, the Science Guy. They needed help to build a southern hemisphere calibration and reference station as part of the EarthDial project by the Planetary Society, alongside the MarsDials soon to be deployed on the twin NASA Mars rovers, Spirit and Opportunity (Read, 2004; Sullivan, 2003).

"Would you be willing to take on this challenge?"

It wasn't the opportunity that struck me; it was the sheer audacity of making something out of this world and then taking my students on this journey. I felt this should belong in a science fiction story or with a university team. Perhaps they had called me at my high school by mistake? Looking back, I can now see why they were keen to tap into our ability to push beyond our comfort zones. They had already seen the work by my class during a live Mag-Net webcast, reenacting the transit

G. Kidman et al., *Knowledge Generation in STEM and STEAM Education*, https://doi.org/10.1007/978-3-032-03247-8

of Venus using a pinhole camera, retort stand, and laptop. We did this whilst recounting Charles Green's astronomical observations of the same transit with the 1768 Endeavour expedition. It also demonstrated that integrating disciplines could foster cognitive transformation.

I responded, "I think we can make it so."

Soon, I was clutching the printed instructions for constructing an Earth-Dial, the companion for the rover Mars Dials. Running back to class, I introduced the idea to my students and wondered how I would frame this. So far, all my students had ever built were cardboard satellite models made of recycled waste that hang by a thread from our classroom ceiling. By the time I reached my classroom door, I had decided this would be a transdisciplinary learning journey. The Year 10 class was a friendly and diverse group of students from Pacifica, Vietnamese, and Cambodian family backgrounds. It was a wonderfully eclectic yet challenging mix of multiculturalism, typical of many Melbourne high schools.

"We'll build a device to help mark time on another planet, Mars."
"I'm not even entirely sure how. Can you help? I cannot do this alone."

My Year 10 students exchanged uncertain glances and whispers of disbelief. I could almost hear the questions of uncertainty forming in their minds as I laid out how we would move beyond the boundary and silo of our subject to participate in an action-based inquiry and off-world investigation.

They asked: "Why us? Why now?"

And then I saw it, a spark, one that I had seen before, that inspired that moment when awe turns to wonder and ignites curiosity. The class suddenly pivoted from a tedious numeracy activity into a fury of questions where science, mathematics, and digital technologies collided with planetary science and applied learning.

One sun, two planets. A student-driven inquiry.

Over the weeks, we worked through calculations on a spreadsheet to customise our equipment for our latitude. We drew pictures, made models, debated different material sciences for our equipment and negotiated where to mount the equipment on the school roof. We secured a rooftop spot, with colleagues welding old desk legs for scaffolding and setting up the outdoor camera and streaming internet connection. In hindsight, it was a genuine design-thinking process involving the class and the broader school community.

My daughter Erika was only 18 months old and learning to walk when we prepared the EarthDial. My father (Johannes, but Opa to Erika) saw me struggling with the power tools while trying to machine a gnomon. "Let me help", he said, stepping in with quiet confidence. He told Erika to pick something from a Christmas tree as the central piece. She came back with a blue bauble. Dad filled it with something, drilled a thread, and fixed it with a coach bolt onto our EarthDial. The blue bauble, mounted at the heart of the ground station, was a quiet reminder of how we can find inspiration from those we love around us.

It nearly didn't work, but our integrative problem-solving abilities came to the fore. I recall rummaging through Hudson Physics Kits with Erika to steal a polarising filter to correct an optical error missed in our calculations that threatened to damage the camera but could be easily fixed. The school threw its weight behind our crazy project, and as the school year wound to a close, we threw the switches and started streaming our EarthDial image into the void just before Christmas.

Our 'experiment' allowed us to intimately compare Mars' time from the back of a rover with the humble sundial we had built on Earth. What had started as a hypothetical exercise suddenly became a STEM project with geographical underpinnings. We had built an EarthDial on the roof of our school, and my students were no longer passive learners but participants in something interplanetary and more significant than we had ever imagined. Their mindset had changed. Our collaborative intuition and insight helped inspire their creativity, persistence, and initiative to go beyond.

As a final touch, we painted a penguin onto the instrument platform, visible during the live broadcast, allowing us to anchor our EarthDial in Melbourne, Victoria. Our learning in that class was more than just accumulating knowledge or mastering a subject. The students had stood at the edge of uncertainty and transformed themselves in a way we had not imagined.

I now humbly realise that the subject was never about siloed numeracy or space science. It wasn't just about the technology but how we holistically integrated our curiosity and experiences to create something significant.

What emerged was greater than all the parts.

This project was more than building a sundial or participating in a NASA mission. It was a testament to the idea that education should become interdisciplinary, inquiry-driven, and engaging. We had done this in a way that no single discipline could have achieved in isolation. Along this journey, we integrated our learnings and problem-solving with cultural perspectives. Our students did not merely learn about science, technology, engineering, the arts, or mathematics; they lived it.

The boundaries of our curriculum had faded and been replaced by something more significant. We had inspired and fuelled our students with the *A*gency to explore, create, and imagine beyond and towards new horizons.

Perhaps the inspiration and invitation wasn't just from the scientists, NASA or Mars.

It was always there all along, waiting for us to say yes.

Read, B. (2004). Astronomy professor leads effort to webcast live views of sundials around the world. *The Chronicle of Higher Education, 50*(18).

Sullivan, W. (2003). Build a sundial and join the worldwide EarthDial project. *Astronomy Education Review, 2*(2), 202–204. https://doi.org/10.3847/AER2003029.

Index

G. Kidman et al., *Knowledge Generation in STEM and STEAM Education*, https://doi.org/10.1007/978-3-032-03247-8

E

F

R

FSC
www.fsc.org
MIX
Papier | Fördert gute Waldnutzung
FSC® C083411

Kolibri 360 GmbH
Ferdinand-Jühlke-Straße 7
99095 Erfurt, Deutschland
produktsicherheit@kolibri360.de